LECTURES IN THEORETICAL PHYSICS

Volume XIII

De Sitter and Conformal Groups and Their Applications

Institute for Theoretical Physics Publications

University of Colorado, Boulder, Colorado 80302

Directors: Asim O. Barut
Wesley E. Brittin

LECTURES IN THEORETICAL PHYSICS

Volume XIII

De Sitter and Conformal Groups and Their Applications

Edited by

Asim O. Barut and Wesley E. Brittin
Department of Physics and Astrophysics
University of Colorado

Colorado Associated University Press
Boulder, Colorado

Colorado Associated University Press
1424 Fifteenth Street
Boulder, Colorado 80302

Library of Congress Catalog Number 59-13034
ISBN 87081-014-6

FOREWORD

This Volume contains the lectures presented at a week-long Conference on De Sitter and Conformal Groups and Their Applications held at the University of Colorado, June 29-July 3, 1970. The lectures were both on the mathematical theory of the representations of a general class of non-compact groups, and also on their various recent uses in many areas of theoretical physics. Correspondingly the Volume is divided in two parts, and each into various Sections. The interest in the representations of many non-compact groups is rapidly increasing among physicists who, unlike mathematicians, need the actual construction of these representations. Perhaps the most important non-compact groups in physics, the Lorentz and Poincaré groups, were the subject matter of a similar Conference here in Boulder about six years ago.† Since then much has been learned about the representations of groups like O(3,2), O(4,1), O(4,2),... and their inhomogeneous and euclidian counterparts. At the same time, the applications of new group theoretical techniques--which go much beyond the customary use compact symmetry groups--have brought new results and new insights to a number of physical problems. We believe therefore that it is timely and useful to bring some of the existing literature, the new results and the unsolved problems in the area of De Sitter and Conformal Groups to the attention of physicists and mathematicians. This is what the present volume intends to do.

I wish to thank the lecturers and the participants for their effort for a lively Conference, and to Mrs. Marion Higa for her invaluable contributions in the organization of the Conference and in the editing of this Volume.

Boulder, December 1970. A. O. Barut

†*Lectures in Theoretical Physics*, Vol. VIIA. The Lorentz Group (Univ. of Colorado Press, 1965).

TABLE OF CONTENTS

INTRODUCTION

INTRODUCTION TO DE SITTER AND CONFORMAL GROUPS AND THEIR PHYSICAL APPLICATIONS†

A. O. Barut

As a prelude to the contributions in these Proceedings we review in some detail the general properties of De Sitter and Conformal Groups and the areas of their physical applications.

Contents

---

†Presented at the Symposium on De Sitter and Conformal Groups, University of Colorado, Summer 1970.

I. The Role of De Sitter and Conformal Groups in Particle Physics

Although the largest exact symmetry group used in particle physics is still the full inhomogeneous Lorentz group, extended with discrete symmetries (e.g. charge conjugation, baryon and lepton numbers), larger simple groups containing the Lorentz group play more and more an important role. Of course, the symmetry group of the general relativity theory is much larger, but there is not yet a theory which combines the gravitational and the strong, electromagnetic and weak interactions of fundamental particles. How do these large groups enter into the theory, and in what sense are we using them? We will answer these questions for the De Sitter and the conformal groups.

There are three major areas of physical applications:

(1) Theories Invariant Under Dilatations and Special Conformal Transformations in flat space, in addition to inhomogeneous Lorentz transformations.

The dilatations are associated with the change of scale of measurements, and the special conformal transformations may be associated, roughly speaking, with the change of scale from point to point. With this interpretation the independence of the physical laws from the scales used should lead to an exact 15-parameter conformal invariance of the theories.† Except for free electromagnetic fields and other wave equations for mass zero particles of arbitrary spin such invariant theories, however, have not been formulated. Instead the usual Lorentz-invariant Lagrangians of interacting massive scalar, spinor and tensor fields with mass terms are obviously not exactly invariant under conformal group; they are only approximately invariant under certain conditions (e.g. at high energies). One speaks then of a "broken scale invariance." This does not mean that we cannot formulate exact conformally invariant theories, e.g. field theories. We shall come back to this question in Section III.1.

(2) Geometrization of Dynamics of Interacting Systems.

The De Sitter and conformal groups are also used as dynamical groups, generalizing the concept of symmetry. If the standard symmetry group gives us the states of a system of a given energy (multiplets), the dynamical group gives all the states of the system (infinite multiplets). It may also be denoted more precisely as the group of the quantum numbers, or group of all rest-frame states. The concept is particularly useful in the relativistic theory of composite systems. In these applications, these groups are interpreted as O(4,1) and O(4,2), respectively: they contain the physical homogeneous Lorentz group O(3,1) as subgroup, but not the translations. In physics, we are not using abstract Lie groups, but groups whose

---

† This is by no means the only interpretation.[1)]

generators have definite physical interpretation. Thus, the same group may occur in entirely distinct situations and distinct interpretations. Details about O(4,1) and O(4,2) dynamical groups are given in Section III.2.

(3) Theories in Curved-Spaces in the Large (and in the Small)

According to Mach's principle, local inertial frames and local isotropy of space (i.e. rotational invariance) are due to the distribution of distant galaxies. If the shape of the distant space is important for phenomena in the small, we should from the beginning start with a curved or closed universe, rather than the flat space-time. Of course, we have to give up the usual energy-momentum vector and the conservation of the total energy-momentum; instead of the translations, we have new displacement operators. It has been conjectured that although the deviations of the space from flatness is very small, it is in principle essential that a consistent theory be formulated in the curved space.[2] Along the same line, the introduction of gravitation, so it is hoped, may supply the necessary cut-off factors for a finite quantum field theory.[3] At any rate, quantum field theory in a curved space-time is one of the important, if not immediate, goals of theoretical physics.[4] (Sec. III.3).

All the above three points of view being valid we can imagine a super-theory in which the conformal group occurs at least three times in three different interpretations: scale-changes, curved space and internal dynamics!

We begin with a review of the necessary mathematics.

II. Mathematical Results

1. Group Properties

We shall be interested in the Lie algebras and in non-compact Lie groups O(3,2), O(4,1) and O(4,2) [O(3,3) and O(5,1)], also in E(3,2), E(4,1) and E(4,2). The notation here is that O(p,q) is the full real non-compact orthogonal group with the invariant form $x_1^2 + x_2^2 + \dots x_p^2 - x_{p+1}^2 - \dots - x_{p+q}^2$, and E(p,q) is the corresponding pseudo-euclidian group, i.e. O(p,q) plus translations.

We assume a knowledge of the Lorentz and Poincaré groups, (O(3,1) and E(3,1)), and their representations.[5]

The group O(p,q) has four pieces, as in the case of the Lorentz group. The notation $SO_0(4,2)$ will be used for the component continuously connected to identity element (i.e. det = +1).

The De Sitter and conformal groups belong to the complex simple groups B_2 and D_3 in Cartan's classification (Table I). Different non-compact groups may have the same complexification. Conversely, for every simple complex group we get a unique compact real group, and a number of distinct real non-compact groups.[6]

Table I. Complex Simple Groups and Their Real Forms

Complex Cartan Group	Real Compact Form	No. of Parameters r	Some Real Non-Compact Forms
B_2	O(5)	10	O(4,1), O(3,2)
D_3	O(6)	15	O(5,1), O(4,2), O(3,3)
C_2	Sp(4)	10	Sp(2,2), Sp_4
A_3	SU(4)	15	SU(3,1), SU(2,2), SL(4,R), Q_2

The isomorphisms between the groups listed in Table I are shown in Table II.

Table II. Isomorphisms Between Non-Compact Groups

Complex Isomorphism	Real Isomorphisms
$B_2 \sim C_2$	O(3,2) ~ Sp(4) ~ Sp(2,R) O(4,1) ~ Sp(2,2) O(5) ~ USp(4)
$D_3 \sim A_3$	O(4,2) ~ SU(2,2) O(3,3) ~ SL(4,R) O(6) ~ SU(4) O(5,1) ~ Q_2 D_6 ~ SU(3,1)

Two groups that have the same Lie algebra are related to each other by $G' = G/D$ where D is a discrete group; i.e. the groups G' and G are locally isomorphic, but not globally. Among all such groups having the same Lie algebra there is only one which is simply connected; it is the universal covering group. In this sense SU(4) is the covering group of SO(6); SU(2,2) is one of the covering groups of SO(4,2), but not the universal covering group. (The explicit correspondence is shown in Sec. IV.)

2. Realizations and Representations (Linear and Nonlinear)

Most groups used in physics may be defined as the groups of transformations, linear or nonlinear, acting on a space with a relatively low dimension. Thus, the groups $O(p,q)$ are defined as the real linear transformation groups on the $p+q = n$-dimensional real space; the groups $U(p,q)$ as the complex linear transformation groups on the $p+q = n$-dimensional complex space. The conformal group has therefore a 6-dimensional real linear representation as $O(4,2)$, and a four-dimensional linear representation, by 4×4-complex matrices, as $SU(2,2)$. There is, however, another possibility: We can realize the conformal group on the 4-dimensional real Minkowski space-time by real, but nonlinear, transformations in such a way that the inhomogeneous Lorentz subgroup is again linearly represented as before. That is besides the usual transformations

$$x_\mu' = \Lambda_\mu^{\ \nu} x_\nu + a_\mu \tag{2.1}$$

we have the new ones

$$x_\mu' = \lambda x_\mu$$

and

$$x_\mu' = \frac{x_\mu - c_\mu x^2}{1 - 2c^\nu x_\nu + c^2 x^2} \tag{2.2}$$

$$(x^2 \equiv x_0^{\ 2} - \underset{\sim}{x}^2)$$

altogether a group with 15 parameters. Eqs. (2.1)-(2.2) represent the most general group which transforms $ds^2 = 0$ into $ds'^2 = 0$ in the Minkowski space (i.e. light cone into light cone).

In quantum theory we need the representations of the group (defined, say, on the Minkowski space) on the linear space of quantum states. For symmetry groups this means the linear unitary representations in the Hilbert space (because of Wigner's theorem[7]). For groups which are not symmetry groups other realizations may also be important.

3. The Lie Algebra

Quite generally, the pseudo-orthogonal groups $O(p,q)$, $p+q=n$, are generated by the $r = \frac{n(n-1)}{2}$ generators which can be written as an antisymmetric tensor $L_{ab} = - L_{ba}$, $a,b = 1,2,\ldots,n$. The group element $e^{i\theta_{ab}L_{ab}}$ generated by L_{ab} is a rotation (or a hyperbolic rotation) by an angle θ_{ab} around an axis perpendicular to the a-b-plane. For this choice of the generators (e.g. choice of basis of the Lie algebra), the group is parametrized by r angles θ_{ab}.

The commutation relations of the generators (i.e. Lie products) are given by

$$[L_{ab}, L_{cd}] = -i[g_{ac}L_{bd} + g_{bd}L_{ac} - g_{bc}L_{ad} - g_{ad}L_{bc}], \qquad (3.1)$$

where $g_{ab} = (\underbrace{++\ldots+}_{p\text{-times}}\ \underbrace{-----}_{q\text{-times}})$.

The quadratic, third and fourth, etc.-order (invariant) Casimir operators of the Lie algebra are (for n = 6, for example)

$$Q_{(2)} = -2\mathrm{Tr}(LGLG) = -(g_{ab}g_{cd} - g_{ad}g_{bc})\, L^{ac}L^{db}$$

$$Q_{(3)} = \epsilon_{abcdef}L^{ab}L^{cd}L^{ef}$$

$$Q_{(4)} = L_{ab}L^{bc}L_{cd}L^{da} \qquad (3.2)$$

Note that lowering and rising of the indices in L^{ab} is carried out with g_{ab}.

3.1 A New Basis

Quite generally for O(p,q) groups, we can define

$$P_A \equiv L_{A,n-1} - L_{An}$$

$$K_A \equiv L_{A,n-1} + L_{An} \quad ; \quad A, B = 1, 2, \ldots n-2$$

$$D \equiv L_{n-1,n} \qquad (3.3)$$

Then the commutation relations (3.1) become

$$[P_A, P_B] = [K_A, K_B] = 0$$

$$[L_{AB}, P_A] = -ig_{AA}P_B;\ [L_{AB}, K_A] = -ig_{AA}K_A;\ [L_{AB}, D] = 0$$

(equation continued)

$$[P_A, K_B] = i(g_{nn} - g_{n-1,n-1}) L_{AB} - 2ig_{AB}D$$

$$[D, P_A] = -i(g_{nn}L_{A,n-1} + g_{n-1,n-1}L_{A,n})$$

$$[D, K_A] = -i(g_{nn}L_{A,n-1} - g_{n-1,n-1}L_{A,n}) \quad , \qquad (3.4)$$

which show that P_A and K_A are (n-2)-vectors whose components commute, and D is a scalar.

As a special case, n = 6, p = 4, q = 2 (i.e. $g_{nn} = -g_{n-1,n-1}$) we get in this basis the two Poincaré subalgebras of the conformal group: $\{L_{\mu\nu}, P_\mu\}$ and $\{L_{\mu\nu}, K_\mu\}$ with the commutation relations:

$$[P_\mu, P_\nu] = [K_\mu, K_\nu] = 0$$

$$[L_{\mu\nu}, P_\mu] = -ig_{\mu\mu}P_\nu, \quad [L_{\mu\nu}, K_\mu] = -ig_{\mu\mu}K_\nu; \quad [L_{\mu\nu}, D] = 0$$

$$[P_\mu, K_\nu] = 2i(L_{\mu\nu} - g_{\mu\nu}D)$$

$$[D, P_\mu] = +i P_\mu,$$

$$[D, K_\mu] = -i K_\mu \qquad (3.5)$$

In the nonlinear realization (2.2), D is the generator of dilatations and K_μ are the generators of special conformal transformations. To see these commutation relations directly, we may use the group law in the x_μ-space and the corresponding composition of representations: For example, consider $x_\mu' = \lambda' x_\mu + a_\mu'$, $x_\mu'' = \lambda'' x_\mu' + a_\mu'' = \lambda''\lambda' x_\mu + \lambda'' a_\mu' + a_\mu'' = \lambda x_\mu + a_\mu$, and correspondingly the representations $G = e^{i(\lambda D + P^\mu a_\mu)}$. Then from the group property $G = G''G'$ and the composition law of the parameters $\lambda = \lambda''\lambda'$, $a_\mu = \lambda'' a_\mu' + a_\mu''$, we obtain the commutation relations $[D, P_\mu] = iP_\mu$. The other commutation relations in (3.5) can be obtained in a similar way.

3.2 Group Contraction

We show here a relation between the simple O(p,q)-groups and the corresponding pseudo-euclidian E(p,q-1) or E(p-1,q) groups. Let us denote, referring to the basis (3.1), the generators $L_{A,n}$ by cP_A, and the remaining generators by L_{AB}, A,B = 1,2...n-1. The commutation relations can be written as

$$[P_A, P_B] = -ig_{nn} \frac{1}{c^2} L_{AB}$$

$$[P_A, L_{AB}] = -ig_{AA} \frac{1}{c} L_{nB} = ig_{AA} P_B \tag{3.6}$$

Now we let $c \to \infty$. In this limit L_{AB} and P_A are generators of the pseudo-euclidian group in (n-1)-dimensions determined by g_{AB}, i.e., L_{AB} generate the homogeneous transformations and P_A, (which is a (n-1)-vector operator) generate the translations. Depending on the sign of g_{nn} we get E(p,q-1) or E(p-1,q). These groups are also written as $T_{n-1} \otimes O(p,q-1)$, a semi-direct product of translations with the orthogonal group, T_{n-1} being an invariant subgroup.

Thus, the contraction of both O(3,2) and O(4,1) can give the inhomogeneous Lorentz group.[8)]

The quantity $P_A P^A$ is an invariant of the contracted inhomogeneous group, but is no longer an invariant in the original group.

3.3 Two Other Parametrizations of the Conformal Group

(a) The U(2,2)-parametrization: We consider the complex four-dimensional space with the invariant form

$$|z_1|^2 + |z_2|^2 - |z_3|^2 - |z_4|^2 = \text{invariant}, \tag{3.7}$$

and parametrize the pseudo unitary infinitesimal transformations as follows

$$z_\mu{}' = z_\mu + (\alpha_\mu{}^\nu + i\beta_\mu{}^\nu)\, z_\nu \tag{3.8}$$

where $\alpha_\mu{}^\nu$ and $\beta_\mu{}^\nu$ are real parameters. The group elements can then be written as

$$G = e^{i(-\frac{1}{2}\alpha^{\mu\nu} M_{\mu\nu} + \frac{1}{2} i\beta^{\mu\nu} K_{\mu\nu})}, \tag{3.9}$$

with

$$M_{\mu\nu} = - M_{\nu\mu}, \quad K_{\mu\nu} = K_{\nu\mu}. \tag{3.10}$$

The commutation relations of these generators are then[9)]

$$[M_{\mu\nu}, M_{\sigma\rho}] = -i(g_{\mu\sigma} M_{\nu\rho} + g_{\nu\rho} M_{\mu\sigma} - g_{\nu\sigma} M_{\mu\rho} - g_{\mu\rho} M_{\nu\sigma})$$

$$[K_{\mu\nu}, K_{\sigma\rho}] = -i(g_{\mu\sigma} K_{\nu\rho} + g_{\nu\rho} K_{\mu\sigma} + g_{\nu\sigma} K_{\mu\rho} + g_{\mu\rho} K_{\nu\sigma})$$

$$[M_{\mu\nu}, K_{\sigma\rho}] = -i(g_{\mu\sigma} K_{\nu\rho} - g_{\nu\rho} K_{\mu\sigma} - g_{\nu\sigma} K_{\mu\rho} + g_{\mu\rho} K_{\nu\sigma}) \qquad (3.11)$$

The $U(2) \otimes U(2)$ subgroup of $U(2,2)$ is generated by $(M_{12}; K_{11}, K_{12}, K_{22}) \oplus (M_{34}; K_{33}, K_{34}, K_{44})$. For $SU(2,2)$ there will a relation between the four-diagonal elements $K_{\mu\mu}$, $\mu = 0,1,2,3$; only three of them will be independent.

(b) Cartan Parametrization: Because the set of generators $K_{\mu\mu}$ commute among themselves, they form the Cartan-subalgebra $\{H_i\}$. Next we define the 12 generalized lowering and raising operators

$$E_{\alpha} \equiv \tfrac{1}{2}(M_{\mu\nu} + i\, K_{\mu\nu})$$

$$E_{-\alpha} \equiv \tfrac{1}{2}(M_{\mu\nu} - i\, K_{\mu\nu})$$

$$\mu \neq \nu; \quad \alpha = 1,2,\ldots 6. \qquad (3.12)$$

In terms of these new generators, the commutation relations are indeed in the canonical form[10)]

$$[E_{\alpha}, E_{\beta}] = N_{\alpha\beta}\, E_{\alpha+\beta}, \qquad \alpha \neq -\beta$$

$$[E_{\alpha}, E_{-\alpha}] = \underset{\sim}{r}(\alpha)\cdot \underset{\sim}{H} \quad . \qquad (3.13)$$

An alternative tensorial form of the canonical basis is also useful: Let

$$Y_{AB} = \begin{pmatrix} Y_{11} & E_{-1} & E_{-2} & E_{-3} \\ E_1 & Y_{22} & E_{-4} & E_{-5} \\ E_2 & E_3 & Y_{33} & E_{-6} \\ E_3 & E_4 & E_5 & Y_{44} \end{pmatrix} \qquad (3.14)$$

$$A,B = 1,2,3,4$$

where Y_{AA} are again linear combinations of $K_{\mu\mu}$'s. Then we have simply

$$[Y_{AB}, Y_{CD}] = g_{BC} Y_{AD} - g_{AD} Y_{BC} , \qquad (3.15)$$

and the Casimir operator becomes

$$Q_{(2)} = Y_{AB} Y^{BA} \qquad (3.16)$$

4. <u>Homomorphism Between SO(4,2) and SU(2,2)</u>

In this section we derive the relation between the six dimensional real coordinates $\eta_1 \ldots \eta_6$ and the four-dimensional complex coordinates $z_1 \ldots z_4$. The discussion parallels that of the well known homomorphisms between SL(2,C) and SO(3,1) [or SO(3) and SU(2)].[11]

Define the anti-symmetric matrix

$$A = \begin{pmatrix} 0 & \eta_1+i\eta_2 & \eta_3+i\eta_4 & \eta_5+i\eta_6 \\ & 0 & \eta_5-i\eta_6 & \eta_3-i\eta_4 \\ & & 0 & \eta_1-i\eta_2 \\ & & & 0 \end{pmatrix} \qquad (4.1)$$

and the metric tensor

$$G = \begin{pmatrix} g_{11} & & & 0 \\ & g_{22} & & \\ & & g_{33} & \\ 0 & & & g_{44} \end{pmatrix} . \qquad (4.2)$$

Then

$$\mathrm{Tr}(AGA^+G) = 2[(g_{11}g_{22}+g_{33}g_{44})(\eta_1^2+\eta_2^2) + (g_{11}g_{33}+g_{22}g_{44})(\eta_3^2+\eta_4^2) + (g_{11}g_{44}+g_{22}g_{33})(\eta_5^2+\eta_6^2)] \qquad (4.3)$$

This expression is invariant under the transformations

$$A \to A' = UAU^+ \qquad (4.4)$$

if $U^+GU = G$, that is, if $U \in SU(p,q)$, $p+q = 6$, then

$$\mathrm{Tr}(A'GA'^{+}G) = \mathrm{Tr}(AGA^{+}G) \tag{4.5}$$

Lemma: If A is antisymmetric and of the form given by Eq. (4.1), then $A' = UAU^{+}$, $U \in SU(p,q)$ is also antisymmetric and of the form (4.1). This follows from the property:

$$A_{AB} = -A_{BA} = \tfrac{1}{2}\epsilon_{ABCD}\overline{(GAG)}^{CD}; \quad A,B,\ldots = 1,2\ldots 4 \ , \tag{4.6}$$

where the bar indicates the complex conjugation. (Example: $A_{12} = g_{33}g_{44}\bar{A}_{34}$). Thus, the matrix A' has the same form (4.1) in terms of six new coordinates $\eta_1', \ldots \eta_6'$. The transformation $A' = UAU^{+}$ induces therefore a transformation in the η-space with the invariant $\mathrm{Tr}(AGA^{+}G)$. It follows from Eq. (4.3) that this invariant is that of (pseudo)-orthogonal transformations in the 6-dimensional space only in two cases:

(a) All g_{AA} have the same sign: then we get the homomorphism $SU(4) \to O(6)$.

(b) Two of the g_{AA} are positive, and the other two negative: then we get the homomorphism $SU(2,2) \to O(4,2)$.

In both cases $\pm U$ correspond to the same $O(p,q)$ transformation.

If we define the antisymmetric tensors Σ_A (generalizing the Pauli matrices),† we can write

$$A = \eta^{A}\Sigma_{A}, \quad A' = \eta'^{A}\Sigma_{A}; \quad \eta'^{A} = O^{A}{}_{B}\,\eta^{B} \tag{4.7}$$

Hence

$$\eta'^{A}\Sigma_{A} = \eta^{B}U\Sigma_{B}U^{+} = O^{A}{}_{B}\,\eta^{B}\Sigma_{A} \tag{4.8}$$

or

$$O^{A}{}_{B} = \tfrac{1}{4}\mathrm{Tr}(U\,\Sigma_{B}\,U^{+}\,\Sigma^{A}) \tag{4.9}$$

The inverse formula is left to the student as an exercise. To my knowledge these formulas are nowhere in the literature.

---

† $$\Sigma_1 = \begin{pmatrix} 0 & 1 & 0 & 0 \\ -1 & 0 & 0 & 0 \\ 0 & 0 & 0 & 0 \\ 0 & 0 & 0 & 0 \end{pmatrix}, \quad \Sigma_2 = \begin{pmatrix} 0 & i & 0 & 0 \\ -i & 0 & 0 & 0 \\ 0 & 0 & 0 & 0 \\ 0 & 0 & 0 & 0 \end{pmatrix}, \quad \text{etc.}$$

$$\Sigma_A^{2} = 1, \quad \mathrm{Tr}(\Sigma_A) = 0$$

5. Representation Theory

5.1 Finite Dimensional Representations

The finite dimensional irreducible representations of the simple group $B_2 \sim C_2$ are also the unitary irreducible representations of the compact group O(5). These same representations are non-unitary for O(4,1) and O(3,2). These representations are well known and can be characterized, as usual, by the weight diagrams (or Gel'fand-Zeitlin patterns).[10),12),13)] The dimensionality of these low-lying finite dimensional irreducible representations and their reduction with respect to the O(4)-subgroup is shown in the following Table:

Dimension of Representations

O(5)	1	4	5	10	14	16
O(4)-reduction	1	2 2	4 1	3 4 3	-	-

(5.1)

There are general formulas for the dimensionality of the finite dimensional representations of classical groups in terms of top weight and roots (Weyl's character and dimension formulas).[14)]

Similarly the finite-dimensional representations of $D_3 \sim A_3$ are the unitary representations of O(6). The irreducible ones and their reduction with respect to the O(5)-subgroup is as follows:

Dimension of Irreducible Representations

O(6)	1	4	6	10	15	20	36
O(5)-Reduction	1	4	5 1		10 5	16 4	20 16

(5.2)

5.2 Explicit Form of the 4-dimensional Representations of "O(5)" and "O(6)" Classes in Terms of Dirac Matrices

This very useful representation is given in the Lie algebra the following matrices

$$L_{ab} \equiv \ell_{ab} = \tfrac{1}{2} i \gamma_a \gamma_b , \qquad a < b; \; a,b = 1,2,3,4,5 \equiv 0,6$$

$$\gamma_a = (\gamma_1 , \gamma_2 , \gamma_3 , -\gamma_5 , \gamma_0 , -iI)$$

$$\gamma^5 = \gamma^0 \gamma^1 \gamma^2 \gamma^3 , \quad \gamma^{0\,2} = 1 , \quad \gamma^{i\,2} = \gamma^{5\,2} = -1 , \quad \gamma^{0+} = \gamma^0$$

$$\gamma^{i+} = -\gamma^i , \quad \gamma^{5+} = -\gamma^5 \qquad (5.3)$$

The matrices ℓ_{ab} satisfy the commutation relations[15] (3.1) with $g_{ab} = (++++--)$. This particular representation is pseudo-unitary with respect to the metric γ^o:

$$\gamma^o \ell_{ab}^{\dagger} = \ell_{ab} \quad . \tag{5.4}$$

There is another inequivalent irreducible 4-dimensional representation of the A_3-groups, namely

$$\bar{L}_{ab} \equiv -\ell_{ab}^{*} \quad . \tag{5.5}$$

If we restrict in the above representation the indices to a,b = 1,2,3,4,5 ≡ 0, we get the four-dimensional representations of $B_2 \sim C_2$-groups, which are also irreducible. We shall see later also infinite dimensional representations of O(4,2) which when restricted to its various subgroup remain irreducible.

In the above representation, the homogeneous Lorentz subgroup SO(3,1) is generated by

$$L_{\mu\nu} = \tfrac{1}{2} i(\gamma_\mu \gamma_\nu - g_{\mu\nu}) \ , \tag{5.6}$$

and with respect to this Lorentz group the remaining nine generators can be grouped into a vector $(\frac{1}{2}\gamma_\mu)$, an axial vector $(-\frac{i}{2}\gamma_5\gamma_\mu)$, and a pseudoscalar $(-\frac{i}{2}\gamma_5)$.

The four-dimensional representation can also be characterized by the representation relation

$$\{\gamma_\mu, \gamma_\nu\} = 2g_{\mu\nu} \tag{5.7}$$

or, by the relation[15]

$$\ell_{34} = -2\,\ell_{06}\,\ell_{12} \quad . \tag{5.8}$$

5.3 <u>Labeling Operators. Choice of Bases</u>

In the following Table we show the enumeration of the complete set of commuting operators (CSCO):

Group	No. of Parameters	No. of Casimir Operators	No. of Cartan Generators	Additional Operators Needed to Label the States Uniquely
$B_2 \sim C_2$ "O(5)"	10	2	2	2
$A_3 \sim D_3$ ("O(6)")	15	3	3	3

For irreducible representations, the choice of the 4 operators for B_2, or the 6-operators for A_3 (beside the Casimir operators) depend on the physical applications (i.e. which quantum numbers are diagonalized and interpreted physically). That the additional operators can always be found is seen by the following general solution for the group O(2n). Consider the chain of subgroups:

$$\underset{(D_n)}{O(2n)} \supset \underset{(B_{n-1})}{O(2n-1)} \supset \underset{(D_{n-1})}{O(2n-2)} \supset \text{------} \tag{5.9}$$

In O(2n) we need besides the n Casimir operators, $n^2 - n$ other operators. If we diagonalize O(2n-1)-subgroups, this one needs precisely $(n^2 - n)$ total labeling operators. We choose the (n-1)-Casimir operators of O(2n-1), the remaining $(n^2 - 2n+1)$ operators are just the total labeling operators of O(2n-2), and so on down the chain.

A similar construction holds for the chain

$$U(n) \supset U(n-1) \supset U(n-2) \supset \text{----}$$

in the case of unitary groups.[16] Thus, for O(4,2)-representations we need 9 labels in general, but for special representations fewer labels are sufficient.

5.4 Representations in Terms of Boson Creation-Annihilation Operators. Tensor Methods.

Consider n pairs of boson creation operators: a_i and a_i^*, i = 1,2,...n.

$$[a_i, a_j^*] = \delta_{ij}.$$

We form the bilinear combinations

$$F_{ij} = a_i a_j \,, \qquad \frac{n(n+1)}{2} \text{ in number}$$

$$G_{ij} = a_i^\dagger a_j + \tfrac{1}{2}\delta_{ij} \,, \qquad n^2 \quad " \quad "$$

$$H_{ij} = a_i^\dagger a_j^\dagger \,, \qquad \frac{n(n+1)}{2} \quad " \quad " \qquad (5.10)$$

These $2n^2+n$ operators span the Lie algebra of C_n: the unitary simplectic group in 2n-dimension which leaves the forms

$$\sum_{i=1}^{\infty} \bar{x}_i y_i \quad \text{and} \quad \sum_{i=1}^{n} (x_i y_{i+n} - x_{i+n} y_i) \qquad (5.11)$$

invariant.

The linear combinations of G_{ij} give the usual basis of the Lie algebra of the unitary groups (A_{n-1} with (n^2-1) parameters). Note that $\sum_{i=1}^{n} a_i^\dagger a_i$ commutes with all the G_{ij}. A subset of the G_{ij} represents the Lie algebra of the orthogonal groups ($B_{\frac{n-1}{2}}$ with $\frac{n(n-1)}{2}$ parameters, or $D_{\frac{n}{2}}$ also with $\frac{n(n-1)}{2}$ parameters), namely the terms $a_i^\dagger a_j$, $i < j$, in the combinations $a_i^\dagger (L_{ab})_{ij}\, a_j$, where $(L_{ab})_{ij}$ are the matrix elements of the n-dimensional representation of the generator L_{ab}. It is interesting that the Lie algebras non-compact groups SO(p,q) and SU(p,q) can also be represented by the bilinear combinations (5.10), even for unitary representations. We first represent the maximal-compact subalgebras $SO(p) \otimes SO(q)$, or $SU(p) \times SU(q) \times U(1)$, by the G_{ij}-terms of (5.10) as outlined above. The generators of these subgroups act on the states

$$a_1^{\dagger\alpha_1} a_2^{\dagger\alpha_2} \ldots a_n^{\dagger\alpha_n} |0\rangle \,, \qquad (5.12)$$

where the powers α_i are real or complex numbers. The remaining so-called noncompact generators are formed by $F_{ij} = a_i a_j$ and $H_{ij} = a_i^\dagger a_j^\dagger$. Because these are made out of two-annihilation or two creation operators, they change the value of the Casimir operator of the compact-subgroup. An example of this method for conformal group can be found in my contribution later in these Proceedings. The method is in fact equivalent to the tensor method of building the higher-dimensional representations out of fundamental representation. The new feature for non-compact groups is the use of formal complex powers in Eq. (5.12).[17)-20)]

5.5 Infinite-dimensional Representations

The classification of the infinite-dimensional representations for the De Sitter and conformal groups is complete only for SO(4,1). For SO(3,2) and SO(4,2) there is not yet a complete list. This is due to the multiplicity theorem, that for SO(n,1) and SU(n,1) every unitary irreducible representation contains each representation of the maximal compact subgroup, SO(n) or SU(n), respectively, only once.[21)] Note that non-unitary representations need not be completely reducible. There are a class of non-unitary representations which are reducible but indecomposable.[22)]

O(4,1)

The continuous unitary representations of SO(4,1) in infinitesimal form were studied by Thomas,[23)] by Newton,[24)] and most rigorously and completely by Dixmier[25)] and others.[26)] These representations are constructed on a basis of the compact subgroup SO(4) which is isomorphic to SO(3) x SO(3) which makes the construction particularly simple. [In fact, all inequivalent continuous unitary representations of SO(n,1) and SU(n,1) have been given.[27)]] One can also relate, by analytic continuation, the representations of SO(4,1) to those of SO(5),[28),29)] as is known from the case of O(2,1)-O(3) analytic continuation.[17),30)] In physical applications, we often need the representations in global form, i.e. the matrix elements of finite group elements. For SO(4,1) these global forms have been given first by Takahashi[31),32)] using the theory of induced representations.[33)] The explicit form of a special class of discrete representations of SO(4,1) in terms of boson-creation and annihilation operators can be found in my contribution in this volume, as a special case of the representations of SO(4,2).

SO(3,2)

Special infinite-dimensional representations have been known for some time,[34),35),16)] but there is no complete list. We refer further to contributions in this volume by A. Böhm and L. Jaffe.

SO(4,2)

The literature on the representations of SO(4,2) is extensive,[18),37)-46)] but again there is no complete list of all unitary irreducible representations, unfortunately. The simplest representations are the so-called most degenerate representations; in the conformal group interpretation these correspond to the states of mass-zero spin j particles, in the dynamical group interepretation to the rest-frame states of composite particles with lowest spin j_o.

These representations have also very remarkable reduction properties with respect to the subgroup SO(4,1) and SO(3,2),[45] and also with respect to the Poincaré and Weyl subgroups.[47] (The Weyl group consists of the Poincaré group plus dilatations (11 parameters); it is also called the causality group, because according to a theorem of Zeeman,[48] it is the largest group of one-to-one mappings of the Minkowski space into itself that preserves the causal order of pairs of vectors.)

The infinitesimal method is used most often in the construction of the representations. Note that it is sufficient to determine the representations of the generators L_{12}, L_{23}, L_{34}, L_{45} and L_{56}; the others are determined by the commutation relations.

Finally, we should like also to mention some results on the most-degenerate representations of groups of the type SO(p,q) and SU(p,q).[27),49)-53]

III. Physical Applications

1. Conformally Invariant Theories and Broken Conformal Symmetry

Historial references to early physical interpretations and applications of the conformal group can be found in the review of Kastrup.[1] We adopt the interpretation that the conformal invariance expresses the change of unit from one frame to another which moreover depends on the space-time point. The physical laws are expected to be invariant under these transformations. The orthodox point of view, however, has been to look at the present Lorentz-invariance theories and see whether they are also conformally invariant. The radical point of view would be to rewrite the theory in such a way that it is conformally invariant. Of course, the concept of mass must also be modified.

In order to compare the physical phenomena at different space-time points a correspondence of units must be established. Thus the physical laws must be invariant under coordinate-dependent transformations of units.[54] Although there should be not much doubt about this point, the problem is to obtain experimentally verifiable and meaningful new consequences of this larger invariance principle. This step apparently has not yet been achieved. Mathematically, conformal invariant wave equations can be written in some cases. Dirac[1] has written such equations for the electromagnetic field (including the current density terms), and for spin-$\frac{1}{2}$ fields. It is interesting that he obtains an additional degeneracy of solutions which has not yet been interpreted physically.

Some properties of the conformal-invariant scattering amplitudes have been given by Castell[1] and Bali et al.[55]

In the orthodox interpretation of conformal invariance, on the other hand, we have first the (trivial) case of the invariance of free wave equations for massless particles of arbitrary spin.[56] The problem of conformal invariance for interacting fields is much more complicated,[57],[58] notably due to renormalization questions.[59]-[60] In this usual interpretation, the hope is that conformal invariance is valid at very high energies when the effect of mass terms are small.[61] The so-called "scaling" phenomenon in the inelastic electron-proton scattering[62] (the fact that the form factors are functions of a dimensionless quantity) has been interpreted in this way.[59]

2. Dynamical Groups

No definite connection is known at the present time between the use of De Sitter and conformal groups as dynamical groups or spectrum generating groups and their use as "space-time-scale" groups. A deeper connection might perhaps exist, because the dynamical groups have been interpreted as the "symmetry" group of the "system + interaction"[63] (for example, a H-atom plus the external electromagnetic field). Perhaps it is not the individual particles or systems but only systems together with the measuring devices that have conformally invariant Hamiltonians. At the moment, however, dynamical groups together with a current operator describe the properties of non-relativistic or relativistic composite quantum systems, such as mass spectrum, form factors, magnetic moments. The conformal group in the O(4,2)-interpretation has been found to describe the Dirac particle, the H-atom and a model for proton interpretable in terms of magnetic charges (see my contribution in these Proceedings and the references given there). The relation of the dynamical groups to other models of strong interaction such as the current algebra framework can be found in a recent review.[64]

3. Theories in Curved Spaces

The invariance requirements give us important information about the possible states of elementary particles and their possible interactions, although these latter are not uniquely determined by these requirements. The origin of these invariance considerations goes back to the isotropy and homogeneity of the space and the time, such as the rotational invariance or more generally the Lorentz invariance. These, according to Mach's principle, are determined by the mass distribution of matter in the universe. [According to the same principle, the inertial mass of a body, or the inertial forces in an accelerated frame, are determined by the distant matter of the universe.]

Thus, we are led to consider the shape of the universe in our considerations. The space-time is not flat according to Einstein's equations, and the simplest curved universes are the uniform universes both in space and time which are the Riemann spaces with constant curvature. These types of 4-dimensional spaces can be imbedded into a five-dimensional space, either of positive curvature (SO(4,1)), or of negative curvature (SO(3,2)) [i.e. spaces on which SO(4,1) and SO(3,2) act transitively]. Little is known about the theory of particles formulated in curved spaces, except the form of the free field equations, and questions like localizability and position operator.[4), 44), 65)] Both De Sitter groups can be contracted to the Poincaré group, so also the field equations to the usual Poincaré invariant equations, as the radius of the universe tends to infinity.

4. Other Applications

There are undoubtedly other applications of these larger noncompact groups. I mention one which is not really an invariance argument. Consider a scattering process involving one or more massless particles. The Hilbert space of one-particle states of the massless particle is also the carrier space of the conformal group, in other words we can perform conformal transformations (scale changes) on massless particles without enlarging the Hilbert space, or introducing new quantum numbers. The S-matrix for the process is an isotropic tensor operator under the Poincaré group. If we make the further requirement[66)] that the S-matrix be an isotropic tensor operator for conformal transformations (scale changes) on the massless particles only --because these transformations act on the same Hilbert space--we are led to specific dependence of the S-matrix on the momenta of massless particles. In particular, the vanishing of the amplitude for mass zero, spin zero particles, as $p_\mu \to 0$ follows from this requirement. (This result is equivalent to "pion gauge condition," or the so-called Adler's self consistency condition.[67)])

There are also some interesting applications of the inhomogeneous De Sitter groups, IO(4,1). This group arises as additional symmetry group of Lorentz-covariant wave equations.[68)] Also, if the mass term in the wave equations is interpreted as the fifth coordinate: $p_5 p^5 = m^2$, the Lorentz invariant wave equations are then formally invariant under IO(4,1).[69)]

References

1. See the reviews of early work in T. Fulton, F. Rohrlich, and L. Witten, Rev. Mod. Phys. 34, 462 (1962);
F. Gürsey, Nuovo Cimento 3, 98 (1956); P. A. M. Dirac, Ann. of Math. 37, 429 (1936) [Wave Equations in Conformal Space];
H. Kastrup, Ann. Phys. 7, 388 (1962);
L. Castell, Nucl. Phys. B4, 343 (1967) and these Proceedings.
M. Flato, J. Simon and D. Sternheimer (Collège de France), to be published.
2. O. Klein, Niels Bohr and the Development of Physics (Pergamon Press, London, 1955); L. Landau, ibid., p. 96; S. Deser, Rev. Mod. Phys. 29, 417 (1957).
3. See, for example, C. J. Isham, Abdus Salam and J. Strathdee, Trieste preprint IC/70/131.
P. Budini and G. Calucci, Nuovo Cimento (to appear).
4. T. O. Phillips and E. P. Wigner, in Group Theory and Its Applications, edited by E. M. Loebl (Academic Press, N. Y., 1968), pp. 631-676.
P. A. M. Dirac, Ann. of Math. 36, 657 (1935) [Electron Wave Equation in De Sitter Space].
P. A. M. Dirac, Max Planck Festschrift, 1958 [The Electron Wave Equation in Riemann Space].
F. Gürsey and T. D. Lee, Proc. Nat. Acad. Sci. 49, 179 (1963).
F. Gürsey, in Group Theoretical Concepts and Methods in Elementary Particle Physics (Gordon and Breach, N.Y., 1964).
5. See, for example, the Proceedings of the Conference on Lorentz Group, in Lectures in Theoretical Physics, Vol. VIIA (Univ. of Colorado Press, 1965).
6. For the classification of real non-compact groups see A. O. Barut and R. Raczka, Proc. Roy. Soc. (London) A287, 519 (1965).
7. E. P. Wigner, Group Theory (Academic Press, N. Y., 1959), App. 20; V. Bargmann, J. Math. Phys. 5, 862 (1964).
8. I. E. Segal, Duke Math. J. 18, 221 (1951).
E. Inönü and E. P. Wigner, Proc. Nat. Acad. Sci. (USA) 39, 510 (1953); 40, 119 (1954).
E. Inönü, in Group Theoretical Concepts and Methods in Elementary Particle Physics, F. Gürsey, ed. (Gordon and Breach, N. Y. 1964), p. 365.
E. J. Saletan, J. Math. Phys. 2, 1 (1961).
9. e. g. A. O. Barut, Journ. Math. Phys. 5, 1652 (1964).
10. G. Racah, Springer Tracts in Modern Physics, Vol. 37 (1965).
11. e. g. A. O. Barut, Electrodynamics and Classical Theory of Fields and Particles (Macmillan, 1964), Ch. I.

12. G. Boerner, Theory of Group Representations (Springer Verlag, 1955).
13. M. Gel'fand and M. L. Zeitlin, Dokl. Akad. Nauk, USSR, 71, 825 and 1017 (1950)
14. H. Weyl, Math. Z. 24, 377 (1926).
15. A. O. Barut, Phys. Rev. 135, B839 (1964); Phys. Rev. Letters 20, 893 (1968).
A. O. Barut, P. Cordero and G. C. Ghirardi, Phys. Rev. D1, 536 (1970).
16. G. Racah, in Group Theoretical Concepts and Methods in Elementary Particle Physics, ed. F. Gürsey (Gordon and Breach, 1964).
W. R. Sharp, Proc. Camb. Phil. Soc. 68, 571 (1970).
17. A. O. Barut and C. Fronsdal, Proc. Roy. Soc. (London) A287, 532 (1965).
18. C. Fronsdal, ibid. A288, 98 and 113 (1965).
19. C. Fronsdal, in High Energy Physics and Elementary Particles (Intern. Atomic Energy Agency, Vienna 1965), p. 585.
20. A. O. Barut, "Unified Algebraic Construction of Representations of Compact and Non-compact Lie Algebras and Lie Groups," in Lectures in Theoretical Physics, Vol. IX (Gordon and Breach, 1967), p. 125.
21. R. Godement, Trans. Am. Math. Soc. 73, 496 (1952).
See also the review E. M. Stein, in High Energy Physics and Elementary Particles (IAEA, Vienna 1965), p. 563.
22. I. M. Gel'fand, V. A. Ponomarev, Russian Math. Survey 23, 2 (1968).
23. L. H. Thomas, Ann. of Math. 42, 113 (1941).
24. T. D. Newton, Ann. of Math. 51, 730 (1950).
25. J. Dixmier, Bull. Soc. Math. France 89, 9 (1961).
26. A. Kihlberg and S. Ström, Ark. Fysik 31, 491 (1966).
27. U. Ottoson, Comm. Math. Phys. 8, 228 (1968) and 10, 114 (1968).
28. J. G. Kuriyan, N. Mukunda and E. C. G. Sudarshan, Comm. Math. Phys. 8, 204 (1968).
29. W. J. Holman, III, Journ. Math. Phys. 10, 1889 (1969).
30. A. O. Barut and E. C. Phillips, Comm. Math. Phys. 8, 52 (1968).
31. R. Takahashi, Bull. Soc. Math. de France 91, 289 (1963).
32. See also S. Ström, these Proceedings.
33. G. Mackey, Induced Representations of Groups and Quantum Mechanics (W. A. Benjamin, NY, 1968).
34. J. B. Ehrman, Proc. Camb. Phil. Soc. 53, 290 (1956) and Ph. D. Thesis, Princeton Univ. (1954).
35. P. A. M. Dirac, Journ. Math. Phys. 4, 901 (1963).

36. N. T. Evans, J. Math. Phys. 8, 170 (1967).
37. Y. Murai, Progr. Theor. Phys. 9, 147 (1953) [Representations in Cartan's Basis].
38. M. I. Graev, Dokl. Akad. Nauk, USSR, 98, 517 (1954).
39. A. Esteve and P. G. Sona, Nuovo Cimento 32, 473 (1964).
40. I. M. Gel'fand and M. I. Graev, Izv. Akad. Nauk, 29, 1329 (1965).
41. A. Kihlberg, V. F. Müller and F. Halbwachs, Comm. Math. Phys. 3, 194 (1966).
42. J. Olszewski, Acta Phys. Polonica 30, 105 (1966).
43. T. Yao, Journ. Math. Phys. 8, 1931 (1967); 9, 1615 (1968).
44. A. Maduemezia, Nuovo Cimento 51A, 320 and 932 (1967).
45. A. O. Barut and A. Böhm, Journ. Math. Phys. 11, 2938 (1970).
46. T. Yao, these Proceedings.
47. J. Mickelson and J. Niederle, Prag preprint 1970.
48. E. C. Zeeman, Journ. Math. Phys. 5, 490 (1964).
49. R. Raczka, N. Limic and J. Niederle, Journ. Math. Phys. 7, 1861 and 2026 (1966).
50. J. Fisher, J. Niederle and R. Raczka, J. Math. Phys. 7, 816 (1966).
51. R. Raczka and J. Fisher, Comm. Math. Phys. 3, 233 (1966).
52. A. Kihlberg, Ark. für Fysik 32, 241 (1966)
53. N. Limic and J. Niederle, Ann. Inst. Henri Poincaré, IX, 327 (1968).
54. For applications of this idea see M. Fierz, Helv. Phys. Acta 29, 128 (1956); R. H. Dicke, Phys. Rev. 125, 2163 (1962).
55. N. F. Bali, D. D. Coon and A. Katz, Journ. Math. Phys. 10, 1939 (1969).
See also C. Garrod, Phys. Rev. D1, 1524 (1970).
56. H. Bateman, Proc. London Math. Soc. 8, 288, 469 (1910).
E. Cunningham, ibid. 10, 77 (1910).
A. J. MacLennan, Nuovo Cimento 3, 1360 (1956); 5, 640 (1957).
J. S. Lomont, ibid. 22, 673 (1961).
L. Gross, Journ. Math. Phys. 5, 687 (1964).
B. Kursunoğlu, J. Math. Phys. 8, 1694 (1967).
57. J. Wess, Nuovo Cimento 18, 1086 (1960).
58. G. Mack, Nucl. Phys. B5, 499 (1968).
G. Mack and Abdus Salam, Ann. Phys. (NY) 53, 174 (1969).
59. See the review P. Carruthers, Broken Scale Invariance in Particle Physics, Nuclear Physics. (to appear).
60. R. Jackiw, these Proceedings and references therein.
61. H. A. Kastrup, Phys. Rev. 142, 1060 (1966); 143, 1041 (1966); 150, 1189 (1966); Nuclear Phys. 58, 561 (1964).

62. B. J. Björken, Phys. Rev. 179, 1547 (1969).
63. A. O. Barut, "Formulation of Quantum Dynamics in Terms of Generalized Symmetries, I." Lecture Notes IC/68/104.
64. A. O. Barut, Review of Hadron Symmetries: Multiplets, Supermultiplets and Infinite-multiplets (Rapporteur's talk given at the XVth International Conference on High Energy Physics, Kiev 1970).
65. See G. Börner, these Proceedings.
66. A. O. Barut, Lett. Nuovo Cimento 2, 94 (1969).
67. S. L. Adler, Phys. Rev. 137, B1022 (1965); 139, B1638 (1965).
68. V. I. Fushchich, Kiev preprint ITF-70-32 (1970).
69. M. M. Bakri, Nuovo Cimento 51A, 864 (1967); J. Math. Phys. 10, 298 (1969); 11, 2015, 2027 (1970).
70. J. A. de Vos and J. Hilgevoord, Nucl. Phys. B1, 494 (1967).
71. M. Humi, J. Math. Phys. 11, 2222 (1970).

PART I: MATHEMATICAL RESULTS

Section A: Groups

ONE-PARAMETER SUBGROUPS OF THE CONFORMAL GROUP OF SPACETIME AND IN GENERAL OF UNITARY GROUPS WITH AN INDEFINITE METRIC†

Johan G. Belinfante
Mathematics Department and Physics Department
Carnegie-Mellon University
Pittsburgh, Pennsylvania

My talk today will be about some preliminary work on the subgroup structure of the conformal group SU(2,2) and related groups. This work was carried out in collaboration with P. Winternitz, who has in his own talk already explained that part of the motivation for this type of study comes from certain applications of the theory of harmonic analysis on noncompact groups.[1] In this, as well as in many other applications of group theory to physics, of course it is useful to know something about the lattice of subgroups of the groups of interest. Here we shall only study the conjugacy classes of one-parameter subgroups of the groups U(p,q) and SU(p,q) as the first step toward finding all the connected analytic subgroups.

The approach considered here may be described as an application of geometric algebra.[2] Our study leads to a complete classification of hermitean operators with respect to an indefinite metric, which may well have some interesting applications to quantum field theory with an indefinite metric. However we must point out that many of the theorems obtained apply only to the case of finite-dimensional vector spaces, whereas the Fock spaces considered in quantum field theory are usually infinite dimensional.

In the special case of the conformal group, we have the local isomorphism $SU(2,2) \approx SO(4,2;R)$. Thus another approach to this case would be through a study of orthogonal groups with indefinite metric. Most of our methods apply equally well to orthogonal groups as to unitary groups. The conjugacy classes of subgroups of orthogonal groups with indefinite metric have been studied previously for special cases, the best known case being the study of the Lorentz group by E. P. Wigner in 1939. His method can be generalized to the case of

†Presented at the Symposium on De Sitter and Conformal Groups, University of Colorado, Summer 1970.

$SO(n,1; R)$, but since it depends on a certain triangle inequality, it does not generalize to groups with index greater than one.[3] The simplest case of an index two group, $SO(2,2; R)$, was studied in a series of papers by H. Zassenhaus. Finally we mention that the case of $SO(3,3; R)$ comes up in connection with the Petrov classification of Einstein spaces.

Since most of the remainder of my talk will lean heavily on general results of vector space theory and the general theory of linear operators and hermitean forms, it may be useful to quickly review some of this material and to agree on the terminology.[4]

When dealing with indefinite metrics, it is important to make a careful distinction between ideas related to linear independence and ideas related to orthogonality. For the moment we shall consider vector spaces in general without any hermitean structure. Subspaces $S_1, \ldots, S_n$ of a vector space V are said to be <u>independent</u> iff for all $\psi_1 \in S_1, \ldots, \psi_n \in S_n$, the only solution of the equation $\psi_1 + \ldots + \psi_n = 0$ is the trivial solution $\psi_1 = \ldots = \psi_n = 0$. In the case $n = 2$ this is equivalent to the condition $S_1 \cap S_2 = 0$, and in general, this is equivalent to a collection of $n - 1$ conditions, a typical one being $(S_1 + \ldots + S_k) \cap S_{k+1} = 0$. A sum of independent subspaces is usually called a <u>direct sum</u>, denoted by $S_1 \oplus \ldots \oplus S_n$.

We use the language of invariant subspaces to describe the structure of a linear operator as exhibited in its Jordan canonical form, in accord with the common view that this theory may be regarded as an application of the theory of torsion modules over a principal ideal domain. We recall that a subspace S of a vector space V is an <u>invariant subspace</u> under the action of a linear operator γ on V iff $\gamma S \subset S$, that is, if $\gamma\psi \in S$ for all $\psi \in S$. Two invariant subspaces S_1 and S_2 are <u>isomorphic</u> iff there exists a one-one and onto linear transformation $\alpha: S_1 \to S_2$ which commutes with γ in the sense that $\alpha \circ \gamma_1 = \gamma_2 \circ \alpha$, where γ_1 and γ_2 are the obvious restrictions of γ to S_1 and S_2 respectively. An <u>indecomposable invariant subspace</u> is defined to be one which cannot be written as the direct sum of two nonzero invariant subspaces. If V is finite-dimensional, then V can be written as the direct sum of a finite set of indecomposable invariant subspaces, the decomposition being unique only up to isomorphism.

The essence of the Jordan canonical form theorem is that we can characterize the indecomposable invariant subspaces as those which are both cyclic and primary. We shall hereafter for simplicity restrict our attention to vector spaces over the complex number field. A <u>generalized eigenvector</u> $\psi \in V$ of a linear operator γ on V is any nonzero vector which satisfies $(\gamma - c\,1)^p \psi = 0$ for some complex number c and some positive integer p. For $p = 1$, this reduces to the ordinary definition of an eigenvector. For any complex number c we may define the primary component $V_\gamma^c \subset V$ to be the subspace consisting of the zero vector and all generalized eigenvectors of γ corresponding to c if any. It is then true that c is an eigenvalue in the ordinary sense iff the corresponding primary component is nonzero. An invariant subspace S is said to be a primary invariant subspace iff all its members are generalized eigenvectors corresponding to a single common complex number c, that is, iff $S \subset V_\gamma^c$ for some value of c. In particular, of course, the primary components are themselves primary invariant subspaces, and V is the direct sum of all of them. An invariant subspace S is said to be a cyclic invariant subspace iff there exists a vector φ in S, called a cyclic vector, such that every vector ψ in S can be written as a polynomial in γ acting on the cyclic vector: $\psi = p(\gamma)\,\varphi$.

We shall be applying these results to hermitean linear operators. In the case of a positive definite metric, of course, the ordinary eigenvector theory suffices, and we are all familiar with this theory from its application to elementary quantum mechanics. What we shall find is that by using the more general concepts, we can set up a parallel theory even in the case of an indefinite metric. Actually, there is one case of an indefinite hermitean metric with which we are all familiar already, namely the Dirac spinor space, which we may characterize as a four-dimensional complex vector space C^4 equipped with the metric (++--). For convenience we shall use the same notation that is used in the Dirac theory, namely $\bar{\varphi}\psi$, for hermitean forms in general. The bar notation will be henceforth reserved for this use, and for clarity we shall use an asterisk to denote the complex conjugate c^* of a scalar c.

To be precise, we define an hermitean form $\bar{\varphi}\psi$ on a complex vector space V to be a complex-valued function of two vector variables, linear in ψ and antilinear in φ, and satisfying the condition

$$(\bar{\varphi}\psi)^* = \bar{\psi}\,\varphi.$$

We shall use the term hermitean space to mean a complex vector space equipped with an hermitean form. If ψ is a vector in an hermitean space, then its norm $\bar{\psi}\psi$ may be positive, negative or zero. It is customary to call a vector with zero norm an isotropic vector. Two vectors φ and ψ are said to be orthogonal iff $\bar{\varphi}\psi = 0$.

The subspaces of a hermitean space can be characterized by their metric signature. Two subspaces S_1 and S_2 of a hermitean space are said to be orthogonal subspaces iff every vector in S_1 is orthogonal to every vector in S_2. Note that orthogonal subspaces need not be independent, and independent subspaces need not be orthogonal. A hermitean space V is said to be the orthogonal direct sum of the subspaces $S_1, \ldots, S_n$, written $V = S_1 \perp \ldots \perp S_n$, iff V is the direct sum of pairwise orthogonal subspaces. Thus the subspaces in an orthogonal direct sum decomposition are required to be both independent and mutually orthogonal. Every subspace of a hermitean space can be written as an orthogonal direct sum of lines: $S = L_1 \perp \ldots \perp L_n$. There are three possible types of lines, because the norms of the nonzero vectors in a one-dimensional subspace all have the same sign, namely either +, -, or 0. Each subspace then has a unique metric signature, which indicates how many lines of each type there are in any decomposition of the subspace as an orthogonal direct sum of lines. Two subspaces S_1 and S_2 of a hermitean space are said to be isometric iff there exists a one-one onto linear mapping $\alpha: S_1 \to S_2$ which preserves the metric in the sense that $(\overline{\alpha\varphi})(\alpha\psi) = \overline{\varphi}\psi$ for all φ, ψ in S_1. Two subspaces are isometric iff they have the same metric signature.

The orthogonal complement $S^\perp$ of a subspace S of a hermitean space V is the subspace consisting of all vectors in V which are orthogonal to every vector in S. A subspace S of a hermitean space is said to be nonsingular iff its radical $S \cap S^\perp$ is zero. Nonsingular subspaces may be characterized as those whose metric signature contains only +'s and -'s, but no 0's. For example, the Dirac spinor space $C^4(++--)$ is a nonsingular hermitean space.

There are several important standard results about nonsingular spaces. If S is a nonsingular subspace of a hermitean space V, then $V = S \perp S^\perp$. If V is nonsingular and if $V = S_1 \perp \ldots \perp S_n$, then $S_1, \ldots, S_n$ are also nonsingular. Finally, if both S and V are nonsingular, then we have $(S^\perp)^\perp = S$.

A subspace S is totally isotropic in the sense that every vector in S is isotropic iff $S \subset S^\perp$. Totally isotropic subspaces may be characterized as those whose metric signatures consist solely of 0's. All maximal totally isotropic subspaces of a given hermitean space V have the same dimension, called the index of V. The index may be computed from the metric signature as the sum of the number of 0's plus the number of (+-) pairs. In determining the possible metric signatures of the subspaces of a given hermitean space, one must observe among other things the requirement that the index of a subspace cannot exceed that of the containing space. Thus, for example, there are fourteen possible types of subspaces of the Dirac spinor space $C^4(++--)$. These are listed below.

dimension ↓ \ index →	0	1	2
0	()		
1	(+), (-)	(0)	
2	(++), (--)	(+-), (+0), (-0)	(00)
3		(++-), (+--)	(+-0)
4			(++--)

The general structure of a hermitean space may also be described in the following way. Any hermitean space may be written as an orthogonal direct sum of its radical, which is totally isotropic, and a nonsingular subspace. Here the radical is of course uniquely determined, but the nonsingular subspace is only determined up to isometry. A subspace is said to be anisotropic iff it does not contain any nonzero isotropic vectors. Anisotropic subspaces may be characterized as those whose metric signature consists solely of +'s or solely of -'s, but not both. A hyperbolic plane is a plane with metric signature (+-). Hyperbolic planes may be characterized also as nonsingular planes which contain at least one nonzero isotropic vector. A hyperbolic space, defined as any orthogonal direct sum of hyperbolic planes, is clearly a space with a balanced metric signature consisting of an equal number of +'s and -'s. A nonsingular hermitean space may be written as the orthogonal direct sum of a hyperbolic subspace and a subspace which is anisotropic, this type of decomposition again being unique only up to isometry.

Unitary and special unitary groups with indefinite metric arise naturally in the study of the geometry of hermitean spaces.[5] An invertible linear operator α on a hermitean space V is said to be a unitary operator iff $(\overline{\alpha\psi})(\alpha\varphi) = \overline{\psi}\varphi$ for all vectors ψ, φ in V. The set of all unitary linear operators on a hermitean space V forms a group called the unitary group on V. The subgroup consisting of all special unitary operators, that is, of unitary operators with determinant one, is a normal subgroup of the unitary group, called the special unitary group on the hermitean space V. In particular, we may then characterize the conformal group SU(2,2) as the special unitary group of Dirac spinor space.[6]

It is often more convenient to deal with the Lie algebras of these groups rather than with the groups themselves. The real Lie algebra of the unitary group on a hermitean space V may be identified with the set of all antihermitean linear operators on V, that is, those linear operators β which satisfy $\overline{(\beta\psi)}\varphi = -\overline{\psi}(\beta\varphi)$ for all ψ, φ in V. Any real linear combination of antihermitean operators is again antihermitean, and if β_1 and β_2 are antihermitean, then their commutator $[\beta_1, \beta_2] = \beta_1 \circ \beta_2 - \beta_2 \circ \beta_1$ is also antihermitean. The Lie algebra of the special unitary group on V may be identified with an ideal in this algebra, consisting of all traceless antihermitean linear operators on V. Instead of working with antihermitean operators, however, we prefer to work instead with hermitean operators, that is, linear operators γ which satisfy the condition $\overline{(\gamma\psi)}\varphi = \overline{\psi}(\gamma\varphi)$ for all ψ, φ. We can obtain hermitean operators from antihermitean operators by multiplying by $i = \sqrt{-1}$. If γ is a nonzero hermitean operator, then the set consisting of all the operators $e^{it\gamma}$, where t is a real number, is a subgroup of the unitary group, called the one-parameter subgroup generated by γ. If two hermitean operators differ by a nonzero real factor, then of course they generate the same one-parameter subgroup. Thus, one-parameter subgroups of a group correspond to one-dimensional subalgebras of the corresponding Lie algebra. More generally, exponentiation of various subalgebras of the Lie algebra yield various connected analytic subgroups of the unitary group.

Two subgroups H_1 and H_2 of a group G are said to be conjugate subgroups iff there exists an element α in the group G such that $H_2 = \alpha H_1 \alpha^{-1}$. If two hermitean operators γ_1 and γ_2 are conjugate in the sense that there exists a unitary operator α such that $\gamma_2 = \alpha \gamma_1 \alpha^{-1}$, then they generate conjugate one-parameter subgroups of the unitary group. If α here can be taken to be a special unitary operator, then we shall speak of special conjugate hermitean operators. To obtain the conjugacy and special conjugacy classes of hermitean operators, we study their invariant subspaces. We are thus led to study also the notion of conjugate subspaces of a hermitean space.

Two subspaces S_1 and S_2 of a hermitean space V are said to be conjugate subspaces iff there exists a unitary operator α on V such that $S_2 = \alpha S_1$, and if this operator can be taken with determinant one, then we speak of special conjugate subspaces. The Witt theorem says that if two subspaces of a nonsingular hermitean space are isometric, then any isometry between them can be extended to a unitary operator, so that the subspaces are also conjugate.[7)] Thus for subspaces of a nonsingular hermitean space, the question of conjugacy can be settled simply by examining the metric signature of the subspaces. An isometry between subspaces of a nonsingular hermitean space can even be extended to a special unitary operator, so that the subspaces are special conjugate, in the case $\dim S + \dim (S \cap S^{\perp}) < \dim V$, which is certainly the case for nonsingular subspaces. The proof of the Witt theorem for general subspaces can be reduced to the simple special case of nonsingular subspaces by using a result known as the hyperbolic enlargement theorem. The hyperbolic enlargement theorem says that if S is a totally isotropic subspace of a nonsingular hermitean space V, then there exists another totally isotropic subspace T with the same dimension as S such that $S \cap T = 0$, and the direct sum $S \oplus T$ is a hyperbolic subspace of V.

We are now ready to study the invariant subspaces of a hermitean operator. If S is an invariant subspace of a hermitean linear operator γ acting on a hermitean space V, then the orthogonal complement $S^{\perp}$ is also an invariant subspace. Sums and intersections of invariant subspaces again yield invariant subspaces.

We can say quite a bit about the primary components of a hermitean operator on a nonsingular hermitean space. If γ is a hermitean linear operator on a nonsingular hermitean space V, then the primary component V_γ^c is nonsingular when c is real. Two primary components $V_\gamma^{c_1}$ and $V_\gamma^{c_2}$ of a hermitean operator are orthogonal to each other if c_1 and c_2 are not complex conjugates of each other. In particular, the primary component V_γ^c is orthogonal to itself, and hence totally isotropic, when c is not real. On the other hand, if c_1 and c_2 are complex conjugates of each other, and if in addition V is nonsingular, then the primary components $V_\gamma^{c_1}$ and $V_\gamma^{c_2}$ have the same dimension. If c is not real and if V is nonsingular, then the direct sum $V_\gamma^c \oplus V_\gamma^{c*}$ is nonsingular, and being the direct sum of two equidimensional totally isotropic subspaces, must be a hyperbolic subspace. All of these assertions follow as corollaries of the following theorem, which generalizes the familiar arguments about orthogonality of eigenvectors of a hermitean operator to the case of an indefinite metric.

Theorem.

If γ is a hermitean linear operator on a finite-dimensional hermitean space V, and if c_1 and c_2 are two complex numbers which are not complex conjugates of each other, then the corresponding primary components $V_\gamma^{c_1}$ and $V_\gamma^{c_2}$ are orthogonal to each other.

Proof.

The kernels of the various powers $(\gamma - c_1\ 1)^p$ for increasing p form an ascending chain of subspaces which eventually terminates so that for sufficiently large p we obtain the primary component $V_\gamma^{c_1}$ as one of these kernels. We shall show by induction on p that the kernel of $(\gamma - c_1\ 1)^p$ is orthogonal to the other primary component $V_\gamma^{c_2}$. For p = 1, the argument is simple because if $(\gamma - c_1\ 1)\psi = 0$ and $(\gamma - c_2\ 1)^q \varphi = 0$, then $0 = \bar{\psi}(\gamma - c_2\ 1)^q \varphi = (c_1{}^* - c_2)^q \bar{\psi}\varphi$, implying that $\bar{\psi}\varphi = 0$. Suppose now that the kernel of $(\gamma - c_1\ 1)^p$ is orthogonal to $V_\gamma^{c_2}$. If ψ belongs to the kernel of $(\gamma - c_1\ 1)^{p+1}$, then $(\gamma - c_1\ 1)\psi$ belongs to the kernel of $(\gamma - c_1\ 1)^p$ and is therefore orthogonal to $V_\gamma^{c_2}$. For any vector φ in $V_\gamma^{c_2}$ there exists a smallest integer q such that $\bar{\psi}(\gamma - c_2\ 1)^q\varphi = 0$. If $q > 0$, then we have $\overline{(\gamma - c_1\ 1)\psi}\,(\gamma - c_2\ 1)^{q-1}\varphi = 0$, and it follows then that $(c_1{*} - c_2)\bar{\psi}(\gamma - c_2\ 1)^{q-1}\varphi = 0$, so that $\bar{\psi}(\gamma - c_2\ 1)^{q-1}\varphi = 0$, in contradiction with the assumed minimal property of q. Hence $q = 0$ and $\bar{\psi}\varphi = 0$, showing that the kernel of $(\gamma - c_1\ 1)^{p+1}$ is orthogonal to the primary component $V_\gamma^{c_2}$ and thereby completing the inductive argument. Q.E.D.

We now introduce the important concept of an elementary invariant subspace, which will bear the same relation to orthogonal direct sums that the concept of an indecomposable invariant subspace bears to ordinary direct sums. An elementary invariant subspace of an hermitean space is an invariant subspace which cannot be decomposed as an orthogonal direct sum of two nonzero invariant subspaces. If V is finite-dimensional, then it can be written as the orthogonal direct sum of a finite set of elementary invariant subspaces, $V = S_1 \perp \ldots \perp S_n$.

This raises the problem of studying elementary hermitean operators, that is, hermitean operators on a nonsingular space V such that the space V is itself elementary. Since we can alter the trace of a hermitean operator by adding a suitable multiple of the identity operator, it is sufficient to study the structure of traceless elementary hermitean operators. If γ is a traceless elementary hermitean operator on a finite-dimensional nonsingular hermitean space V, then either $V = V_\gamma^0$ or else $V = V_\gamma^{is} \oplus V_\gamma^{-is}$ for some real $s \neq 0$. Hence such operators are either nilpotent, that is, $\gamma^n = 0$ for some integer n, or else they satisfy the equation $(\gamma^2 + s^2)^p = 0$ for some integer p. At this point it is appropriate to state some relevant theorems about nilpotent hermitean operators. For the moment we may drop the assumption that they be elementary.

Lemma.

If a hermitean operator γ on a hermitean space V is nilpotent with $\gamma^m = 0$, then the cyclic invariant subspace S generated by any vector ψ in V is a nonsingular m-dimensional subspace iff $\bar{\psi}\gamma^{m-1}\psi \neq 0$.

Proof.

Since S is spanned by $\psi, \gamma\psi, \ldots, \gamma^{m-1}\psi$, any vector φ in the radical $S \cap S^\perp$ can be written as a linear combination $\varphi = (c_0 + c_1\gamma + \ldots + c_{m-1}\gamma^{m-1})\psi$, and we have $\bar{\psi}\gamma^k\varphi = 0$ for all k. Since $\gamma^m = 0$, we have $0 = \bar{\psi}\gamma^{m-1}\varphi = c_0\bar{\psi}\gamma^{m-1}\psi$, and if $\bar{\psi}\gamma^{m-1}\psi \neq 0$, then we have $c_0 = 0$. By a similar argument we can also show that $c_1 = \ldots = c_{m-1} = 0$, so that $\varphi = 0$ and S is nonsingular. Moreover, this same argument shows that the vectors $\psi, \gamma\psi, \ldots, \gamma^{m-1}\psi$ are linearly independent, and hence S has dimension m.

Conversely, if $\bar{\psi}\gamma^{m-1}\psi = 0$, then any vector φ in S can be written as some polynomial in γ acting on ψ and hence satisfies $\bar{\psi}\gamma^{m-1}\varphi = c_0\bar{\psi}\gamma^{m-1}\psi = 0$, and hence it follows that $\gamma^{m-1}\psi$ belongs to the radical $S \cap S^{\perp}$. If S were nonsingular, this would imply that $\gamma^{m-1}\psi = 0$, and hence S would have a dimension less than m. Q.E.D.

The above lemma allows us to prove the following theorem, which has as a corollary the result that elementary nilpotent hermitean operators are indecomposable.

Theorem.

If γ is a nilpotent hermitean operator on a finite-dimensional nonsingular hermitean space V, then V is the orthogonal direct sum of a finite set of nonsingular cyclic invariant subspaces.

Proof.

Since γ is nilpotent, there exists a smallest integer m such that $\gamma^m = 0$. If $m = 1$, then $\gamma = 0$, and every subspace of V is invariant, and hence any orthogonal direct sum decomposition of V into nonsingular lines does the trick. If $m > 1$, then by the assumed minimal property of m, we have $\gamma^{m-1} \neq 0$. If $\bar{\psi}\gamma^{m-1}\psi = 0$ for all ψ in V, then by polarization, we obtain $\bar{\psi}\gamma^{m-1}\varphi = 0$ for all ψ, φ in V, and since V is nonsingular, this would imply that $\gamma^{m-1} = 0$, a contradiction. Hence there is a vector ψ in V such that $\bar{\psi}\gamma^{m-1}\psi \neq 0$, and by the lemma it follows that the cyclic invariant subspace S generated by ψ is a nonzero nonsingular subspace. Then we may write $V = S \perp S^{\perp}$, and the argument may be repeated with $S^{\perp}$ replacing V. This process must finally stop somewhere because V is finite-dimensional. Q.E.D.

We may prove a similar type of theorem for the case of a hermitean operator γ when $\gamma^2 + s^2$ is nilpotent.

Theorem.

If γ is a hermitean operator on a finite-dimensional nonsingular hermitean space, and if $\gamma^2 + s^2$ is nilpotent for some real $s \neq 0$, then there exist nonzero totally isotropic indecomposable subspaces $S_{\pm}$ such that $S_+ \oplus S_-$ is a nonsingular invariant subspace.

Proof.

Since the nonsingular space V is the direct sum of the totally isotropic primary components V_γ^{is} and V_γ^{-is}, it follows that for every nonzero vector in V_γ^{is}, there exists another vector in V_γ^{-is} such that these two vectors are not orthogonal. If there is a vector φ_+ in V_γ^{is} such that $(\gamma - is)^m \varphi_+ \neq 0$, for some integer m, then there is a vector φ_- in V_γ^{-is} such that $\bar{\varphi}_-(\gamma - is)^m \varphi_+ \neq 0$, and hence $(\gamma + is)^m \varphi_- \neq 0$. Now let p be the smallest integer such that $(\gamma^2 + s^2)^p = 0$. If $(\gamma - is)^{p-1}$ is zero on V_γ^{is}, then $(\gamma + is)^{p-1}$ is zero on V_γ^{-is}, and hence $(\gamma^2 + s^2)^{p-1} = 0$, a contradiction. Hence there must exist vectors $\varphi_\pm$ in $V_\gamma^{\pm is}$ such that $\bar{\varphi}_-(\gamma - is)^{p-1}\varphi_+ \neq 0$. The cyclic subspaces $S_\pm$ generated by $\varphi_\pm$ are totally isotropic and primary since $S_\pm \subset V_\gamma^{\pm is}$, and are hence indecomposable. Finally one may verify that the radical of the direct sum $S_+ \oplus S_-$ is zero by using the fact that $\varphi_\pm$ satisfy $\bar{\varphi}_-(\gamma - is)^{p-1}\varphi_+ \neq 0$. Q.E.D.

These results show that there is a rather simple relation between elementary invariant subspaces and indecomposable ones. An elementary invariant subspace of a hermitean operator on a nonsingular hermitean space is either a nonsingular indecomposable invariant subspace, or else is the ordinary direct sum of a pair of totally isotropic invariant subspaces which are indecomposable and have equal dimensions.

We next study the possible metric signatures for elementary invariant subspaces. A nonsingular hermitean space is said to have maximal index iff the number of +'s and -'s in its metric signature are either the same, or else differ only by one. An even-dimensional maximal index nonsingular space is just a hyperbolic space.

Theorem.

Elementary invariant subspaces of a hermitean operator on a nonsingular hermitean space are maximal index nonsingular subspaces.

Proof.

If γ is an elementary hermitean operator on a nonsingular space V, then V is either itself indecomposable, or is the direct sum of two totally isotropic indecomposable subspaces. In the latter case V is hyperbolic, while in the former case we can subtract a real multiple of the identity from γ to obtain a nilpotent operator. If γ is a nilpotent hermitean operator, and if ψ is a cyclic vector for V, then every other cyclic vector is of the form $p(\gamma)\psi$, where $p(\gamma)$ is a polynomial in γ whose constant term is nonzero. By explicitly constructing a suitable polynomial, it is possible to show that there exists a cyclic vector ψ in V such that $\bar{\psi}\gamma^k\psi = 0$ for all k except for $k = n - 1$, where $n = \dim V$. We could multiply ψ by a suitable factor to make $\bar{\psi}\gamma^{n-1}\psi = \pm 1$. If $m = [n/2]$ is the largest integer not exceeding $n/2$, then the vectors $\psi, \gamma\psi, \ldots, \gamma^{m-1}\psi$ span an m-dimensional totally isotropic subspace of V, and it follows that V has maximal index. Q.E.D.

These results have a direct bearing on the problem of finding conjugate classes of hermitean operators. Let $V = S_1 \perp \ldots \perp S_n$ an elementary decomposition of a nonsingular space V with respect to a hermitean operator γ. Suppose that $V = T_1 \perp \ldots \perp T_n$ is any orthogonal direct sum decomposition of V into maximal index subspaces such that S_i is isometric with T_i for each $i = 1, \ldots, n$. Then by the Witt theorem, there exists a special unitary operator α on V such that $T_i = \alpha S_i$ and $V = T_1 \perp \ldots \perp T_n$ is an elementary decomposition of the hermitean operator $\alpha\gamma\alpha^{-1}$. The classification of the special conjugacy classes of hermitean operators thus reduces to two problems. The first problem is to find the possible ways of decomposing a given nonsingular hermitean space as an orthogonal direct sum of maximal index nonsingular subspaces. The second problem is to find the special conjugacy classes of elementary hermitean operators on a given maximal index nonsingular space.

The maximal index decompositions of a nonsingular hermitean space are readily found in each case by inspection. For example, there are six such breakups for the Dirac spinor space $C^4(++--)$. These six possible maximal index decompositions of Dirac spinor space are the following:

Case 1. (++--)

2. (++-)(-)

3. (+)(+--)

4. (+-)(+-)

5. (+-)(+)(-)

6. (+)(+)(-)(-)

Our approach to the second problem was based on the expectation that special conjugacy for hermitean operators would reduce pretty much to the determination of the metric signatures and Jordan canonical forms arising in their elementary subspace decompositions. By the use of explicit canonical forms we find that it is almost but not quite true that hermitean operators are special conjugate iff their elementary invariant subspaces are isometric and isomorphic.

To obtain the canonical forms we make use of the theorems stated earlier. For the case of a nilpotent elementary hermitean operator γ, we obtain a canonical form by using the vectors ψ, $\gamma\psi$, ..., $\gamma^{n-1}\psi$ as a basis, where ψ is the cyclic vector found in the proof of the last theorem, normalized so that $\bar{\psi}\gamma^{n-1}\psi = \epsilon$, where $\epsilon = \pm 1$. In this basis, the matrix of the hermitean operator γ is fairly simple, but the hermitean metric form is not diagonal. It is therefore convenient to introduce a slightly different canonical form by using another basis, chosen to make the metric diagonal so that its signature can be easily read off. The final canonical forms obtained are slightly different for the cases of even and odd dimensional spaces, so that in all we obtain four canonical forms for nilpotent elementary hermitean operators. By restoring a real multiple of the identity, we thus also get four cases for any elementary hermitean operator with a single real eigenvalue. In the odd-dimensional case, the quantity ϵ just determines whether the metric has one more + or - sign. In the even-dimensional case, however, the two cases with $\epsilon = +1$ and $\epsilon = -1$ represent examples of hermitean operators whose elementary invariant subspaces are both isometric and isomorphic, but still not special conjugate.

A similar procedure is used to obtain canonical forms for elementary hermitean operators with a pair of complex conjugate eigenvalues, that is, the case $V = V_\gamma^c \oplus V_\gamma^{c^*}$. In this case we can find a pair of vectors $\varphi_\pm$ in V such that $\bar{\varphi}_-(\gamma - c\,1)^k \varphi_+ = 0$ unless $k = p-1$ where dim $V = 2p$. We can normalize these two vectors so that $\bar{\varphi}_-(\gamma - c\,1)^{p-1} \varphi_+ = 1$, so that no ϵ is necessary here, and we get just one more canonical form for even-dimensional maximal index spaces.

We thus get a total of five cases in all for canonical forms.

Case	Metric	Eigenvalues	Other Parameters
I	$C^{2p+1}(p+1, p)$	r	$\epsilon = +1$
II	$C^{2p+1}(p, p+1)$	r	$\epsilon = -1$
III	$C^{2p}(p, p)$	r	$\epsilon = +1$
IV	$C^{2p}(p, p)$	r	$\epsilon = -1$
V	$C^{2p}(p, p)$	r ± is	...

Canonical Elementary Hermitean Matrix With A Single Real Eigenvalue r In The Odd Dimensional Case

$$
\begin{bmatrix}
r & 1 & 0 & & & & & & & & 0 & -1 & 0 \\
1 & r & 1 & & & & & & & & -1 & 0 & 1 \\
0 & 1 & r & \cdot & & & & & & \cdot & 0 & 1 & 0 \\
 & & \cdot & \cdot & \cdot & & & & \cdot & \cdot & \cdot & & \\
 & & & \cdot & r & 1 & 0 & -1 & 0 & \cdot & & & \\
 & & & & 1 & r & \epsilon & 0 & 1 & & & & \\
 & & & & 0 & 1 & r & 1 & 0 & & & & \\
 & & & & 1 & 0 & -\epsilon & r & 1 & & & & \\
 & & & \cdot & 0 & -1 & 0 & 1 & r & \cdot & & & \\
 & & \cdot & \cdot & \cdot & & & & \cdot & \cdot & \cdot & & \\
0 & 1 & 0 & \cdot & & & & & & \cdot & r & 1 & 0 \\
1 & 0 & -1 & & & & & & & & 1 & r & 1 \\
0 & -1 & 0 & & & & & & & & 0 & 1 & r
\end{bmatrix}
$$

Canonical Elementary Hermitean Matrix With A Single Real Eigenvalue r In The Even Dimensional Case

$$
\begin{bmatrix}
r & 1 & 0 & & & & & & & & & 0 & -1 & 0 \\
1 & r & 1 & & & & & & & & & -1 & 0 & 1 \\
0 & 1 & r & \cdot & & & & & & & \cdot & 0 & 1 & 0 \\
 & & \cdot & \cdot & \cdot & & & & & \cdot & \cdot & \cdot & & \\
 & & & \cdot & r & 1 & 0 & 0 & -1 & 0 & \cdot & & & \\
 & & & & 1 & r & 1 & -1 & 0 & 1 & & & & \\
 & & & & 0 & 1 & r+\epsilon & +\epsilon & 1 & 0 & & & & \\
 & & & & 0 & 1 & -\epsilon & r-\epsilon & 1 & 0 & & & & \\
 & & & & 1 & 0 & -1 & 1 & r & 1 & & & & \\
 & & & \cdot & 0 & -1 & 0 & 0 & 1 & r & \cdot & & & \\
 & & \cdot & \cdot & \cdot & & & & & \cdot & \cdot & \cdot & & \\
0 & 1 & 0 & \cdot & & & & & & & \cdot & r & 1 & 0 \\
1 & 0 & -1 & & & & & & & & & 1 & r & 1 \\
0 & -1 & 0 & & & & & & & & & 0 & 1 & r
\end{bmatrix}
$$

Canonical Elementary Hermitean Matrix With A Complex Conjugate Pair Of Nonreal Eigenvalues $r \pm is$

$$
\begin{bmatrix}
r & 1 & 0 & & & & & & & & & 0 & -1 & is \\
1 & r & 1 & & & & & & & & & -1 & is & 1 \\
0 & 1 & r & \cdot & & & & & & & \cdot & is & 1 & 0 \\
 & & \cdot & \cdot & \cdot & & & & & \cdot & \cdot & \cdot & & \\
 & & & \cdot & r & 1 & 0 & 0 & -1 & is & \cdot & & & \\
 & & & & 1 & r & 1 & -1 & is & 1 & & & & \\
 & & & & 0 & 1 & r & is & 1 & 0 & & & & \\
 & & & & 0 & 1 & is & r & 1 & 0 & & & & \\
 & & & & 1 & is & -1 & 1 & r & 1 & & & & \\
 & & & \cdot & is & -1 & 0 & 0 & 1 & r & \cdot & & & \\
 & & \cdot & \cdot & \cdot & & & & & \cdot & \cdot & \cdot & & \\
0 & 1 & is & \cdot & & & & & & & \cdot & r & 1 & 0 \\
1 & is & -1 & & & & & & & & & 1 & r & 1 \\
is & -1 & 0 & & & & & & & & & 0 & 1 & r
\end{bmatrix}
$$

In the three tables listing the actual canonical matrices for the five possible cases, we have combined those cases which differ only by the value of the parameter ϵ. The significance of these canonical matrices may be summarized as follows. A set of vectors $\psi_1, \ldots, \psi_n$ is said to be an orthonormal basis for the nonsingular hermitean space $C^n(p,q)$ iff $\bar{\psi}_i \psi_j$ is 0 when $i \neq j$, +1 when $i = j \leq p$, and -1 when $i = j > p$. If γ is an elementary hermitean operator on a nonsingular hermitean space V, then there exists an orthonormal basis for V such that the matrix of γ with respect to this basis has one of the five canonical forms listed. These canonical forms actually represent nonconjugate classes of hermitean operators, except that in the fifth case, the canonical forms differing only by the sign of the parameter s are special conjugate to each other. Thus we obtain distinct special conjugacy classes of elementary hermitean operators only if we restrict the parameter s to be positive, (say).

It is instructive to illustrate the general theory with some special cases. The simplest illustration is the application of the general theory to the case of hermitean operators on the two-dimensional nonsingular hermitean space $C^2(+-)$. In this case, the most general hermitean matrix is of the form

$$\begin{bmatrix} A + D & B + iC \\ -B + iC & A - D \end{bmatrix} ,$$

where A, B, C, D are real numbers. The eigenvalues of this matrix are given by $A \pm \sqrt{D^2 - B^2 - C^2}$. There are thus three spectral cases, depending on whether $D^2 - B^2 - C^2$ is positive, negative or zero, corresponding respectively to a pair of distinct real eigenvalues, a complex conjugate pair of nonreal eigenvalues, or the degenerate case of a single real eigenvalue.

There are only two possible breakups of the hermitean space $C^2(+-)$ as an orthogonal direct sum of maximal index subspaces: (+-) and (+)(-). The detailed correspondence between the three possible spectral cases and the two maximal index subspace breakups is clearly as follows: The spectral case of a real pair of distinct eigenvalues can only correspond to the maximal index subspace breakup (+)(-), while the spectral case of a complex conjugate pair of nonreal eigenvalues can only correspond to the breakup (+-). On the other hand, the degenerate spectral case of a single real eigenvalue could correspond to either the breakup (+-) or the breakup (+)(-). In terms of the parameters A, B, C, D, the degenerate spectral case occurs whenever $B^2 + C^2 = D^2$, but this leads to a maximal index breakup of the type (+)(-) only in the very special case B = C = D = 0, and to (+-) otherwise.

In all cases here we can give prescriptions for writing down the canonical form corresponding to the original hermitean matrix in terms of the parameters A, B, C, D. There are six such prescriptions, depending on the values of the parameters. These six cases can also be given a geometrical description in terms of the cone $B^2 + C^2 = D^2$ in (B, C, D) space. (See following page.)

In determining the conjugacy classes of one-parameter subgroups of the group U(1,1), we have to remember that two matrices which differ by a nonzero real factor generate the same subgroup. We then obtain the following conjugacy classes of one-parameter subgroups of U(1,1), listed by generator:

$$\begin{bmatrix} r+1 & 0 \\ 0 & r-1 \end{bmatrix} \qquad -\infty < r < \infty$$

$$\begin{bmatrix} 1 & 0 \\ 0 & 1 \end{bmatrix}$$

$$\begin{bmatrix} r & i \\ i & r \end{bmatrix} \qquad -\infty < r < \infty$$

$$\begin{bmatrix} a+1 & 1 \\ -1 & a-1 \end{bmatrix} \qquad a = 0, \pm 1 .$$

For the subgroup SU(1,1), only traceless generators are permitted, and thus we obtain only three special conjugacy classes of one-parameter subgroups for SU(1,1), generated respectively by the three matrices:

$$\begin{bmatrix} 1 & 0 \\ 0 & -1 \end{bmatrix}, \qquad \begin{bmatrix} 0 & i \\ i & 0 \end{bmatrix}, \qquad \begin{bmatrix} 1 & 1 \\ -1 & -1 \end{bmatrix}.$$

Prescriptions For Writing Down The Canonical Forms Of Hermitean Operators On The Hermitean Space $C^2(+-)$

1. Inside the Upper Cone

$D > \sqrt{B^2 + C^2} > 0$

$$\begin{bmatrix} A + \sqrt{D^2 - B^2 - C^2} & 0 \\ 0 & A - \sqrt{D^2 - B^2 - C^2} \end{bmatrix} \quad (+)(-)$$

2. Inside the Lower Cone

$D < -\sqrt{B^2 + C^2} < 0$

$$\begin{bmatrix} A - \sqrt{D^2 - B^2 - C^2} & 0 \\ 0 & A + \sqrt{D^2 - B^2 - C^2} \end{bmatrix} \quad (+)(-)$$

3. Vertex of the Cone

$D = B = C = 0$

$$\begin{bmatrix} A & 0 \\ 0 & A \end{bmatrix} \quad (+)(-)$$

4. Surface of Upper Cone

$D = \sqrt{B^2 + C^2} > 0$

$$\begin{bmatrix} A + 1 & 1 \\ -1 & A - 1 \end{bmatrix} \quad (+-)$$

5. Surface of Lower Cone

$D = -\sqrt{B^2 + C^2} < 0$

$$\begin{bmatrix} A - 1 & -1 \\ 1 & A + 1 \end{bmatrix} \quad (+-)$$

6. Exterior of the Cone

$D^2 < B^2 + C^2$

$$\begin{bmatrix} A & i\sqrt{B^2 + C^2 - D^2} \\ i\sqrt{B^2 + C^2 - D^2} & A \end{bmatrix} \quad (+-)$$

References

1. The related subgroup problem for the proper Lorentz group was studied in the same connection in I. Friš and P. Winternitz, Sov. J. Nucl. Phys. 1, 636 (1965), and for SU(2,1) by W. Montgomery, L. O'Raifeartaigh and P. Winternitz, Nucl. Phys. B11, 39 (1969).
These authors use an approach to harmonic analysis which is based on the method of horocycles, described for instance in I. M. Gel'fand, M. I. Graev and N. Ya. Vilenkin, Generalized Functions, Vol. 5, Integral Geometry and Representation Theory (Academic Press, N.Y., 1966).
Another approach to harmonic analysis on connected semisimple Lie groups makes use of the Iwasawa decomposition: Harish-Chandra, Bull. Amer. Math. Soc. 76, 529 (1970).
In particular, harmonic analysis on the groups SU(p,q) has been studied recently: T. Hirai, J. Math. Soc. Japan 22, 134 (1970).
2. A useful general reference for geometric algebra is Artin's book, which, however, deals only with the orthogonal and symplectic groups: E. Artin, Geometric Algebra (Interscience, N.Y., 1957).
Another standard reference here is known to the French: N. Bourbaki, Sesquilinear Forms and Quadratic Forms [French] (Hermann, Paris, 1959).
3. The local isomorphism between SU(2,2) and SO(4,2; R) is one of several special isomorphisms existing among low rank Lie groups. For a more complete discussion we refer to Helgason's book, S. Helgason, Differential Geometry and Symmetric Spaces (Academic Press, N.Y., 1962), p. 352.
For the conjugacy classes of the Lorentz group there is a considerable literature:
E. P. Wigner, Ann. of Math. 40, 149 (1939).
G. Abraham, Proc. Indian Acad. Sci. 28, 87 (1948).
J. S. Lomont, Bull. Calcutta Math. Soc. 55, 51 (1963).
The complete subgroup lattice of the Lorentz group is also known. Besides the paper of Friš and Winternitz quoted earlier, we may also mention the following:
J. Segercrantz, Ann. Acad. Sci. Fennicae A. VI. 233, 1 (1967).
R. Shaw, Quart. J. Math. Oxford (2) 21, 101 (1970).
The classes of the De Sitter groups are discussed in:
T. O. Philips and E. P. Wigner, in Group Theory and Its Applications, edited by E. M. Loebl (Academic Press, N.Y., 1968), p. 631-676.
The group SO(2,2; R) and its generalizations over arbitrary fields have been extensively studied. H. Zassenhaus, Canad. Math. Bull. 1, 31, 101, 183 (1958).

The curvature tensor on a Riemann space with tangent space V of signature (p,q) may be regarded as a symmetric operator on the Grassmann product space $V \wedge V$ which has an induced signature (p', q') where $p' = \frac{1}{2}[p(p-1) + q(q-1)]$ and $q' = pq$. In particular, if $p = 3$ and $q = 1$, then $p' = q' = 3$. A. Z. Petrov, Einstein Spaces (Pergamon Press, N.Y., 1969).

4. We could refer here to any reasonably advanced linear algebra text, such as K. Hoffman and R. Kunze, Linear Algebra (Prentice-Hall, N.J., 1961).
   Our own terminology has been influenced by S. Warner, Modern Algebra, Vol. 2 (Prentice-Hall, N.J., 1965).
5. This approach to unitary groups with indefinite metric is standard. J. Dieudonné, The Geometry of the Classical Groups, 2nd ed. [French] (Springer-Verlag, Berlin, 1963).
6. The relation between the conformal group and the geometry of Dirac spinor space has long been known. O. Veblen, Proc. Natl. Acad. Sci. 19, 503 (1933).
7. The Witt theorem for orthogonal and symplectic geometry is discussed in Artin's book cited earlier. For hermitean geometry there is a good discussion in N. Jacobson, Lectures in Abstract Algebra, Vol. 2, Linear Algebra (Van Nostrand, Princeton, 1953).

Section B: O(4,1)

CLASSIFICATION OF THE IRREDUCIBLE REPRESENTATIONS OF THE O(4,1) DE SITTER GROUP†

Fritz Schwarz
Center for Particle Theory
Department of Physics
The University of Texas
Austin, Texas 78712

Abstract

By M^5 we denote the vector space of real 5-tupels $(x_1 \ldots x_5)$ in which the quadratic form $x_1^2 + \ldots + x_4^2 - x_5^2$ is given. The De Sitter group O(4,1) is the group of linear homogeneous transformations of the M^5 which leave this form invariant. At first we determine the irreducible representations of the identity component. These irreducible representations are extended in all possible ways to inequivalent representations of the whole group.

I. Introduction

Let M^5 be the 5-dimensional vector space whose elements are the real 5-tupels $(x_1 \ldots x_5)$ and in which the quadratic form $x_1^2 + \ldots + x_4^2 - x_5^2$ is given. By O(4,1) we denote the group of linear homogeneous transformations of the M^5 which leave this form invariant, i.e., the elements of O(4,1) are the real 5 by 5 matrices g which obey the equation $g^t G g = G$. Here G is the diagonal matrix with the nonzero elements $\{+1, +1, +1, +1, -1\}$. The group O(4,1) consists of four disconnected pieces, which are characterized by det $g = \pm 1$ and $g_{55} \geqq +1$ or $g_{55} \leqq -1$.[1)] We choose the following notation:

†Presented at the Symposium on De Sitter and Conformal Groups, University of Colorado, Summer 1970.

Notation	det g	sign g_{55}	Representative
$O^{++}(4,1)$	+1	+	$E_5 = \begin{pmatrix} +1 & 0 & 0 & 0 & 0 \\ 0 & +1 & 0 & 0 & 0 \\ 0 & 0 & +1 & 0 & 0 \\ 0 & 0 & 0 & +1 & 0 \\ 0 & 0 & 0 & 0 & +1 \end{pmatrix}$
$O^{-+}(4,1)$	+1	-	$R = \begin{pmatrix} +1 & 0 & 0 & 0 & 0 \\ 0 & +1 & 0 & 0 & 0 \\ 0 & 0 & +1 & 0 & 0 \\ 0 & 0 & 0 & -1 & 0 \\ 0 & 0 & 0 & 0 & -1 \end{pmatrix}$
$O^{+-}(4,1)$	-1	+	$S = \begin{pmatrix} +1 & 0 & 0 & 0 & 0 \\ 0 & +1 & 0 & 0 & 0 \\ 0 & 0 & +1 & 0 & 0 \\ 0 & 0 & 0 & -1 & 0 \\ 0 & 0 & 0 & 0 & +1 \end{pmatrix}$
$O^{--}(4,1)$	-1	-	$T = \begin{pmatrix} +1 & 0 & 0 & 0 & 0 \\ 0 & +1 & 0 & 0 & 0 \\ 0 & 0 & +1 & 0 & 0 \\ 0 & 0 & 0 & +1 & 0 \\ 0 & 0 & 0 & 0 & -1 \end{pmatrix}$

Evidently $O^{++}(4,1)$ denotes the identity component of $O(4,1)$ and the other three pieces are the cosets with respect to it. There are three other subgroups which we will use frequently:

$$O^{+}(4,1) = O^{++}(4,1) \cup RO^{++}(4,1) = O^{++}(4,1) \cup O^{-+}(4,1) \qquad (1a)$$

$$O^{1}(4,1) = O^{++}(4,1) \cup SO^{++}(4,1) = O^{++}(4,1) \cup O^{+-}(4,1) \qquad (1b)$$

$$O^{2}(4,1) = O^{++}(4,1) \cup TO^{++}(4,1) = O^{++}(4,1) \cup O^{--}(4,1) \qquad (1c)$$

The identity component is a normal subgroup of index 4 in the whole group $O(4,1)$ and a normal subgroup of index 2 in each one of the subgroups defined by Eqs. (1). Further, each one of the subgroups defined by Eqs. (1) is a normal subgroup of index 2 in the whole group $O(4,1)$.

The group O(4,1) is not simply connected. The universal covering group of the identity component $O^{++}(4,1)$ is isomorphic to the group SL(2,Q).[2] This group consists of the 2 by 2 matrices, whose elements are quaternions, and which leave the form $\bar{x}x - \bar{y}y$ in the 2-dimensional vector space of quaternionic 2-tupels invariant. Let φ be the homomorphism from SL(2,Q) onto $O^{++}(4,1)$ and e be the unit element of SL(2,Q). Then we have $\varphi(\pm e) = E_5$. The situation is more complicated for the group O(4,1); for this discussion see Ref. 3. Let r, s and t be the elements of a covering group which correspond to R, S and T respectively, i.e., $\varphi(\pm r) = R$, $\varphi(\pm s) = S$ and $\varphi(\pm t) = T$. Then a covering group of O(4,1) is uniquely determined as soon as the square of a representative of each coset is fixed. If we take as representatives the elements r, s and t we may have $r^2 = \pm e$, $s^2 = \pm e$ and $t^2 = \pm e$. According to the different choices of the signs, there are 8 different covering groups of O(4,1), which we denote by $C_jO(4,1)$ with $1 \leqq j \leqq 8$. We fix the notation in the following way:

$C_jO(4,1)$	r^2	s^2	t^2	-: r,s,t commute +: r,s,t anticommute
j = 1	+e	+e	+e	-
j = 2	+e	+e	-e	+
j = 3	+e	-e	+e	+
j = 4	+e	-e	-e	-
j = 5	-e	+e	+e	+
j = 6	-e	+e	-e	-
j = 7	-e	-e	+e	-
j = 8	-e	-e	-e	+

The elements ±e, ±r, ±s and ±t combine differently in the different groups $C_jO(4,1)$; their multiplication schemes are given in Table 1.

The identity component has a basic set of 10 one-parameter subgroups. For 6 of them we take the rotations in the x_i-x_j-coordinates planes with $1 \leq i < j \leq 4$. The rotation in the x_1-x_2-plane, for example, is described by a matrix of the form

$$g_{12}(\alpha) = \begin{pmatrix} \cos\alpha & \sin\alpha & 0 & 0 & 0 \\ -\sin\alpha & \cos\alpha & 0 & 0 & 0 \\ 0 & 0 & 1 & 0 & 0 \\ 0 & 0 & 0 & 1 & 0 \\ 0 & 0 & 0 & 0 & 1 \end{pmatrix} \qquad (2)$$

The matrices $g_{ij}(\alpha)$, which describe the rotations in the other coordinate planes, are similar. The remaining 4 one-parameter subgroups are the pseudorotations in the x_i-x_5-coordinate planes with $1 \leqq i \leqq 4$. The first one of them is described by the matrix

$$g_{15}(\alpha) = \begin{pmatrix} \cosh\alpha & 0 & 0 & 0 & \sinh\alpha \\ 0 & 1 & 0 & 0 & 0 \\ 0 & 0 & 1 & 0 & 0 \\ 0 & 0 & & 1 & 0 \\ \sinh\alpha & 0 & 0 & 0 & \cosh\alpha \end{pmatrix} \tag{3}$$

The matrices for the other pseudorotations are obtained from Eq. (3) in an obvious fashion. We denote the Lie algebra of the identity component by so(4,1). The basis elements A_{ij} with $1 \leqq i < j \leqq 5$ are defined by

$$A_{ij} = \frac{d}{d\alpha} g_{ij}(\alpha)\Big|_{\alpha=0} \tag{4}$$

A simple calculation shows that they obey the following commutation relations:

$$[A_{ij}, A_{kl}] = g_{jk}A_{il} + g_{il}A_{jk} - g_{ik}A_{il} - g_{jl}A_{ik} \tag{5}$$

with $g_{ii} = +1$ for $1 \leqq i \leqq 4$, $g_{55} = -1$ and $g_{ij} = 0$ for $i \neq j$. Later we will need also the commutation relations of the infinitesimal generators $A_{i,i+1}$ for $1 \leqq i \leqq 4$ with the elements S, T and R, and therefore we give them here. It is easy to see that they have the following form:

$$[A_{12}, R]_- = [A_{23}, R]_- = [A_{34}, R]_+ = [A_{45}, R]_- = 0 \tag{6}$$

$$[A_{12}, S]_- = [A_{23}, S]_- = [A_{34}, S]_+ = [A_{45}, S]_+ = 0 \tag{7}$$

$$[A_{12}, T]_- = [A_{23}, T]_- = [A_{34}, T]_- = [A_{45}, T]_+ = 0 \tag{8}$$

where we used the notation $[X,Y]_\pm = XY \pm YX$. To classify the irreducible representations (IR's) of the group O(4,1) we proceed as follows: At first we determine in Section II the IR's of the identity component. Then we extend these representations in all possible ways, which lead to inequivalent representations, to the group O(4,1) or its covering groups. This is done in two steps. In the first step the IR's of the identity component are extended to the groups defined by Eqs. (1) or its covering groups. In the second step the IR's of these subgroups are extended to all four pieces. This has the advantage that we need only the connection between the representations of a group and those of a normal subgroup of index 2 contained in it. The general case, where the normal subgroup is not restricted to the index 2, has been treated by A. H. Clifford.[4] However, the special case we need is considerably simpler and therefore we describe it in Section III. In Sections IV and V these results are applied to the De Sitter group for m_{51} integer or halfinteger respectively. The label m_{51} will be introduced in the next section.

II. The Irreducible Representations of the Identity Component

In this section we classify the IR's of the identity component. We do this by determining a set of irreducible matrices which obey the commutation relations (5). As a special case we recover the unitary representations, which are already known (see Refs. 5 and 6). We denote a matrix, which represents the infinitesimal generator A_{ij}, by $D^{(m_{51},z_{52})}(A_{ij})$. The labels m_{51} and z_{52} will be explained later in this section.

It is easy to see that a representation is completely determined if one knows the matrices which represent the generators $A_{i,i+1}$ for $1 \leq i \leq 5$, because the other generators can be expressed through them with the commutation relations (5). We define a new set of generators B_{ij} by

$$B_{ij} = \sqrt{g_{ii}}\sqrt{g_{jj}}\ A_{ij} \tag{9}$$

Putting these new generators into Eq. (5) one sees easily that they obey the commutation relations of the Lie algebra of the 5-dimensional rotation group, i.e., an equation, which results from Eq. (5) by replacing the g_{ij} by the Kronecker symbol δ_{ij}. In the reduction $SO(2) \subset SO(3) \subset SO(4) \subset O^{++}(4,1)$ a vector within a representation space is completely specified by the labels m_{21}, m_{31}, m_{41} and m_{42}. The matrices $D^{(m_{51},z_{52})}(A_{i,i+1})$ act in the following way on these vectors:

$$D^{(m_{51},z_{52})}(A_{12})|m_{41},m_{42},m_{31},m_{21}\rangle = im_{21}|m_{41},m_{42},m_{31},m_{21}\rangle \tag{10}$$

$$D^{(m_{51},z_{52})}(A_{23})|m_{41},m_{42},m_{31},m_{21}\rangle =$$

$$= A(m_{21})|m_{41},m_{42},m_{31},m_{21}+1\rangle - A(m_{21}-1)|m_{41},m_{42},m_{31},m_{21}-1\rangle \tag{11}$$

$$D^{(m_{41},z_{52})}(A_{34})|m_{41},m_{42},m_{31},m_{21}\rangle = iC_2|m_{41},m_{42},m_{31},m_{21}\rangle + B(m_{31})|m_{41},m_{42},m_{31}+1,m_{21}\rangle - B(m_{31}-1)|m_{41},m_{42},m_{31}-1,m_{21}\rangle \tag{12}$$

$$D^{(m_{51},z_{52})}(A_{45})|m_{41},m_{42},m_{31},m_{21}\rangle = = A(m_{41})|m_{41}+1,m_{42},m_{31},m_{21}\rangle - A(m_{41}-1)|m_{41}-1,m_{42},m_{31},m_{21}\rangle + A(m_{42})|m_{41},m_{42}+1,m_{31},m_{21}\rangle - A(m_{42}-1)|m_{41},m_{42}-1,m_{31},m_{21}\rangle \tag{13}$$

From the commutation relations it follows for the matrix elements

$$A(m_{21}) = \tfrac{1}{2}\sqrt{(m_{31}+\tfrac{1}{2})^2 - (m_{21}+\tfrac{1}{2})^2} \tag{14}$$

$$A(m_{41}) = \tfrac{1}{2}\sqrt{(m_{41}+\tfrac{1}{2})^2 - (m_{31}+\tfrac{1}{2})^2}\;\sqrt{\frac{[(z_{51}+\tfrac{1}{2})^2 - (m_{41}+\tfrac{1}{2})^2][(z_{52}+\tfrac{3}{2})^2 - (m_{41}+\tfrac{1}{2})^2]}{[(m_{42}+1)^2 - m_{41}^2][(m_{42}+1)^2 - (m_{41}+1)^2]}} \tag{15}$$

$$A(m_{42}) = \tfrac{1}{2}\sqrt{(m_{42}+\tfrac{3}{2})^2 - (m_{31}+\tfrac{1}{2})^2}\;\sqrt{\frac{[(z_{51}+\tfrac{1}{2})^2 - (m_{42}+\tfrac{3}{3})^2][(z_{52}+\tfrac{3}{2})^2 - (m_{42}+\tfrac{3}{2})^2]}{[m_{41}^2 - (m_{42}+1)^2][m_{41}^2 - (m_{42}+2)^2]}} \tag{16}$$

$$B(m_{31}) = \sqrt{m_{21}^2 - (m_{31}+1)^2}\;\sqrt{\frac{[m_{41}^2-(m_{31}+1)^2][(m_{42}+1)^2-(m_{31}+1)^2]}{(m_{31}+1)^2[4(m_{31}+1)^2-1]}} \tag{17}$$

$$C_2 = \frac{m_{21}\; m_{41}\;(m_{42}+1)}{m_{31}\;(m_{31}+1)} \tag{18}$$

These expressions are taken from the appendix of Ref. 6. There also the occurrence of the complex labels z_{51} and z_{52} is explained. The labels m_{41}, m_{42}, m_{31} and m_{21} are integer or halfinteger at the same time and obey the inequalities

$$|m_{41}| \leqq m_{31} \leqq m_{42} \tag{19a}$$

$$|m_{21}| \leqq m_{31} \tag{19b}$$

The complex constants z_{51} and z_{52} are restricted by the requirement that the SO(4) labels m_{41} and m_{42} obey the condition (19a) and that the representation specified through them is irreducible. We determine now the restrictions for z_{51} and z_{52} which follow from these requirements.

Eq. (19a)means that $m_{41} \leqq m_{41}^{max} = m_{42}^{min} \leqq m_{42}$. For this to be true the following equation must be valid:

$$A(m_{41}^{max}) = A(m_{42}^{min} - 1) = 0 \tag{20}$$

From this equation we get the condition

$$(z_{51}+\tfrac{1}{2})^2 = (m_{41}^{max}+\tfrac{1}{2})^2 = (m_{42}^{min}+\tfrac{1}{2})^2 \tag{21}$$

The solution of this equation is $z_{51} = m_{51} = m_{41}^{max} = m_{42}^{min}$ with the condition

$$|m_{41}| \leqq m_{51} \leqq m_{42} \tag{22}$$

If the SO(4) labels are integer, the range of m_{51} is $0,1,2,\ldots$; if they are halfinteger m_{51} has the range $\frac{1}{2}, \frac{3}{2}, \frac{5}{2}, \ldots$.

The irreducibility requires that there is no subspace which is invariant under the action of the operator $D^{(m_{51}, z_{52})}(A_{45})$. If the SO(4) labels are integer, this is certainly the case if $z_{52}=x_{52}+iy_{52}$ is an arbitrary complex number except an integer m_{52} which does not fulfill $m_{51} = m_{52}+1$. To avoid having the same representation occur more than once, we restrict the imaginary part of z_{52} by $0 \leqq y_{52}$. If m_{52} is an integer together with m_{51} and $m_{51} \neq m_{52}+1$, the representation splits into the direct sum of two IR's, which differ only by the SO(4) content. For $m_{51} \leqq m_{52}$ the two IR's are

$$|m_{41}| \leqq m_{51} \leqq m_{42} \leqq m_{52} \tag{23a}$$

$$|m_{41}| \leqq m_{51}, \quad m_{52}+1 \leqq m_{42} \tag{23b}$$

For $0 \leqq m_{52}+1 < m_{51}$ there are 3 possibilities, one of which is already contained in (23b):

$$|m_{41}| \leqq m_{52}+1, \quad m_{51} \leqq m_{42} \tag{24a}$$

$$m_{52}+2 \leqq +m_{41} \leqq m_{51} \leqq m_{42} \tag{24b}$$

$$m_{52}+2 \leqq -m_{41} \leqq m_{51} \leqq m_{42} \tag{24c}$$

Analogous considerations can be made for the case where the discrete label m_{51} is halfinteger and we do not repeat them here. The results for m_{51} integer or halfinteger are collected in Tables 2 and 3 respectively. It is easy to derive the additional conditions which the labels m_{51} and z_{52} must obey so that these IR's are unitary. However, they were already derived in Ref. 6 and we include the results in Tables 2 and 3 without further discussion.

III. Some Results From A Paper by A. H. Clifford

In this section we describe the results of A. H. Clifford[4)] for the special case where the normal subgroup is of index 2; see also Refs. 1 and 7. We give only the results we need. The interested reader can find the proofs in these references. Let G be a group, $H \subset G$ a normal subgroup of index 2. By h, h_i we always denote elements of H, by g, g_i elements of G which are not necessarily in H. However, let always $g_o \notin H$, then $g_o H$ is the coset with respect to H and we have $G = H + g_o H$. If $D(h)$ is a representation of H, then also $D(g^{-1} hg) = D^*(h)$ with fixed $g \in G$ is a representation of H, because always $g^{-1} hg \in H$. The representation $D^*(h)$ is called a representation conjugate to $D(h)$. It may happen that the representation $D^*(h)$ is equivalent to $D(h)$ for a subset of G, in this case it is called selfconjugate in this subset. Trivially this is the case for $g \in H$, because then $D(g^{-1} hg) = D(g^{-1})D(h)D(g)$. However, in general the subset of G for which a given representation of H is selfconjugate, may be larger. It can be shown that this subset is always a subgroup of G, called the little group of the representation $D(h)$. If H is of index 2 in G, the little group of an arbitrary representation of H is either H itself or the whole group G.

Let $\bar{D}(g)$ be an IR of G. If $\bar{D}(g)$ is restricted to H there are two possibilities which can occur. If $\bar{D}(g)$ remains irreducible, the representation $D(h)$, subduced by $\bar{D}(g)$, is selfconjugate in G. The other possibility is that $D(h)$ is reducible. In this case the little group of H is H itself. $\bar{D}(g)$ splits into the direct sum of two IR's $D_1(h)$ and $D_2(h)$ of H which are conjugate to each other.

We want to describe now how the IR's $D(h)$ of H, which are supposed to be known, can be extended to those of G. Such an extension is determined if we know the operator which corresponds to one representative g_o of the coset. At first we consider the case where the representation $D(h)$ of H is selfconjugate in G. There exists an operator C with

$$D(g_o^{-1} hg) = C^{-1} D(h) C \tag{25}$$

for all $h \in H$, and consequently

$$D(g_o^{-2} h g_o^2) = D(g_o^{-2}) D(h) D(g_o^2) = C^{-2} D(h) C^2 \tag{26}$$

because $g_o^2 \in H$. It follows that $D(g_o^2) = \alpha^2 C^2$, i.e., $D(g_o)$ is determined up to a constant and we have

$$D(g_o) = \pm \alpha C \tag{27}$$

where α is an arbitrary complex constant. It is fixed by the requirement that $D(g_o^2)$ has a prescribed value. The two possibilities of $D(h)$ corresponding to the different signs at the right hand side of (27) give two inequivalent representations of G.

The other possibility is that the little group of $D(h)$ is the group H itself. In this case the extension of $D(h)$ to an IR of G can be induced from $D(h)$. We take the unit element e and the element g_o as representatives of H and the coset respectively. The representation $\bar{D}(g)$, induced by $D(h)$, is irreducible and given by

$$\bar{D}(g) = \begin{pmatrix} D(g) & D(gg_o) \\ D(g_o^{-1}) & D(g_o^{-1} g g_o) \end{pmatrix} \quad (D(g') = 0 \text{ if } g' \notin H,\ g' = g,\ gg_o,\ g_o^{-1} g \text{ or } g_o^{-1} g g_o \text{ respectively}) \tag{28}$$

where g is an arbitrary element of G. For $g \in H$ it follows from (28)

$$\bar{D}(g) = \begin{pmatrix} D(g) & 0 \\ 0 & D(g_o^{-1} g g_o) \end{pmatrix} \tag{29}$$

and for $g \notin H$

$$\bar{D}(g) = \begin{pmatrix} 0 & D(gg_o) \\ D(g_o^{-1} g) & 0 \end{pmatrix} \tag{30}$$

From (29) and (30) one sees that the representation space of $\bar{D}(g)$ is a system of imprimitivity for G. For $g = g_o$ one gets from (30)

$$\bar{D}(g) = \begin{pmatrix} 0 & D(g_o^2) \\ D(e) & 0 \end{pmatrix} \tag{31}$$

The extensions of the IR's of H to the whole group G, described in this section, exhaust all possibilities which lead to inequivalent representations of G.

IV. Extension to the Whole Group If m_{51} Is Integer

If the discrete label m_{51} is integer, the results of Section III can be applied to the group O(4,1) itself. At first we extend the IR's of the identity component $O^{++}(4,1)$ to the three subgroups $O^{+}(4,1)$, $O^{1}(4,1)$ and $O^{2}(4,1)$.

We begin with $O^{+}(4,1)$. As representatives of the identity component and the coset we take the elements E_5 and R respectively. If an IR of $O^{++}(4,1)$ is selfconjugate in $O^{+}(4,1)$, there exists an operator B with

$$B^{-1} D^{(m_{51}, z_{52})}(A_{i,i+1})B = D^{(m_{51}, z_{52})}(R^{-1} A_{i,i+1} R) = = D^{(m_{51}, z_{52})}(A_{i,i+1}) \tag{32a}$$

for i = 1,2 and 4, and

$$B^{-1} D^{(m_{51}, z_{52})}(A_{34})B = D^{(m_{51}, z_{52})}(R^{-1} A_{34} R) = -D^{(m_{51}, z_{52})}(A_{34}) \tag{32b}$$

The calculations which lead to the operator B are similar to those for O(5) and can be found in Ref. 8. Therefore we do not repeat them here. Eqs. (32) determine B up to a complex constant α with the result

$$B|m_{41}, m_{42}, m_{31}, m_{21}\rangle = \alpha(-1)^{m_{31}}(-1)^{m_{41}}|-m_{41}, m_{42}, m_{31}, m_{21}\rangle \tag{33}$$

There are no conditions for the labels m_{51} and z_{52}. However, from Table 2 one sees that the operator B does not exist for the representations of the classes IVa and IVb. B exists only in a representation which is the direct sum of two IR's of the identity component, one of which is from class IVa, the other from class IVb, and for which the labels m_{51} and z_{52} have the same values. We call the representations of this type simply the class IV. The requirement $D^{(m_{51}, z_{52})}(R^2) = D^{(m_{51}, z_{52})}(E_5)$ fixes the value of the constant α in Eq. (33) to $\alpha = 1$. According to Eq. (27) the two possible extensions are

$$D^{(m_{51}, z_{52})}(R)|m_{41}, m_{42}, m_{31}, m_{21}\rangle = \pm(-1)^{m_{31}}(-1)^{m_{41}}|-m_{41}, m_{42}, m_{31}, m_{21}\rangle \tag{34}$$

The extensions to $O^{1}(4,1)$ and $O^{2}(4,1)$ follow from similar considerations and we do not repeat them here. The results are as follows: The IR's of the classes I...IV are selfconjugate in $O^{1}(4,1)$. The two inequivalent extensions are

$$D^{(m_{51},z_{52})}(S)|m_{41},m_{42},m_{31},m_{21}\rangle =$$

$$= (-1)^{m_{31}}(-1)^{m_{42}}|-m_{41},m_{42},m_{31},m_{21}\rangle \quad (35)$$

Contrary to these results, all IR's of the classes I...III, IVa and IVb are selfconjugate in $O^2(4,1)$ and the two inequivalent extensions by $D^{(m_{51},z_{52})}(T)$ are

$$D^{(m_{51},z_{52})}(T)|m_{41},m_{42},m_{31},m_{21}\rangle =$$

$$= (-1)^{m_{41}}(-1)^{m_{42}}|m_{41},m_{42},m_{31},m_{21}\rangle \quad (36)$$

Now we determine the IR's of the whole group O(4,1). To do this we can start with one of the subgroups $O^+(4,1)$, $O^1(4,1)$ or $O^2(4,1)$ and extend their IR's in all possible ways to the whole group O(4,1). Clearly the results are the same in all three cases, and we choose $O^+(4,1)$. As representative of $O^+(4,1)$ we take the unit element E_5 and as representative of the coset with respect to $O^+(4,1)$ the element S. If an IR of $O^+(4,1)$ is selfconjugate in O(4,1), there exists an operator C which obeys the following commutation relations:

$$C^{-1}D^{(m_{51},z_{52})}(A_{i,i+1})C = D^{(m_{51},z_{52})}(S^{-1}A_{i,i+1}S) =$$

$$D^{(m_{51},z_{52})}(A_{i,i+1}) \quad (37a)$$

for i = 1 and 2,

$$C^{-1}D^{(m_{51},z_{52})}(A_{i,i+1})C = D^{(m_{51},z_{52})}(S^{-1}A_{i,i+1}S) =$$

$$-D^{(m_{51},z_{52})}(A_{i,i+1}) \quad (37b)$$

for i = 3 and 4, and

$$C^{-1}D^{(m_{51},z_{52})}(R)C = D^{(m_{51},z_{52})}(S^{-1}RS) = D^{(m_{51},z_{52})}(R) \quad (37c)$$

From Eqs. (37a) and (37b) it follows that C has the form

$$C|m_{41},m_{42},m_{31},m_{21}\rangle = \alpha(-1)^{m_{31}}(-1)^{m_{42}}|-m_{41},m_{42},m_{31},m_{21}\rangle \quad (38)$$

A simple calculation shows that (37c) is automatically fulfilled. The condition $D^{(m_{51},z_{52})}(S^2) = D^{(m_{51},z_{52})}(E_5)$ leads to $\alpha = 1$ and the two inequivalent extensions are again given by Eq. (35). The operator, which represents the element T is determined by the equation

$$D^{(m_{51},z_{52})}(T) = D^{(m_{51},z_{52})}(R)\ D^{(m_{51},z_{52})} \qquad (39)$$

Let us summarize the results which we have found in this section for m_{51} integer. Each one of the IR's of the classes I...III given in Table 2 can be extended to an IR of the whole group O(4,1). However this is not true for the IR's of the classes IVa and IVb separately. Only the representations of the class IV, which were defined previously, can be extended to the whole group O(4,1). In each one of these representations the elements R and S are represented by the operators (34) and (35) respectively. The operator, which represents the element T is determined by Eq. (39). The representations of O(4,1), which belong to the classes I...III are irreducible under restriction to $O^+(4,1)$, $O^1(4,1)$, $O^2(4,1)$ and the identity component itself. The representations of the class IV remain irreducible under restriction to $O^+(4,1)$ and $O^1(4,1)$. However, restricted to $O^2(4,1)$ or $O^{++}(4,1)$ they are reducible and decompose into the direct sum of two inequivalent representations.

V. Extension to the Whole Group If m_{51} Is Halfinteger

If the discrete label m_{51} is halfinteger, the extension is more complicated. This is a consequence of the fact that the group O(4,1) has eight different covering groups $C_jO(4,1)$. That means if m_{51} is halfinteger we have to determine the IR's of these eight covering groups. We proceed in a similar way as in the case where m_{51} is integer, i.e., at first we extend the IR's of the identity component to the groups $C_jO^+(4,1)$, $C_jO^1(4,1)$ and $C_jO^2(4,1)$.

We begin with $C_jO^+(4,1)$. If an IR of the identity component is selfconjugate in $C_jO^+(4,1)$ there exists an operator B with

$$B^{-1} D^{(m_{51},z_{52})}(A_{i,i+1})B = D^{(m_{51},z_{52})}(r^{-1}A_{i,i+1}r) =$$

$$= D^{(m_{51},z_{52})}(A_{i,i+1}) \qquad (40a)$$

for i = 1, 2 and 4, and

$$B^{-1} D^{(m_{51},z_{52})}(A_{34})B = D^{(m_{51},z_{52})}(r^{-1}A_{34}r) = -D^{(m_{51},z_{52})}(A_{34}) \quad (40b)$$

From these equations it follows that B exists in the classes I...IV and is given by Eq. (33). The constant α is fixed by the requirement that $D^{(m_{51},z_{52})}(r^2)$ has a prescribed value. It follows that the two inequivalent extensions for $r^2 = +e$ are

$$D^{(m_{51},z_{52})}(r)|m_{41},m_{42},m_{31},m_{21}\rangle =$$
$$= \pm i(-1)^{m_{31}}(-1)^{m_{41}}|-m_{41},m_{42},m_{31},m_{21}\rangle \quad (41)$$

and for $r^2 = -e$

$$D^{(m_{51},z_{52})}(r)|m_{41},m_{42},m_{31},m_{21}\rangle =$$
$$= \pm(-1)^{m_{31}}(-1)^{m_{41}}|-m_{41},m_{42},m_{31},m_{21}\rangle \quad (42)$$

For the extensions to $C_jO^1(4,1)$ and $C_jO^2(4,1)$ we give only the results, because the calculations are completely analogous. The IR's of the classes I...IV are selfconjugate in $C_jO^1(4,1)$. If $s^2 = +e$ the two extensions are

$$D^{(m_{51},z_{52})}(s)|m_{41},m_{42},m_{31},m_{21}\rangle =$$
$$= \pm(-1)^{m_{31}}(-1)^{m_{42}}|-m_{41},m_{42},m_{31},m_{21}\rangle \quad (43)$$

and for $s^2 = -e$

$$D^{(m_{51},z_{52})}(s)|m_{41},m_{42},m_{31},m_{21}\rangle =$$
$$= i(-1)^{m_{31}}(-1)^{m_{42}}|-m_{41},m_{42},m_{31},m_{21}\rangle \quad (44)$$

The IR's of the classes I...III and IVa and IVb separately are selfconjugate in $C_jO^2(4,1)$ with the extensions

$$D^{(m_{51},z_{52})}(t)|m_{41},m_{42},m_{31},m_{21}\rangle =$$
$$(-1)^{m_{41}}(-1)^{m_{42}}|m_{41},m_{42},m_{31},m_{21}\rangle \quad (45)$$

$$D^{(m_{51},z_{52})}(t)|m_{41},m_{42},m_{31},m_{21}\rangle =$$
$$i(-1)^{m_{41}}(-1)^{m_{42}}|m_{41},m_{42},m_{31},m_{21}\rangle \quad (46)$$

for $t^2 = +e$ or $t^2 = -e$ respectively.

Finally we determine the representations of the groups $C_jO(4,1)$. As in the case where m_{51} is integer, we can start with IR's of each one of the subgroups $C_jO^+(4,1)$, $C_jO^1(4,1)$ or $C_jO^2(4,1)$, and we choose $C_jO^+(4,1)$. As representative of the normal subgroup, which is now $C_jO^+(4,1)$, we take the element e, and as representative of the coset with respect to it the element s. If an IR of

$C_jO^+(4,1)$ is selfconjugate in $C_jO(4,1)$, there exists an operator C with

$$C^{-1} D^{(m_{51}, z_{52})}(A_{i,i+1})C = D^{(m_{51}, z_{52})}(s^{-1} A_{i,i+1} s) = $$
$$= D^{(m_{51}, z_{52})}(A_{i,i+1}) \tag{47a}$$

for i = 1 and 2,

$$C^{-1} D^{(m_{51}, z_{52})}(A_{i,i+1})C = D^{(m_{51}, z_{52})}(s^{-1} A_{i,i+1} s) = $$
$$= -D^{(m_{51}, z_{52})}(A_{i,i+1}) \tag{47b}$$

for i = 3 and 4, and

$$C^{-1} D^{(m_{51}, z_{52})}(r)C = D^{(m_{51}, z_{52})}(s^{-1} rs) = \pm D^{(m_{51}, z_{52})}(r) \tag{47c}$$

The +-sign at the right hand side of Eq. (47c) is valid if r and s commute, the --sign if they anticommute. From Eqs. (47a) and (47b) it follows that C is given by Eq. (38). A simple calculation shows that Eq. (47c) is automatically fulfilled if r and s anticommute. In this case the operator which represents the element s is given by Eqs. (43) and (44) for $s^2 = +e$ or $s^2 = -e$ respectively. From the discussion in the introduction we see that r and s anticommute in the covering groups $C_jO(4,1)$ with j = 2, 3, 5 or 8. That means that the IR's of the classes I...IV can be extended on the same representation space to an IR of these covering groups. The operators which represent the elements r, s and t in these representations are given in Table 4.

The representations of the covering groups $C_jO(4,1)$ with j = 1, 2, 4 and 7 have to be induced, because for these groups there exists no operator C which obeys the commutation relations (47). We denote this induced representation by $\overline{D}^{(m_{51}, z_{52})}(A)$. For the matrices which represent the infinitesimal generators we get according to Section III

$$\overline{D}^{(m_{51}, z_{52})}(A_{i,i+1}) = \begin{pmatrix} D^{(m_{51}, z_{52})}(A_{i,i+1}) & 0 \\ 0 & D^{(m_{51}, z_{52})}(A_{i,i+1}) \end{pmatrix} \tag{48}$$

for i = 1 and 2, and

$$\bar{D}^{(m_{51},z_{52})}(A_{i,i+1}) = \begin{pmatrix} D^{(m_{51},z_{52})}(A_{i,i+1}) & 0 \\ 0 & -D^{(m_{51},z_{52})}(A_{i,i+1}) \end{pmatrix} \quad (49)$$

for i = 3 and 4. The elements r and s are represented by the matrices

$$\bar{D}^{(m_{51},z_{52})}(r) = \begin{pmatrix} D^{(m_{51},z_{52})}(r) & 0 \\ 0 & D^{(m_{51},z_{52})}(r) \end{pmatrix} \quad (50)$$

$$\bar{D}^{(m_{51},z_{52})}(s) = \begin{pmatrix} D^{(m} & 0 & \pm D^{(m_{51},z_{52})}(e) \\ D^{(m_{51},z_{52})}(e) & & 0 \end{pmatrix} \quad (51)$$

The +-sign at the right hand side of Eq. (51) is valid if $s^2 = +e$, the --sign, if $s^2 = -e$. In the former case we transform $\bar{D}^{(m_{51},z_{52})}(A)$ into an equivalent representation $\bar{\bar{D}}^{(m_{51},z_{52})}(A)$ with the matrix $\begin{pmatrix} E & 0 \\ 0 & B \end{pmatrix}$, in the second case with the matrix $\begin{pmatrix} E & 0 \\ 0 & -iB \end{pmatrix}$, where B is defined by Eq. (33) with $\alpha = 1$. The result has the following form: The operators which represent the infinitesimal generators $A_{i,i+1}$ are always given by

$$\bar{\bar{D}}^{(m_{51},z_{52})}(A_{i,i+1}) = \begin{pmatrix} D^{(m_{51},z_{52})}(A_{i,i+1}) & 0 \\ 0 & D^{(m_{51},z_{52})}(A_{i,i+1}) \end{pmatrix} \quad (52)$$

The matrices which represent the elements r, s and t in the representations $\bar{D}^{(m_{51},z_{52})}(A)$ and $\bar{\bar{D}}^{(m_{51},z_{52})}(A)$ are given in Table 5.

At the end of this section let us again summarize the results which we have found for m_{51} halfinteger. We distinguish two cases.

1) An arbitrary IR of the classes I...IV, which are given in Table 3, can be extended to an IR of the covering groups $C_jO(4,1)$ for

j = 2, 3, 5 and 8 on the same representation space. The operators, which represent the elements r, s and t, are given in Table 4. If the IR of the identity component belongs to the classes I, II or III, the corresponding representation of the whole group remains irreducible under restriction to each one of the subgroups $C_jO^+(4,1)$, $C_jO^1(4,1)$, $C_jO^2(4,1)$ and the identity component itself. If it belongs to the class IV the representation of $C_j O(4,1)$ remains irreducible under restriction to $C_jO^+(4,1)$ and $C_jO^1(4,1)$. However, under restriction to $C_jO^2(4,1)$ or $C_jO^{++}(4,1)$ it is reducible and decomposes into the direct sum of two inequivalent representations.

2) The IR's of the groups $C_jO(4,1)$ for j = 1, 4, 6 and 7 cannot be constructed on the representation space of a single IR of the identity component. They are induced from a subgroup. The matrices, which represent the elements r, s and t, are given in Table 5. Under restriction to $C_jO^+(4,1)$ or $C_jO^1(4,1)$ these representations are reducible and decompose into the direct sum of two equivalent representations. Under restriction to $C_jO^2(4,1)$ or $C_jO^{++}(4,1)$ they are also reducible, and they decompose either into two equivalent representations of one of the classes I, II or III, or into four representations, two of which are from class IVa and two from class IVb.

Acknowledgment

The author thanks Prof. J. Ehlers and Dr. S. Strom for some helpful discussions, and Dr. Strom also for the reading of the manuscript.

Table 1

$C_1O(4.1)$

e	-e	r	-r	s	-s	t	-t
-e	e	-r	r	-s	s	-t	t
r	-r	e	-e	t	-t	s	-s
-r	r	-e	e	-t	t	-s	s
s	-s	t	-t	e	-e	r	-r
-s	s	-t	t	-e	e	-r	r
t	-t	s	-s	r	-r	e	-e
-t	t	-s	s	-r	r	-e	e

$C_2O(4,1)$

e	-e	r	-r	s	-s	t	-t
-e	e	-r	r	-s	s	-t	t
r	-r	e	-e	t	-t	s	-s
-r	r	-e	e	-t	t	-s	s
s	-s	-t	t	e	-e	-r	r
-s	s	t	-t	-e	e	r	-r
t	-t	-s	s	r	-r	-e	e
-t	t	s	-s	-r	r	e	-e

$C_3O(4,1)$

e	-e	r	-r	s	-s	t	-t
-e	e	-r	r	-s	s	-t	t
r	-r	e	-e	-t	t	-s	s
-r	r	-e	e	t	-t	s	-s
s	-s	t	-t	-e	e	-r	r
-s	s	-t	t	e	-e	r	-r
t	-t	s	-s	r	-r	e	-e
-t	t	-s	s	-r	r	-e	e

$C_4O(4,1)$

e	-e	r	-r	s	-s	t	-t
-e	e	-r	r	-s	s	-t	t
r	-r	e	-e	t	-t	s	-s
-r	r	-e	e	-t	t	-s	s
s	-s	t	-t	-e	e	-r	r
-s	s	-t	t	e	-e	r	-r
t	-t	s	-s	-r	r	-e	e
-t	t	-s	s	r	-r	e	-e

$C_5O(4,1)$

e	-e	r	-r	s	-s	t	-t
-e	e	-r	r	-s	s	-t	t
r	-r	-e	e	t	-t	-s	s
-r	r	e	-e	-t	t	s	-s
s	-s	-t	t	e	-e	-r	r
-s	s	t	-t	-e	e	r	-r
t	-t	s	-s	r	-r	e	-e
-t	t	-s	s	-r	r	-e	e

$C_6O(4,1)$

e	-e	r	-r	s	-s	t	-t
-e	e	-r	r	-s	s	-t	t
r	-r	-e	e	t	-t	-s	s
-r	r	e	-e	-t	t	s	-s
s	-s	t	-t	e	-e	r	-r
-s	s	-t	t	-e	e	-r	r
t	-t	-s	s	r	-r	-e	e
-t	t	s	-s	-r	r	e	-e

$C_7O(4,1)$

e	-e	r	-r	s	-s	t	-t
-e	e	-r	r	-s	s	-t	t
r	-r	-e	e	-t	t	s	-s
-r	r	e	-e	t	-t	-s	s
s	-s	-t	t	-e	e	r	-r
-s	s	t	-t	e	-e	-r	r
t	-t	s	-s	r	-r	e	-e
-t	t	-s	s	-r	r	-e	e

$C_8O(4,1)$

e	-e	r	-r	s	-s	t	-t
-e	e	-r	r	-s	s	-t	t
r	-r	-e	e	t	-t	-s	s
-r	r	e	-e	-t	t	s	-s
s	-s	-t	t	-e	e	r	-r
-s	s	t	-t	e	-e	-r	r
t	-t	s	-s	-r	r	-e	e
-t	t	-s	s	r	-r	e	-e

Table 1. The multiplication tables for the elements ±e, ±r, ±s and ±t in the covering groups $C_jO(4,1)$.

Table 2

Class	Conditions for m_{51} and z_{52}	SO(4) Content
I	$0 \leq m_{51}$; z_{52} complex number, not integer, $0 \leq y_{52}$ In the following 3 cases these IR's are unitary: a.) $0 \leq m_{51}$; $z_{52} = iy_{52}$; $0 < y_{52}$ b.) $1 \leq m_{51}$; $z_{52} = x_{52}$; $0 \leq x_{52}+\frac{3}{2} < \frac{1}{2}$ c.) $0 = m_{51}$; $z_{52} = x_{52}$; $0 \leq x_{52}+\frac{3}{2} < \frac{3}{2}$	$\lvert m_{41}\rvert \leq m_{51} \leq m_{42}$ The same The same $m_{41} = 0,\ m_{51} \leq m_{42}$
II	$0 \leq m_{51}$; $z_{52} = m_{52}$ integer with $m_{51} \leq m_{52}$ These IR's are finite dimensional and therefore not unitary	$\lvert m_{41}\rvert \leq m_{51} \leq m_{42} \leq m_{52}$
III	$0 \leq m_{51}$; $z_{52} = m_{52}$ integer with $0 \leq m_{52}+1 \leq m_{51}$ These IR's are unitary if the following condition is fulfilled $1 \leq m_{51}$; $m_{52}+1 = 0$	$\lvert m_{41}\rvert \leq m_{52}+1 \leq m_{51} \leq m_{42}$ $m_{41} = 0,\ m_{51} \leq m_{42}$
IVa	$1 \leq m_{51}$; $z_{52} = m_{52}$ integer with $1 \leq m_{52}+2 \leq m_{51}$ These IR's are all unitary	$m_{52}+2 \leq +m_{41} \leq m_{51} \leq m_{42}$
IVb	$1 \leq m_{51}$; $z_{52} = m_{52}$ integer with $1 \leq m_{52}+2 \leq m_{51}$ These IR's are all unitary	$m_{52}+2 \leq -m_{41} \leq m_{51} \leq m_{42}$

Table 2. Classification of the IR's of the identity component if m_{51} is integer.

Table 3

Class	Conditions for m_{51} and z_{52}	SO(4) Content
I	$\frac{1}{2} \leq m_{51}$; z_{52} complex number, not halfinteger, $0 \leq y_{52}$ These IR's are unitary if the following condition is fulfilled: $z_{52} = iy_{52}$; $0 < y_{52}$	$\lvert m_{41}\rvert \leq m_{51} \leq m_{42}$ The same
II	$\frac{1}{2} \leq m_{51}$; $z_{52} = m_{52}$ halfinteger with $m_{51} \leq m_{52}$ These IR's are finite dimensional and therefore not unitary	$\lvert m_{41}\rvert \leq m_{51} \leq m_{42} \leq m_{52}$
III	$\frac{1}{2} \leq m_{51}$; $z_{52} = m_{52}$ halfinteger with $\frac{1}{2} \leq m_{52}+1 \leq m_{51}$ None of these IR's is unitary	$\lvert m_{41}\rvert \leq m_{52}+1 \leq m_{51} \leq m_{42}$
IVa	$\frac{1}{2} \leq m_{51}$; $z_{52} = m_{52}$ halfinteger with $\frac{1}{2} \leq m_{52}+2 \leq m_{51}$ These IR's are all unitary	$m_{52}+2 \leq +m_{41} \leq m_{51} \leq m_{42}$
IVb	$\frac{1}{2} \leq m_{51}$; $z_{52} = m_{52}$ halfinteger with $\frac{1}{2} \leq m_{52}+2 \leq m_{51}$ These IR's are all unitary	$m_{52}+2 \leq -m_{41} \leq m_{51} \leq m_{42}$

Table 3. Classification of the IR's of the identity component if m_{51} is halfinteger.

Table 4

$C_jO(4,1)$	$D^{(m_{51},z_{52})}(r)$	$D^{(m_{51},z_{52})}(s)$	$D^{(m_{51},z_{52})}(r)D^{(m_{51},z_{52})}(s)$
j = 2	$\pm i(-1)^{m_{31}}(-1)^{m_{41}}$	$\pm(-1)^{m_{31}}(-1)^{m_{42}}$	$+D^{(m_{51},z_{52})}(t)$
j = 3	$\pm i(-1)^{m_{31}}(-1)^{m_{41}}$	$\pm i(-1)^{m_{31}}(-1)^{m_{42}}$	$-D^{(m_{51},z_{52})}(t)$
j = 5	$\pm(-1)^{m_{31}}(-1)^{m_{41}}$	$\pm(-1)^{m_{31}}(-1)^{m_{42}}$	$+D^{(m_{51},z_{52})}(t)$
j = 8	$\pm(-1)^{m_{31}}(-1)^{m_{41}}$	$\pm i(-1)^{m_{31}}(-1)^{m_{42}}$	$+D^{(m_{51},z_{52})}(t)$

Table 4. The operators which represent the elements r, s and t in an IR of the covering groups $C_jO(4,1)$ for j = 2, 3, 5 and 8. The expressions in the second and third column have the following meaning: $D^{(m_{51},z_{52})}(r)$ and $D^{(m_{51},z_{52})}(s)$, applied on a vector $|m_{41},m_{42},m_{31},m_{21}\rangle$, give a new vector $|-m_{41},m_{42},m_{31},m_{21}\rangle$ times a scalar factor. This factor is given in the respective columns. In the last column the right hand side of the equation $D^{(m_{51},z_{52})}(r)D^{(m_{51},z_{52})}(s) = \pm D^{(m_{51},z_{52})}(t)$ is given.

Table 5

$C_jO(4,1)$	$\bar{D}^{(m_{51},z_{52})}(r)$	$\bar{D}^{(m_{51},z_{52})}(s)$	$\bar{D}^{(m_{51},z_{52})}(t)$	$\bar{\bar{D}}^{(m_{51},z_{52})}(r)$	$\bar{\bar{D}}^{(m_{51},z_{52})}(s)$	$\bar{\bar{D}}^{(m_{51},z_{52})}(t)$
j = 1	$\begin{pmatrix} iB & 0 \\ 0 & iB \end{pmatrix}$	$\begin{pmatrix} 0 & E \\ E & \end{pmatrix}$	$\begin{pmatrix} 0 & iB \\ iB & 0 \end{pmatrix}$	$\begin{pmatrix} iB & 0 \\ 0 & iB \end{pmatrix}$	$\begin{pmatrix} 0 & B \\ -B & 0 \end{pmatrix}$	$\begin{pmatrix} 0 & -iE \\ iE & 0 \end{pmatrix}$
j = 4	$\begin{pmatrix} iB & 0 \\ 0 & iB \end{pmatrix}$	$\begin{pmatrix} 0 & -E \\ E & 0 \end{pmatrix}$	$\begin{pmatrix} 0 & -iB \\ iB & 0 \end{pmatrix}$	$\begin{pmatrix} iB & 0 \\ 0 & iB \end{pmatrix}$	$\begin{pmatrix} 0 & iB \\ -iB & 0 \end{pmatrix}$	$\begin{pmatrix} 0 & E \\ -E & 0 \end{pmatrix}$
j = 6	$\begin{pmatrix} B & 0 \\ 0 & B \end{pmatrix}$	$\begin{pmatrix} 0 & E \\ E & 0 \end{pmatrix}$	$\begin{pmatrix} 0 & B \\ B & 0 \end{pmatrix}$	$\begin{pmatrix} B & 0 \\ 0 & B \end{pmatrix}$	$\begin{pmatrix} 0 & B \\ -B & 0 \end{pmatrix}$	$\begin{pmatrix} 0 & -E \\ E & 0 \end{pmatrix}$
j = 7	$\begin{pmatrix} B & 0 \\ 0 & B \end{pmatrix}$	$\begin{pmatrix} 0 & -E \\ E & 0 \end{pmatrix}$	$\begin{pmatrix} 0 & B \\ -B & 0 \end{pmatrix}$	$\begin{pmatrix} B & 0 \\ 0 & B \end{pmatrix}$	$\begin{pmatrix} 0 & iB \\ -iB & 0 \end{pmatrix}$	$\begin{pmatrix} 0 & iE \\ -iE & 0 \end{pmatrix}$

Table 5. The operators which represent the elements r, s and t in the IR's $\bar{D}^{(m_{51},z_{52})}(A)$ and $\bar{\bar{D}}^{(m_{51},z_{52})}(A)$ of the covering groups $C_jO(4,1)$ for j = 1, 4, 6 and 7.

References

1. H. Boerner, Darstellungen von Gruppen (Springer, Berlin, 1967).
2. R. Takahashi, Bull. Soc. Math. France **91**, 289 (1963).
3. G. Krafft, Mitt. Math. Sem. Giessen **53**, 1 (1955).
4. A. H. Clifford, Ann. Math. **37**, 533 (1937).
5. L. H. Thomas, Ann. Math. **42**, 113 (1941); T. D. Newton, Ann. Math. **51**, 730 (1950); J. Dixmier, Bull. Soc. Math. France **89**, 9 (1961); S. Ström, Ark. Fysik **30**, 455 (1965).
6. F. Schwarz, in these Proceedings.
7. H. Weyl, The Classical Groups (Princeton University Press, Princeton, 1946).
8. F. Schwarz, University of Texas preprint, CPT-71.

THE MATRIX ELEMENTS OF FINITE TRANSFORMATIONS IN THE DE SITTER GROUP Sp(2,2)†

Wayne J. Holman, III
Theoretical Physics Department
Middle East Technical University
Ankara, Turkey

Abstract

Spinor basis states for the irreducible representations of Sp(2,2), the spinor covering group of the de Sitter group SO(4,1), are obtained by analytic continuation from those of the compact group Sp(4). An algorithm is established for the determination of matrix elements of finite transformations from the form of these basis states, and they are written down explicitly for all the unitary irreducible representations of Sp(2,2).

I. Introduction

In previous papers[1),2)] the basis states of irreducible representations of Sp(4) and the matrix elements of its finite transformations were determined. Also, the analytic continuation of these basis states to those of the irreducible representations of Sp(2,2), a complex extension of Sp(4) was performed. Ref. 2 contains a short bibliography of previous work on the representation theory of Sp(2,2). In the present work we shall consider the analytic continuation of the representation functions of Sp(4) to those of Sp(2,2). It is evident that such a method is possible, i.e., that the representation functions of Sp(2,2) may be obtained from those of Sp(4) by suitable analytic continuation in the coordinates of the group manifold and in one of the two parameters which label the irreducible representations of Sp(4), since the matrix elements of the Sp(2,2) generators may be obtained from those of the Sp(4) generators by such an analytic continuation. The matrix elements of finite transformations are obtained simply by exponentiation of the generators, hence the representation matrices of Sp(2,2) must be analytic continuations of those of Sp(4).

The representation functions of Sp(2,2) are of interest in particle physics even if we do not make the physical assumption of a de Sitter universe, since the representation functions of the Poincaré group are all asymptotic forms of those of Sp(2,2). E.g., the matrix

†Presented at the Symposium on de Sitter and Conformal Groups, University of Colorado, Summer 1970.

elements of a Lorentz transformation along the z-axis in de Sitter space is an analytic function of the invariant which we here denote as Φ. The matrix elements of this transformation in the Poincaré group may be found for all irreducible representations as different asymptotic forms of the de Sitter function. The different representation functions of the Poincaré group cannot be obtained from one another by analytic continuation in the complex planes of mass and spin, but they may be found as different asymptotic limits of a single analytic function, the matrix element of the Lorentz transformation in the de Sitter group. Thus this latter function constitutes an "analytic medium" within which all the irreducible representation functions of the Poincaré group may be connected by analytic continuation and contraction. Ström[3] has investigated the Lie algebras of both the de Sitter and Poincaré groups and found which representations of the former yield any given representation of the latter under Wigner-Inönü contraction and how the appropriate asymptotic limit is to be taken. In the limit of the contraction the SO(4) subgroup of the de Sitter group passes over to the E(3) subgroup of the Poincaré group, and so we obtain representation functions of the Poincaré group in infinite-dimensional matrix form, with states labeled by momentum and helicity.

In crossed-channel momentum-helicity amplitudes of which we perform a harmonic analysis by means of these infinite dimensional representation functions of the Poincaré group, a pole in the complex mass plane will be given as the asymptotic limit of a pole in the plane of the complex parameter Φ which labels the irreducible representations of the de Sitter group Sp(2,2). Ström[4] has also found the form of the matrix which transforms an irreducible representation of Sp(2,2) from the basis in its maximal compact subgroup Sp(2) $\times$ Sp(2) to a basis in one of its Lorentz subgroups SL(2,c). In the limit of the contraction, then, this matrix will transform an irreducible representation of the Poincaré group from its E(3) to its SO(3,1) basis and show how a pole in the complex mass plane of the Poincaré group generates a family of Lorentz poles which in turn generate families of Regge poles. We shall then be able to study the behaviour of these Lorentz poles for scattering through arbitrary angles. The possibility of such a formulation is sufficient to motivate the study of the explicit structure of the Sp(2,2) representation functions.

In Section II we shall review the construction of the representation functions of Sp(4), and we shall obtain two forms for these functions, Eqs. (II.14) and (II.16). Eq. (II.14) provides us with the most convenient expression both for the analytic continuation to the Sp(2,2) case and for the subsequent contraction to the matrix elements of time translations in the Poincaré group. The analytic continuation of (II.16) is extremely difficult since prescriptions must be found which specify a unique continuation of the (9-j) symbol, and such prescriptions do not immediately suggest themselves. In order to obtain such a continuation we shall presumably have to obtain the representation functions of Sp(2,2) in the form (II.14), then apply the Burchnall-Chaundy identities to obtain the form (II.16).

We first perform the continuation of the spinor basis states for irreducible representations of Sp(4) to those of Sp(2,2). In so doing we are given the parametrization of these basis states in Sp(2,2). We obtain two equivalent forms of this parametrization (Eqs. (III.9) and (III.10)), which are given by the two linearly independent analytic continuations of the hypergeometric function which occurs in the semi-maximal basis state. These two forms of the Sp(2,2) basis state are related by unitary equivalence. We then establish an algorithm for the determination of matrix elements of finite transformations from the explicit expression for the basis states: We apply an operator of a finite transformation to the basis state, expand the resulting expression in powers of the elementary spinors a^i_j, then perform the commutations with the spinors of the final state as if the inner product were being taken between Sp(4) states. At this point we perform the analytic continuation in Φ and obtain representation functions of Sp(2,2). We must verify that in the limit of infinitesimal transformations we obtain the correct matrix elements of the generators. Having done so we have verified the parametrization of the spinor basis states. The parametrization of the remaining structural units of the general representation function, i.e., the four Wigner rotation functions in (II.14), is then uniquely determined by the condition of regularity at the origin. The results are expressed in Eqs. (III.18), (III.20), and (III.22), which give the matrix elements of finite Lorentz transformations along the 4-axis in the continuous, positive discrete, and negative discrete irreducible representations, respectively.

II. The Basis States and Representation Functions of Sp(4)

In Ref. 1 the basis states and representation functions of Sp(4) were derived by a lowering operator method in the realization of the Sp(4) irreducible representations in terms of a calculus of boson operators. The generators of the group are:

$$
\begin{array}{ll}
\frac{1}{2}(E_{11} - E_{22}) & \frac{1}{2}(L_{12} + L_{34}) \\
E_{12} & \frac{1}{2}[(L_{14} + L_{23}) + i(L_{31} + L_{24})] \\
E_{21} & \frac{1}{2}[(L_{14} + L_{23}) - i(L_{31} + L_{24})] \\
\frac{1}{2}(E_{33} - E_{44}) & \frac{1}{2}(L_{12} - L_{34}) \\
-E_{34} & \frac{1}{2}[(L_{23} - L_{14}) + i(L_{31} - L_{24})] \\
-E_{43} & \frac{1}{2}[(L_{23} - L_{14}) - i(L_{31} - L_{24})] \\
\sqrt{\tfrac{1}{2}}(E_{14} + E_{32}) & \sqrt{\tfrac{1}{2}}(L_{52} + iL_{15}) \\
\sqrt{\tfrac{1}{2}}(E_{41} + E_{23}) & \sqrt{\tfrac{1}{2}}(L_{52} - iL_{15}) \\
\sqrt{\tfrac{1}{2}}(E_{13} - E_{42}) & \sqrt{\tfrac{1}{2}}(L_{45} + iL_{53}) \\
\sqrt{\tfrac{1}{2}}(E_{31} - E_{24}) & \sqrt{\tfrac{1}{2}}(L_{45} - iL_{53}) \ ,
\end{array}
\qquad (II.1)
$$

where E_{ij} represents a four-dimensional matrix with unity in the (ij) place and zeros elsewhere. The corresponding generators of the locally isomorphic group SO(5) are given in the right column.

The irreducible representations of Sp(4) are labeled, in accordance with Cartan's Main Theorems, by the maximal eigenvalues of the generators of the Cartan subalgebra, $\frac{1}{2}(E_{11} - E_{22})$ and $\frac{1}{2}(E_{33} - E_{44})$, which we denote J_m and Λ_m, respectively. We may take $J_m - \Lambda_m$ to be half of a non-negative integer without loss of generality. The space of the irreducible representation (J_m, Λ_m) is reduced by the representations of its Sp(2) x Sp(2) subgroup generated by the first six generators in the table above. The representations of the Sp(2) subgroup generated by $(\frac{1}{2}(E_{11} - E_{22}), E_{12}, E_{21})$ are labeled by the angular momentum states $|J, M_J\rangle$, and those of the subgroup generated by $(\frac{1}{2}(E_{33} - E_{44}), -E_{34}, -E_{43})$ by the states $|\Lambda, M_\Lambda\rangle$. The states of the Sp(2) x Sp(2) subgroup which occur in the irreducible representation (J_m, Λ_m) are those which satisfy the conditions

$$J_m + \Lambda_m \geqq J + \Lambda \geqq J_m - \Lambda_m \geqq |J - \Lambda| \quad , \qquad (II.2)$$

where these quantities are all either non-negative integers or half-integers.

We realize the generators E_{ij} in terms of the boson operators

$$E_{ij} \to \sum_{p=1,2} a_i^p \bar{a}_j^p \quad , \tag{II.3}$$

where

$$[\bar{a}_j^i , a_\ell^k] = \delta_{ik}\, \delta_{j\ell} \, , \tag{II.4}$$

and we define the vacuum states

$$\bar{a}_j^i \,|0\rangle = \langle 0|\, a_\ell^k = 0 \ . \tag{II.5}$$

In terms of these boson operators the general semimaximal state of an Sp(4) irreducible representation may be written as

$$|J_m, \Lambda_m; J, J; \Lambda, \Lambda\rangle = \left[\frac{(2J_m-2\Lambda_m+1)!\,(J_m+\Lambda_m+J-\Lambda+1)!\,(J_m+\Lambda_m+J+\Lambda+2)!\,(2\Lambda+1)!}{(J_m-\Lambda_m-J+\Lambda)!\,(J_m+\Lambda_m-J-\Lambda)!\,(2J_m+2\Lambda_m+2)!\,(J_m+\Lambda_m-J+\Lambda+1)!} \times \right.$$

$$\left. \frac{1}{(J_m-\Lambda_m+J+\Lambda+1)!\,(J_m-\Lambda_m+J-\Lambda)!\,(J+\Lambda-J_m+\Lambda_m)!\,(2J+1)!}\right]^{\frac{1}{2}} \times$$

$${}_2F_1(J+\Lambda-J_m-\Lambda_m, J-\Lambda-J_m-\Lambda_m-1\,|\,2J+2\,|-\frac{a_{12}}{a_{34}})(a_{34})^{J_m+\Lambda_m-J-\Lambda}(a_{13})^{J+\Lambda-J_m+\Lambda_m}$$

$$(a_1)^{J_m-\Lambda_m+J-\Lambda}(a_3)^{J_m-\Lambda_m-J+\Lambda}\,|0\rangle \ , \tag{II.6}$$

in which

$$a_i \equiv a_i^1$$

$$a_{ij} \equiv a_i^1 a_j^2 - a_j^1 a_i^2 \ . \tag{II.7}$$

We may obtain the most general state by operating on (II.6) with the normalized lowering operators:

$$|J_m, \Lambda_m; J, M_J; \Lambda, M_\Lambda\rangle = \left[\frac{(J+M_J)!\,(\Lambda+M_\Lambda)!}{(2J)!\,(J-M_J)!\,(2\Lambda)!\,(\Lambda-M_\Lambda)!}\right]^{\frac{1}{2}} (E_{21})^{J-M_J}(-E_{43})^{\Lambda-M_\Lambda} \times$$

$$\times\, |J_m, \Lambda_m; J, J; \Lambda, \Lambda\rangle \quad , \tag{II.8}$$

and we may then project out the angular momentum state of the subgroup of rotations in the dimensions (1,2,3) by means of the standard coupling:

$$\sum_{M_J+M_\Lambda=M} C^{J\ \ \Lambda\ \ L}_{M_J M_\Lambda M}\,|J_m,\Lambda_m;J,M_J;\Lambda,M_\Lambda\rangle = |J_m,\Lambda_m;J,\Lambda;L,M\rangle\ . \tag{II.9}$$

The result may be written in the form

$$|J_m,\Lambda_m;J,M_J;\Lambda,M_\Lambda\rangle = \sum_{\substack{j_1+\lambda_1=J_m\\ j_2+\lambda_2=\Lambda_m}}\ \sum_{\substack{m_1+m_2=M_J\\ \mu_1+\mu_2=m_\Lambda}} (-1)^{2j_2}\Bigg[\frac{(2J_m-2\Lambda_m+1)!(J_m+\Lambda_m+J-\Lambda+1)!}{(J_m-\Lambda_m-J+\Lambda)!(2J_m+2\Lambda_m+2)!}$$

$$\times\frac{(J_m-\Lambda_m+J+\Lambda+1)!(J+\Lambda-J_m+\Lambda_m)!(J_m+\Lambda_m+J+\Lambda+2)!(J_m+\Lambda_m-J-\Lambda)!(J_m+\Lambda_m-J+\Lambda+1)!}{(J_m-\Lambda_m+J-\Lambda)!(2J+1)!(2\Lambda+1)!}\times$$

$$\frac{(J+j_1-j_2)!(\Lambda+\lambda_1-\lambda_2)!}{(J+j_2-j_1)!(j_1+j_2-J)!(j_1+j_2+J+1)!(\Lambda+\lambda_2-\lambda_1)!(\lambda_1+\lambda_2-\Lambda)!(\lambda_1+\lambda_2+\Lambda+1)!}\Bigg]^{\frac{1}{2}}(-1)^{\Lambda-M_\Lambda}$$

$$C^{j_1\ j_2\ J}_{m_1 m_2 M_J}\,C^{\lambda_1\lambda_2\ \Lambda}_{\mu_1\mu_2 M_\Lambda}\left(\frac{(a^1_1)^{j_1+m_1}(a^1_2)^{j_1-m_1}}{[(j_1+m_1)!(j_1-m_1)!]^{\frac{1}{2}}}\right)\left(\frac{(a^2_1)^{j_2+m_2}(a^2_2)^{j_2-m_3}}{[(j_2+m_2)!(j_2-m_2)!]^{\frac{1}{2}}}\right)$$

$$\left(\frac{(a^1_3)^{\lambda_1+\mu_1}(a^1_4)^{\lambda_1-\mu_1}}{[(\lambda_1+\mu_1)!(\lambda_1-\mu_1)!]^{\frac{1}{2}}}\right)\left(\frac{(a^2_3)^{\lambda_2+\mu_2}(a^2_4)^{\lambda_2-\mu_2}}{[(\lambda_2+\mu_2)!(\lambda_2-\mu_2)!]^{\frac{1}{2}}}\right)|0\rangle \equiv$$

$$(-1)^{\Lambda-M_\Lambda}\sum_{\substack{j_1+\lambda_1=J_m\\ j_2+\lambda_2=\Lambda_m}}\ \sum_{\substack{m_1+m_2=M_J\\ \mu_1+\mu_2=M_\Lambda}} \mathfrak{F}(J_m\Lambda_m;J\Lambda;j_1 j_2)\,C^{j_1\ j_2\ J}_{m_1 m_2 M_J}\,C^{\lambda_1\lambda_2\ \Lambda}_{\mu_1\mu_2 M_\Lambda}$$

$$\left(\frac{(a^1_1)^{j_1+m_1}(a^1_2)^{j_1-m_1}}{[(j_1+m_1)!(j_1-m_1)!]^{\frac{1}{2}}}\right)\left(\frac{(a^2_1)^{j_2+m_2}(a^2_2)^{j_2-m_2}}{[(j_2+m_2)!(j_2-m_2)!]^{\frac{1}{2}}}\right)\left(\frac{(a^1_3)^{\lambda_1+\mu_1}(a^1_4)^{\lambda_1-\mu_1}}{[(\lambda_1+\mu_1)!(\lambda_1-\mu_1)!]^{\frac{1}{2}}}\right)$$

$$\left(\frac{(a^2_3)^{\lambda_2+\mu_2}(a^2_4)^{\lambda_2-\mu_2}}{[(\lambda_2+\mu_2)!(\lambda_2-\mu_2)!]^{\frac{1}{2}}}\right)|0\rangle\ . \tag{II.10}$$

In order to find the matrix elements of finite rotations we first note that we may parametrize the general rotation operator as

$$(R)\, e^{iL_{45}\theta}\, (R) \;; \qquad \text{(II.11)}$$

where (R) denotes a rotation in the manifold of the Sp(2) x Sp(2) subgroup (locally isomorphic to the SO(4) subgroup of rotations in the dimensions (1,2,3,4)). Hence we need investigate only the matrix elements

$$\langle J_m, \Lambda_m; J', M_J'; \Lambda', M_\Lambda' |\, e^{iL_{45}\theta}\, | J_m, \Lambda_m; J, M_J; \Lambda, M_\Lambda \rangle \qquad \text{(II.12)}$$

where

$$e^{iL_{45}\theta} = e^{\frac{i}{2}\theta(E_{13}+E_{31})}\; e^{-\frac{i}{2}\theta(E_{24}+E_{42})} \;. \qquad \text{(II.13)}$$

This relation prescribes that we treat the bosons in (II.10) as forming angular momentum states in the pairs ($a_1^p\, a_3^p$) and $a_2^p\, a_4^p$). The result is that we may express (II.12) as

$$\sum_{\substack{j_1+\lambda_1=J_m \\ j_2+\lambda_2=\Lambda_m}} \; \sum_{\substack{j_1'+\lambda_1'=J_m \\ j_2'+\lambda_2'=\Lambda_m}} \; \sum_{\substack{m_1+m_2=M_J \\ \mu_1+\mu_2=M_\Lambda}} \; \sum_{\substack{m_1'+m_2'=M_J' \\ \mu_1'+\mu_2'=M_\Lambda'}} (-1)^{\Lambda'-M_\Lambda'} (-1)^{\Lambda-M_\Lambda} \times$$

$$\mathfrak{F}(J_m \Lambda_m ; J' \Lambda' ; j_1' j_2')\; C^{j_1'\; j_2'\; J'}_{m_1' m_2' M_J'}\; C^{\lambda_1' \lambda_2' \Lambda'}_{\mu_1' \mu_2' M_\Lambda'}\; (-1)^{j_1'-j_1+j_2'-j_2}$$

$$d^{\frac{1}{2}(j_1+\lambda_1+m_1+\mu_1)}_{\frac{1}{2}(j_1'-\lambda_1'+m_1'-\mu_1')\, \frac{1}{2}(j_1-\lambda_1+m_1-\mu_1)}(\theta)\; d^{\frac{1}{2}(j_1+\lambda_1-m_1-\mu_1)}_{\frac{1}{2}(j_1'-\lambda_1'-m_1'+\mu_1')\, \frac{1}{2}(j_1-\lambda_1-m_1+\mu_1)}(-\theta)$$

$$d^{\frac{1}{2}(j_2+\lambda_2+m_2+\mu_2)}_{\frac{1}{2}(j_2'-\lambda_2'+m_2'-\mu_2')\, \frac{1}{2}(j_2-\lambda_2+m_2-\mu_2)}(\theta)\; d^{\frac{1}{2}(j_2+\lambda_2-m_2-\mu_2)}_{\frac{1}{2}(j_2'-\lambda_2'-m_2'+\mu_2')\, \frac{1}{2}(j_2-\lambda_2-m_2+\mu_2)}(-\theta)$$

$$C^{j_1\; j_2\; J}_{m_1 m_2 M_J}\; C^{\lambda_1 \lambda_2 \Lambda}_{\mu_1 \mu_2 M_\Lambda}\; \mathfrak{F}(J_m, \Lambda_m ; J\Lambda ; j_1 j_2)\; \delta_{m_i+\mu_i,\, m_i'+\mu_i'}\,, \quad i=1,2. \qquad \text{(II.14)}$$

We now note that

$$d^{\frac{1}{2}(j_i+\lambda_i+m_i+\mu_i)}_{\frac{1}{2}(j_i'-\lambda_i'+m_i'-\mu_i')\;\frac{1}{2}(j_i-\lambda_i+m_i-\mu_i)}(\theta)\; d^{\frac{1}{2}(j_i+\lambda_i-m_i-\mu_i)}_{\frac{1}{2}(j_i'-\lambda_i'-m_i'+\mu_i')\;\frac{1}{2}(j_i-\lambda_i-m_i+\mu_i)}(-\theta)$$

$$= (-1)^{j_i'-j_i-m_i'+m_i} \sum_{K_i} C^{\frac{1}{2}(j_i+\lambda_i+m_i+\mu_i)\;\frac{1}{2}(j_i+\lambda_i-m_i-\mu_i)\;K_i}_{\frac{1}{2}(j_i'-\lambda_i'+m_i'-\mu_i')\;\frac{1}{2}(j_i'-\lambda_i'\;m_i'+\mu_i')\;j_i'-\lambda_i'} \times$$

$$\times\, d^{K_i}_{j_i'-\lambda_i'\;j_i-\lambda_i}(\theta)\; C^{\frac{1}{2}(j_i+\lambda_i+m_i+\mu_i)\;\frac{1}{2}(j_i+\lambda_i-m_i-\mu_i)\;K_i}_{\frac{1}{2}(j_i-\lambda_i+m_i-\mu_i)\;\frac{1}{2}(j_i-\lambda_i-m_i+\mu_i)\;j_i-\lambda_i}$$

$$= (-1)^{j_i'-m_i'-j_i+m_i} \sum_{K_i} C^{j_i'\lambda_i'\;K_i}_{m_i'\mu_i'\;m_i'+\mu_i'}\; d^{K_i}_{j_i'-\lambda_i'\;j_i-\lambda_i}(\theta)\; C^{j_i\lambda_i\;K_i}_{m_i\mu_i\;m_i+\mu_i}\,, \qquad \text{(II.15)}$$

where we have applied the Regge symmetries of the Sp(2) Wigner coefficients. Projecting out the angular momentum state of the subgroup of rotations in dimensions (1,2,3) and combining the Wigner coefficients into (9-j) symbols we obtain:

$$\langle J_m,\Lambda_m;J',\Lambda';L',M'|\,e^{iL_{45}\theta}\,|J_m,\Lambda_m;J,\Lambda;L,M\rangle \equiv \mathcal{D}^{J_m\Lambda_m}_{J'\Lambda';J\Lambda;L}(\theta)$$

$$= (-1)^{\Lambda-\Lambda'} \sum_{\substack{j_1+\lambda_1=J_m\\ j_2+\lambda_2=\Lambda_m}} \;\sum_{\substack{j_1'+\lambda_1'=J_m\\ j_2'+\lambda_2'=\Lambda_m}} \;\sum_{K_1,K_2} (2K_1+1)(2K_2+1)\,\mathfrak{F}(J_m\Lambda_m;J'\Lambda';j_1'j_2')$$

$$[(2J'+1)(2\Lambda'+1)]^{\frac{1}{2}} \begin{Bmatrix} j_1' & \lambda_1' & K_1 \\ j_2' & \lambda_2' & K_2 \\ J' & \Lambda' & L \end{Bmatrix} d^{K_1}_{j_1'-\lambda_1'\;j_1-\lambda_1}(\theta)\; d^{K_2}_{j_2'-\lambda_2'\;j_2-\lambda_2}(-\theta)\;(-1)^{2j_2'-2j_2}$$

$$\mathfrak{F}(J_m\Lambda_m;J\Lambda;j_1j_2)[(2J+1)(2\Lambda+1)]^{\frac{1}{2}} \begin{Bmatrix} j_1 & \lambda_1 & K_1 \\ j_2 & \lambda_2 & K_2 \\ J & \Lambda & L \end{Bmatrix} \delta_{MM'}\,\delta_{LL'}\,, \qquad \text{(II.16)}$$

where the phase $(-1)^{2j_2'-2j_2}$ is canceled by similar factors contained in the monomials $\mathfrak{F}$. It is interesting to compare (II.16) with the matrix element of a finite rotation in the (2,4) plane:

$$\langle J,\Lambda;L',M'|e^{iL_{24}\theta}|J,\Lambda;L,M\rangle = \sum_{\substack{M_J+M_\Lambda=M\\ M'_J+M'_\Lambda=M'}} C^{J\ \ \Lambda\ \ L'}_{M'_J\ M'_\Lambda\ M'}$$

$$d^{J}_{M'_J\ M_J}(\theta)\, d^{\Lambda}_{M'_\Lambda\ M_\Lambda}(-\theta)\ C^{J\ \ \Lambda\ \ L}_{M_J\ M_\Lambda\ M}\ . \tag{II.17}$$

In the matrix element (II.16) we also have a factor of the form $d^{j_1}(\theta)d^{j_2}(-\theta)$, but the coupling is accomplished not by means of the Wigner coefficient but by means of the (9-j) symbol. That is, the (9-j) symbol acts as a generalization of the angular momentum vector coupling coefficient, in which one column denotes the angular momenta being coupled and the differences between the other two columns denote the magnetic quantum numbers. This is a phenomenon of Regge symmetry and has been observed in connection with the construction of SU(3) Wigner coefficients.[5] In fact, Jucys and his collaborators and R. T. Sharp[5),6),7)] have found the following remarkable expressions for doubly and singly stretched (9-j) symbols which bring out their analogy to the (3-j) symbol quite clearly:

$$\begin{Bmatrix} a & b & c\\ d & e & f\\ a+d & b+e & j\end{Bmatrix} = \left[\frac{(2a)!(2b)!(2d)!(2e)!(a+d+b+e-j)!(a+d+b+e+j+1)!}{(2j+1)(a+b-c)!(d+e-f)!(a+c+b+1)!(d+f+e+1)!}\ \times\right.$$

$$\left.\times\ \frac{1}{(2a+2d+1)!(2b+2e+1)!}\right]^{\frac{1}{2}} C^{c\ \ \ f\ \ \ \ j}_{a-b\ d-e\ a+d-b-e}\ , \tag{II.18}$$

$$\begin{Bmatrix} a & b & c\\ d & e & f\\ h & b+e & j\end{Bmatrix} = \left[\frac{(2b)!(2e)!(a+c-b)!(d+f-e)!(a+d-h)!(b+e+h-j)!}{(2j+1)(a+b-c)!(b+c-a)!(a+b+c+1)!(d+e-f)!}\ \times\right.$$

$$\left.\times\ \frac{(b+e+h+j+1)!}{(e+f-d)!(d+e+f+1)!(d+h-a)!(a+h-d)!(a+d+h+1)!(2b+2e+1)!}\right]^{\frac{1}{2}}\ \times$$

$$\sum_x (-1)^x\ \frac{(d+h-a+x)!(2a-x)!}{x!(a+d-h-x)!}\ \left[\frac{(b+c-a+x)!(a+e+f-h-x)!}{(a+c-b-x)!(f+h-a-e+x)!}\right]^{\frac{1}{2}}$$

$$C^{\ \ \ c\ \ \ \ \ \ \ \ \ \ f\ \ \ \ \ \ \ \ \ \ j}_{a-b-x\ \ h-a-e+x\ \ h-b-e}\ . \tag{II.19}$$

When we consider the fact that the (9-j) and (3-j) symbols have isomorphic symmetry groups of 72 elements, it becomes evident that the analogy between them is quite profound.

We note that the function $\mathcal{D}^{J_m \Lambda_m}_{J'\Lambda';J\Lambda;L}(\theta)$ is pure real (pure imaginary) if $\Lambda-\Lambda'$ is integral (half an odd integer). It has the following symmetry properties:

$$\mathcal{D}^{J_m \Lambda_m}_{J'\Lambda';J\Lambda;L}(\theta) = \mathcal{D}^{J_m \Lambda_m}_{\Lambda'J';\Lambda J;L}(\theta) = \mathcal{D}^{J_m \Lambda_m}_{J\Lambda;J'\Lambda';L}(\theta) =$$

$$= \mathcal{D}^{J_m \Lambda_m *}_{J'\Lambda';J\Lambda;L}(-\theta) = (-1)^{2J'-2J}\, \mathcal{D}^{J_m \Lambda_m}_{J'\Lambda';J\Lambda;L}$$

$$= (-1)^{2\Lambda'-2\Lambda}\, \mathcal{D}^{J_m \Lambda_m *}_{J'\Lambda';J\Lambda;L} \quad , \tag{II.20}$$

and the orthogonality property

$$\frac{3}{4}\int_0^{\pi} \sin^3\theta d\theta\, \mathcal{D}^{J_m \Lambda_m}_{J'\Lambda';J\Lambda;L}(\theta)\, \mathcal{D}^{J'_m \Lambda'_m *}_{J'\Lambda';J\Lambda;L}(\theta)$$

$$= \frac{\delta_{J'_m J_m}\,\delta_{\Lambda'_m \Lambda_m}}{\frac{1}{6}(2J_m+2\Lambda_m+3)(2J_m-2\Lambda_m+1)(2\Lambda_m+1)(2J_m+2)} \quad . \tag{II.21}$$

III. The Analytic Continuation of Sp(4) Basis States and Representation Matrices to Those of the de Sitter Group Sp(2,2)

We now adopt the following notation for the state labels and invariants of the irreducible representations of Sp(4):

$$\begin{aligned}
J_m+\Lambda_m+1 &\equiv \Phi \\
J_m-\Lambda_m &\equiv \Delta \\
J+\Lambda+1 &\equiv \varphi \\
J-\Lambda &\equiv \delta \\
J-M_J &\equiv P \\
\Lambda-M_\Lambda &\equiv Q \quad .
\end{aligned} \tag{III.1}$$

We may now obtain the spinor basis states for all the irreducible representations of the de Sitter group Sp(2,2), which is the spinor covering group of SO(4,1), by the analytic continuation of (II.10) in the complex plane of the parameter Φ. This is essentially the method of master analytic representations discussed by Kuriyan, Mukunda, and Sudarshan,[8] except that we are now treating an explicit realization of the spinor basis states rather than their abstract representation.

In order to treat (II.10) as a spinor basis state for representations of a complex extension of the Lie algebra of Sp(4), we omit the vacuum state $|0\rangle$ and regard the operators a_j^i as abstract spinors, i.e., as arbitrary complex numbers. Simultaneously we express the generators E_{ij} as

$$E_{ij} \rightarrow \sum_{p=1}^{2} a_i^p \frac{\partial}{\partial a_j^p} \qquad \text{(III.2)}$$

in place of (II.3). We must also establish an algorithm for the evaluation of inner products of our new basis states as analytic continuations of the inner products of states of the form (II.10), which we can evaluate by means of the commutation of boson operators. Our method will be essentially the following: We shall first determine a parametrization of the states (II.10) such that we obtain the correct matrix elements of the generators of Sp(2,2). We perform the commutations of the spinors a_j^i in Sp(4) according to the rule

$$((a_j^i)^n, (a_{j'}^{i'})^{n'}) = (n)!\delta_{nn'}\delta_{ii'}\delta_{jj'} \quad , \qquad \text{(III.3)}$$

then perform the analytic continuation in those parameters which are to be continued, then perform the internal summations which occur in the matrix element, and obtain the monomial result. When we have found a parametrization of the basis state (II.10) which under this algorithm produces the correct matrix elements of the Sp(2,2) generators, then we have also found the parametrization of the basis state which provides us with the analytic continuation of the representation functions (II.14) to those of Sp(2,2). That is, the analytic continuation of the $_2F_1$ functions contained in the Wigner rotation functions is uniquely specified by the requirement of the regularity of the total function at the identity element of the group manifold, and that of all other structural units is prescribed by the parametrization of the basis states. In this manner we may uniquely specify that analytic continuation of (II.14) which transforms this function into the matrix element of a Lorentz transformation along the 4-axis in an irreducible representation of the de Sitter group Sp(2,2).

We note that we may obtain the finite-dimensional non-unitary irreducible representations of Sp(2,2) immediately from (II.14) or (II.16) simply by the continuation $\theta \to i\zeta$, where ζ is real, $-\infty < \zeta < +\infty$. Then ζ is simply the hyperbolic angle which parametrizes a Lorentz transformation along the 4-axis in de Sitter space. In order to obtain matrix elements of irreducible unitary representations we must perform a further analytic continuation in the Φ plane.

We give now the matrix elements of the generators of rotations in the (i5) plane, $1 \leqq i \leqq 4$:

$$
\begin{aligned}
(E_{41}+E_{23})|\Phi\Delta;JM_J;\Lambda M_\Lambda\rangle = & -A_{J\Lambda} C^{J\ \frac{1}{2}\ J-\frac{1}{2}}_{M_J\ -\frac{1}{2}\ M-\frac{1}{2}} C^{\Lambda\ \frac{1}{2}\ \Lambda-\frac{1}{2}}_{M_\Lambda\ -\frac{1}{2}\ M_\Lambda-\frac{1}{2}} \\
& |\Phi\Delta;J-\tfrac{1}{2},M_J-\tfrac{1}{2};\Lambda-\tfrac{1}{2},M_\Lambda-\tfrac{1}{2}\rangle \\
& -B_{J\Lambda} C^{J\ \frac{1}{2}\ J-\frac{1}{2}}_{M_J\ -\frac{1}{2}\ M_J-\frac{1}{2}} C^{\Lambda\ \frac{1}{2}\ \Lambda+\frac{1}{2}}_{M_\Lambda\ -\frac{1}{2}\ M_\Lambda-\frac{1}{2}} \\
& |\Phi\Delta;J-\tfrac{1}{2},M_J-\tfrac{1}{2};\Lambda+\tfrac{1}{2},M_\Lambda-\tfrac{1}{2}\rangle \\
& +C_{J\Lambda} C^{J\ \frac{1}{2}\ J+\frac{1}{2}}_{M_J\ -\frac{1}{2}\ M_J-\frac{1}{2}} C^{\Lambda\ \frac{1}{2}\ \Lambda-\frac{1}{2}}_{M_\Lambda\ -\frac{1}{2}\ M_\Lambda-\frac{1}{2}} \\
& |\Phi\Delta;J+\tfrac{1}{2},M_J-\tfrac{1}{2};\Lambda-\tfrac{1}{2},M_\Lambda-\tfrac{1}{2}\rangle \\
& -D_{J\Lambda} C^{J\ \frac{1}{2}\ J+\frac{1}{2}}_{M_J\ -\frac{1}{2}\ M_J-\frac{1}{2}} C^{\Lambda\ \frac{1}{2}\ \Lambda+\frac{1}{2}}_{M_\Lambda\ -\frac{1}{2}\ M_\Lambda-\frac{1}{2}} \\
& |\Phi\Delta;J+\tfrac{1}{2},M_J-\tfrac{1}{2};\Lambda+\tfrac{1}{2},M_\Lambda-\tfrac{1}{2}\rangle \quad ,
\end{aligned}
$$

$$
\begin{aligned}
(E_{14}+E_{32})|\Phi\Delta;JM_J;\Lambda M_\Lambda\rangle = & -A_{J\Lambda} C^{J\ \frac{1}{2}\ J-\frac{1}{2}}_{M_J\ \frac{1}{2}\ M_J+\frac{1}{2}} C^{\Lambda\ \frac{1}{2}\ \Lambda-\frac{1}{2}}_{M_\Lambda\ \frac{1}{2}\ M_\Lambda+\frac{1}{2}} \\
& |\Phi\Delta;J-\tfrac{1}{2},M_J+\tfrac{1}{2};\Lambda-\tfrac{1}{2},M_\Lambda+\tfrac{1}{2}\rangle
\end{aligned}
$$

(equation continued)

$$-B_{J\Lambda} C^{J\ \frac{1}{2}\ J-\frac{1}{2}}_{M_J\ \frac{1}{2}\ M_J+\frac{1}{2}} C^{\Lambda\ \frac{1}{2}\ \Lambda+\frac{1}{2}}_{M_\Lambda\ \frac{1}{2}\ M_\Lambda+\frac{1}{2}}$$

$$|\Phi\Delta; J-\tfrac{1}{2}, M_J+\tfrac{1}{2}; \Lambda+\tfrac{1}{2}, M_\Lambda+\tfrac{1}{2}\rangle$$

$$+C_{J\Lambda} C^{J\ \frac{1}{2}\ J+\frac{1}{2}}_{M_J\ \frac{1}{2}\ M_J+\frac{1}{2}} C^{\Lambda\ \frac{1}{2}\ \Lambda-\frac{1}{2}}_{M_\Lambda\ \frac{1}{2}\ M_\Lambda+\frac{1}{2}}$$

$$|\Phi\Delta; J+\tfrac{1}{2}, M_J+\tfrac{1}{2}; \Lambda-\tfrac{1}{2}, M_\Lambda+\tfrac{1}{2}\rangle$$

$$-D_{J\Lambda} C^{J\ \frac{1}{2}\ J+\frac{1}{2}}_{M_J\ \frac{1}{2}\ M_J+\frac{1}{2}} C^{\Lambda\ \frac{1}{2}\ \Lambda+\frac{1}{2}}_{M_\Lambda\ \frac{1}{2}\ M_\Lambda+\frac{1}{2}}$$

$$|\Phi\Delta; J+\tfrac{1}{2}, M_J+\tfrac{1}{2}; \Lambda+\tfrac{1}{2}, M_\Lambda+\tfrac{1}{2}\rangle$$

$$(E_{13}-E_{42})|\Phi\Delta; JM_J; \Lambda M_\Lambda\rangle = -A_{J\Lambda} C^{J\ \frac{1}{2}\ J-\frac{1}{2}}_{M_J\ \frac{1}{2}\ M_J+\frac{1}{2}} C^{\Lambda\ \frac{1}{2}\ \Lambda-\frac{1}{2}}_{M_\Lambda\ -\frac{1}{2}\ M_\Lambda-\frac{1}{2}}$$

$$|\Phi\Delta; J-\tfrac{1}{2}, M_J+\tfrac{1}{2}; \Lambda-\tfrac{1}{2}, M_\Lambda-\tfrac{1}{2}\rangle$$

$$-B_{J\Lambda} C^{J\ \frac{1}{2}\ J-\frac{1}{2}}_{M_J\ \frac{1}{2}\ M_J+\frac{1}{2}} C^{\Lambda\ \frac{1}{2}\ \Lambda+\frac{1}{2}}_{M_\Lambda\ -\frac{1}{2}\ M_\Lambda-\frac{1}{2}}$$

$$|\Phi\Delta; J-\tfrac{1}{2}, M_J+\tfrac{1}{2}; \Lambda+\tfrac{1}{2}, M_\Lambda-\tfrac{1}{2}\rangle$$

$$+C_{J\Lambda} C^{J\ \frac{1}{2}\ J+\frac{1}{2}}_{M_J\ \frac{1}{2}\ M_J+\frac{1}{2}} C^{\Lambda\ \frac{1}{2}\ \Lambda-\frac{1}{2}}_{M_\Lambda\ -\frac{1}{2}\ M_\Lambda-\frac{1}{2}}$$

$$|\Phi\Delta; J+\tfrac{1}{2}, M_J+\tfrac{1}{2}; \Lambda-\tfrac{1}{2}, M_\Lambda-\tfrac{1}{2}\rangle$$

$$-D_{J\Lambda} C^{J\ \frac{1}{2}\ J+\frac{1}{2}}_{M_J\ \frac{1}{2}\ M_J+\frac{1}{2}} C^{\Lambda\ \frac{1}{2}\ \Lambda+\frac{1}{2}}_{M_\Lambda\ -\frac{1}{2}\ M_\Lambda-\frac{1}{2}}$$

$$|\Phi\Delta; J+\tfrac{1}{2}, M_J+\tfrac{1}{2}; \Lambda+\tfrac{1}{2}, M_\Lambda-\tfrac{1}{2}\rangle$$

(equation continued)

$$
\begin{aligned}
(E_{31} - E_{24})|\Phi\Delta;JM_J;\Lambda M_\Lambda\rangle = {} & +A_{J\Lambda} C^{J\ \frac{1}{2}\ J-\frac{1}{2}}_{M_J\ -\frac{1}{2}\ M_J-\frac{1}{2}} C^{\Lambda\ \frac{1}{2}\ \Lambda-\frac{1}{2}}_{M_\Lambda\ \frac{1}{2}\ M_\Lambda+\frac{1}{2}} \\
& |\Phi\Delta;J-\tfrac{1}{2},M_J-\tfrac{1}{2};\Lambda-\tfrac{1}{2},M_\Lambda+\tfrac{1}{2}\rangle \\
& +B_{J\Lambda} C^{J\ \frac{1}{2}\ J-\frac{1}{2}}_{M_J\ -\frac{1}{2}\ M_J-\frac{1}{2}} C^{\Lambda\ \frac{1}{2}\ \Lambda+\frac{1}{2}}_{M_\Lambda\ \frac{1}{2}\ M_\Lambda+\frac{1}{2}} \\
& |\Phi\Delta;J-\tfrac{1}{2},M_J-\tfrac{1}{2};\Lambda+\tfrac{1}{2},M_\Lambda+\tfrac{1}{2}\rangle \\
& -C_{J\Lambda} C^{J\ \frac{1}{2}\ J+\frac{1}{2}}_{M_J\ -\frac{1}{2}\ M_J-\frac{1}{2}} C^{\Lambda\ \frac{1}{2}\ \Lambda-\frac{1}{2}}_{M_\Lambda\ \frac{1}{2}\ M_\Lambda+\frac{1}{2}} \\
& |\Phi\Delta;J+\tfrac{1}{2},M_J-\tfrac{1}{2};\Lambda-\tfrac{1}{2},M_\Lambda+\tfrac{1}{2}\rangle \\
& +D_{J\Lambda} C^{J\ \frac{1}{2}\ J+\frac{1}{2}}_{M_J\ -\frac{1}{2}\ M_J-\frac{1}{2}} C^{\Lambda\ \frac{1}{2}\ \Lambda+\frac{1}{2}}_{M_\Lambda\ \frac{1}{2}\ M_\Lambda+\frac{1}{2}} \\
& |\Phi\Delta;J+\tfrac{1}{2},M_J-\tfrac{1}{2};\Lambda+\tfrac{1}{2},M_\Lambda+\tfrac{1}{2}\rangle \qquad \text{(III.4)}
\end{aligned}
$$

where

$$
\begin{aligned}
A_{J\Lambda} &= \left[\frac{(\Phi-\varphi+1)(\varphi+\Delta)(\varphi-\Delta-1)(\Phi+\varphi)}{(\varphi+\delta-1)(\varphi-\delta-1)}\right]^{\frac{1}{2}} \\
B_{J\Lambda} &= \left[\frac{(\Delta-\delta+1)(\Phi-\delta+1)(\Phi+\delta)(\Delta+\delta)}{(\varphi+\delta-1)(\varphi-\delta+1)}\right]^{\frac{1}{2}} \\
C_{J\Lambda} &= \left[\frac{(\Delta-\delta)(\Phi+\delta+1)(\Phi-\delta)(\Delta+\delta+1)}{(\varphi+\delta+1)(\varphi-\delta-1)}\right]^{\frac{1}{2}} \\
D_{J\Lambda} &= \left[\frac{(\Phi-\varphi)(\varphi+\Delta+1)(\Phi+\varphi+1)(\varphi-\Delta)}{(\varphi+\delta+1)(\varphi-\delta+1)}\right]^{\frac{1}{2}} \qquad \text{(III.5)}
\end{aligned}
$$

We have given here the matrix elements of the generators in Sp(4). We obtain the matrix elements of the Sp(2,2) generators by multiplying each of the four relations (III.4) by $i = \sqrt{-1}$, i.e., we multiply the last four generators in (III.1) by $\sqrt{-1}$, and we identify

$$iL_{\kappa 5} = L_{\kappa 0}, \quad 1 \leq \kappa \leq 4 \qquad \text{(III.6)}$$

as the generator of a Lorentz transformation along the κ-axis. That is, the 5-axis becomes the time axis in the space of SO(4,1). The four quantities (III.5) are then multiplied by $\sqrt{-1}$, and we obtain unitary irreducible representations of Sp(2,2) for the following values of (Φ,Δ):

I. The continuous class:
(a) $\Delta = 1,2,3,\ldots\ ;\ -\Phi(\Phi+1) > 0.$
(b) $\Delta = \frac{1}{2}, \frac{3}{2}, \frac{5}{2},\ldots\ ;\ -\Phi(\Phi+1) > \frac{1}{4}.$
(c) $\Delta = 0;\ -\Phi(\Phi+1) > -2.$

II. The discrete class:
(a) $\Delta = \frac{1}{2},1,\frac{3}{2},2,\frac{5}{2},\ldots;\ \Phi = \Delta-1,\Delta-2,\ldots,0$ or $-\frac{1}{2}$;
$\Delta \geq -\delta \geq \Phi + 1$
(b) $\Delta = \frac{1}{2},1,\frac{3}{2},2,\frac{5}{2},\ldots;\ \Phi = \Delta-1,\Delta-2,\ldots,0$ or $-\frac{1}{2}$;
$\Delta \geq +\delta \geq \Phi+1$
(c) $\Delta = 1,2,3,\ldots\ ;\ \Phi = 0$ (III.7)

The representations II.(a) are designated as positive discrete by Ström[3] and II.(b) as negative discrete.

We must now perform the analytic continuation of the spinor basis states and verify that our parametrization of this continuation yields the correct matrix elements of the generators. Let us consider first the semimaximal state (II.6):

$$\left[\frac{(2\Delta+1)!\,\Gamma(\Phi-\delta+1)\Gamma(\Phi+\varphi+1)(\varphi+\delta)!}{(\Delta-\delta)!\,\Gamma(\Phi-\varphi+1)\Gamma(2\Phi+1)\Gamma(\Phi+\delta+1)(\varphi+\Delta)!(\Delta+\delta)!(\varphi-\Delta-1)!(\varphi-\delta)!}\right]^{\frac{1}{2}} \times$$

$$(-1)^{\varphi-\Delta}\ {}_2F_1\!\left(-\Phi+\varphi,-\Phi-\delta\,\middle|\,\varphi-\delta+1\,\middle|\,-\frac{a_{34}}{a_{12}}\right)(a_{12})^{\Phi-\varphi}(a_{13})^{\varphi-\Delta-1}(a_1)^{\Delta+\delta}(a_3)^{\Delta-\delta}. \qquad \text{(III.8)}$$

We note that we may continue (III.8) in the Φ plane in either of two ways. We may use the expression (II.6), in which the ${}_2F_1$ function has the form

$${}_2F_1\!\left(-\Phi+\varphi,\ -\Phi+\delta\,\middle|\,\varphi+\delta+1\,\middle|\,-\frac{a_{12}}{a_{34}}\right), \qquad \text{(III.9)}$$

or we may reverse the series before performing the analytic continuation in Φ away from its discrete lattice on the positive real axis. In this latter case we must replace (III.9) with the expression

$$\frac{\Gamma(\Phi-\delta+1)(\varphi+\delta)!}{\Gamma(\Phi+\delta+1)(\varphi-\delta)!}\left(-\frac{a_{12}}{a_{34}}\right)^{\Phi-\varphi} {}_2F_1\left(-\Phi+\varphi,-\Phi-\delta\,|\,\varphi-\delta+1\,|-\frac{a_{34}}{a_{12}}\right) . \tag{III.10}$$

It is more convenient to use the form (III.10) and discard an invariant phase $(-1)^{\Phi-\Delta}$ before the analytic continuation. This alternative gives us the basis state (III.8). If we were to perform our continuation from the expression (III.9) we would find that the matrix elements of the generators, evaluated for arbitrary complex Φ, would include a meaningless factor of the sort

$$\left[\frac{\sin\pi(\Phi-\delta)}{\sin\pi(\varphi-\delta)}\right]^2 , \tag{III.11}$$

which is simply unity for Φ on its initial discrete lattice but undefined for Φ at an arbitrary point in its complex plane. Hence we use the expression (III.10) and discard the invariant phase in advance. The resulting expression for the general semimaximal state is given by (III.8).

In order to obtain the general state of the Sp(2,2) representation (Φ,Δ) we apply the normalized lowering operators

$$\left[\frac{(\varphi+\delta-1-P)!(\varphi-\delta-1-Q)!}{(\varphi+\delta-1)!P!(\varphi-\delta-1)!Q!}\right]^{\frac{1}{2}} (E_{21})^P(-E_{43})^Q \tag{III.12}$$

to the state (III.8). We note that this operator commutes with the double spinors a_{12} and a_{34}, so that we have

$$\left[\frac{(\varphi+\delta-1-P)!(\varphi-\delta-1-Q)!}{(\varphi+\delta-1)!P!(\varphi-\delta-1)!Q!}\right]^{\frac{1}{2}} (E_{21})^P(-E_{43})^Q(a_{13})^{\varphi-\Delta-1}(a_1)^{\Delta+\delta}(a_3)^{\Delta-\delta} \longrightarrow$$

$$(-1)^Q\left[\frac{(\varphi+\delta-1-P)!(\varphi-\delta-1-Q)!}{(\varphi+\delta-1)!P!(\varphi-\delta-1)!Q!}\right]^{\frac{1}{2}} \sum_{r,x,y}(-1)^r\binom{\varphi-\Delta-1}{r}\binom{P}{x}\binom{Q}{y}$$

$$\frac{(\varphi+\delta-1-r)!r!(\Delta-\delta+r)!(\varphi-\Delta-1-r)!}{(\varphi+\delta-1-r-P+x)!(r-x)!(\Delta-\delta+r-Q+y)!}\;\frac{1}{(\varphi-\Delta-1-r-y)!}$$

$$(a_1^1)^{\varphi+\delta-1-r-P+x}(a_2^1)^{P-x}(a_1^2)^{r-x}(a_2^2)^x(a_3^1)^{\Delta-\delta+r-Q+y}(a_4^1)^{Q-y}$$

$$(a_3^2)^{\varphi-\Delta-1-r-y}(a_4^2)^y , \tag{III.13}$$

whereupon we expand the double spinors by means of the binomial theorem:

$$(a_{12})^{\Phi-\varphi-\ell} = \sum_{\zeta} (-1)^{\zeta} \binom{\Phi-\varphi-\ell}{\zeta} (a_1^1)^{\Phi-\varphi-\ell-\zeta} (a_2^1)^{\zeta} (a_1^2)^{\zeta} (a_2^2)^{\Phi-\varphi-\ell-\zeta}$$

$$(a_{34})^{\ell} = \sum_{\eta} (-1)^{\eta} \binom{\ell}{\eta} (a_3^1)^{\ell-\eta} (a_4^1)^{\eta} (a_3^2)^{\eta} (a_4^2)^{\ell-\eta} \quad . \qquad \text{(III.14)}$$

The resulting spinor basis state may be written as:

$$|\Phi\Delta;JM_J;\Lambda M_\Lambda\rangle = |\Phi\Delta;\varphi\delta;PQ\rangle = \left[\frac{(2\Delta+1)!\Gamma(\Phi-\varphi+1)\Gamma(\Phi-\delta+1)\Gamma(\Phi+\varphi+1)}{(\Delta-\delta)!(\Delta+\delta)!(\varphi+\Delta)!}\right.$$

$$\left.\frac{\Gamma(\Phi+\delta+1)(\varphi-\Delta-1)!(\varphi+\delta)(\varphi-\delta)(\varphi+\delta-1-P)!(\varphi-\delta-1-Q)!}{\Gamma(2\Phi+1)P!Q!}\right]^{\frac{1}{2}} (-1)^{\varphi-\Delta+Q}$$

$$\sum_{\substack{\ell r\\ \zeta\eta}} \frac{(-1)^{\ell+r+\zeta+\eta} \qquad (\varphi+\delta-1-r)!(\Delta-\delta+r)!}{(\ell-\eta)!(\Phi-\varphi-\ell-\zeta)!(\varphi-\delta+\ell)!(\Phi+\delta-\ell)!\zeta!\eta!r!(\varphi+\delta-1-P-r)!(\Delta-\delta-Q+r)!}$$

$$\frac{1}{(\varphi-\Delta-1-r)!} \quad {}_3F_2(-P,-r,-\Phi+\varphi+\ell+\zeta\,|\,\zeta+1,\varphi+\delta-P-r\,|\,1)$$

$${}_3F_2(-Q,-\varphi+\Delta+1+r,\ -\ell+\eta\,|\,\eta+1,\ \Delta-\delta-Q+r+1\,|\,1)$$

$$(a_1^1)^{\Phi+\delta-1-P-\ell-r-\zeta} (a_2^1)^{\zeta+P} (a_1^2)^{r+\zeta} (a_2^2)^{\Phi-\varphi-\ell-\zeta}$$

$$(a_3^1)^{\Delta-\delta-Q+\ell+r-\eta} (a_4^1)^{\eta+Q} (a_3^2)^{\varphi-\Delta-1-r+\eta} (a_4^2)^{\ell-\eta} \quad . \qquad \text{(III.15)}$$

This expression, then, is the parametrization of (II.10) which we shall use for the analytic continuation in the complex Φ plane. We must establish an algorithm for the evaluation of inner products of the type

$$\langle\Phi\Delta;\varphi'\delta';P'Q'|F|\Phi\Delta;\varphi\delta;PQ\rangle \quad , \qquad \text{(III.16)}$$

where F is some polynomial or transcendental function of the de Sitter group generators (III.2), and Φ is taken at an arbitrary point in its complex plane. We do so by first performing the operation $F|\Phi\Delta;\varphi\delta;PQ\rangle$, writing the result in an expansion in powers of single spinors a^i_j, then commuting the spinors of the initial and final states in (III.16) as if Φ had its Sp(4) values, i.e., according to the rule (III.3), then performing the analytic continuation in Φ back to our chosen point in its complex plane, then summing the internal series which remain after we have eliminated the Kronecker delta functions $\delta_{nn'}$ which result from commutation of the spinors and which are indicated in (III.3). In case the power n in (III.3) involves $+\Phi$, as is the case for the powers of a^1_1 and a^2_2 in (III.15), then the Kronecker delta $\delta_{nn'}$ in (III.3) is to be interpreted as $\delta_{n-\Phi,n'-\Phi}$. We shall form inner products of the type (III.16) only when both initial and final states have the same values of the invariants Φ and Δ.

When we perform the summations which remain after the elimination of the Kronecker deltas, we will in general find that some will be divergent. These may always be regularized by application of the identity

$$\frac{\Gamma(p)\,\Gamma(q)}{\Gamma(p+q)} = \int_0^1 dt\; t^{p-1}(1-t)^{q-1} \quad , \qquad \text{(III.17)}$$

with substitution of the appropriate Pochhammer contour for the line integral on the right, then performance of the summation before the integration.

With this algorithm we find that we obtain correct matrix elements for the generators of the de Sitter group; hence, necessarily, the correct matrix elements for finite transformations. The condition that we obtain correct matrix elements of the generators determines the admissibility of a parametrization of the basis state (III.15), and the parametrization of the basis state determines a unique analytic continuation of the matrix elements of finite transformations, since we must apply the condition of regularity at the origin to these matrix elements.

We obtain, then, the matrix elements of Lorentz transformations along the 4-axis:

$$\langle\Phi\Delta;\varphi'\delta';P'Q'|e^{iL_{04}\zeta}|\Phi\Delta;\varphi\delta;PQ\rangle = (-1)^{\delta'-\varphi'-\delta+\varphi}$$

$$\left[\frac{(2\Delta+1)!\,\Gamma(\Phi-\varphi'+1)\,\Gamma(\Phi-\delta'+1)}{(\Delta-\delta')!\,(\Delta+\delta')!} \times\right.$$

$$\left.\frac{\Gamma(\Phi+\varphi'+1)\,\Gamma(\Phi+\delta'+1)\,(\varphi'-\Delta-1)!\,(\varphi'+\delta')\,(\varphi'-\delta')\,(\varphi'+\delta'-1-P')!\,(\varphi'-\delta'-1-Q')!}{(\varphi'+\Delta)!\,\Gamma(2\Phi+1)\,P'!\,Q'!}\right]^{\frac{1}{2}}$$

(equation continued)

$$\sum_{\substack{\ell r x z \\ \ell' r' x'}} \frac{(-1)^{r'}(\varphi'+\delta'-1-r')!(\Delta-\delta'+r')!}{x'!(\ell'-x')!\Gamma(\Phi-\varphi-\ell'-z+x'+1)(\varphi'-\delta'+\ell')!\Gamma(\Phi+\delta'-\ell'+1)(\varphi-\varphi'+z-x')!r'!}$$

$$\frac{1}{(\varphi'+\delta'-1-P'-r')!(\Delta-\delta'-Q'+r')!(\varphi'-\Delta-1-r')!}$$

$${}_3F_2(-P',-r',-\Phi+\varphi+\ell'+z-x'|\varphi-\varphi'+z-x'+1,\ \varphi'+\delta'-P'-r'|1)$$

$${}_3F_2(-Q',-\varphi'+\Delta+1+r',\ -\ell'+x'|x'+1,\ \Delta-\delta'-Q'+r'+1|1)$$

$$\frac{\Gamma(\Phi+\delta'+\varphi'-\varphi-P'-\ell'-r'-z+x')(\Delta-\delta-Q+\ell+r-x)!(\varphi-\varphi'-P'+z-x')!(x+Q)!(\varphi-\varphi'+r'+z-x')!}{(Q'+\delta'-Q-\delta-\ell'-r'+x'+\ell+r-x)!(Q-Q'+x-x')!(\varphi-\varphi'-r+x+r'-x')!(\ell-x-\ell'+x')!}$$

$$(\varphi-\Delta-1-r+x)!\Gamma(\Phi-\varphi-\ell'-z+x'+1)(\ell-x)!(\cosh\frac{\zeta}{2})^{2\Phi+\delta'+\delta-\varphi'-\varphi-2\ell'-2\ell}$$

$$(\sinh\frac{\zeta}{2})^{\delta'-\varphi'-\delta+\varphi-2\ell'+2\ell}$$

$${}_2F_1(-\Delta+\delta'+Q'-\ell'-r'+x',\Phi+\delta'-P'+\varphi'-\varphi-\ell'-r'-z+x'|$$
$$|\delta'+Q'-\delta-Q-\ell'-r'+x'+\ell+r-x+1|-\sinh^2\frac{\zeta}{2})$$

$${}_2F_1(-x'-Q',\varphi-\varphi'+P'+z-x'+1|Q-Q'+x-x'+1|-\sinh^2\frac{\zeta}{2})$$

$${}_2\Gamma_1(-\varphi'+\Delta+1+r'-x',\varphi-\varphi'+r'+z-x'+1|\varphi-\varphi'+r'-x'-r+x+1|-\sinh^2\frac{\zeta}{2})$$

$${}_2F_1(-\ell'+x',\Phi-\varphi-\ell'-z+x'+1|\ell-\ell'-x+x'+1|-\sinh^2\frac{\zeta}{2})$$

$$(-1)^r\frac{(\varphi+\delta-1-r)!(\Delta-\delta+r)!}{x!(\ell-x)!\Gamma(\Phi-\varphi-\ell-z+x+1)(\varphi-\delta+\ell)!\Gamma(\Phi+\delta-\ell+1)(z-x)!r!(\varphi+\delta-1-P-r)!}$$

$$\frac{1}{(\Delta-\delta-Q+r)!(\varphi-\Delta-1-r)!}\ {}_3F_2(-P,-r,-\Phi+\varphi+\ell+z-x|z-x+1,\varphi+\delta-P-r|1)$$

$${}_3F_2(-Q,-\varphi+\Delta+1+r,-\ell+x|x+1,\Delta-\delta-Q+r+1|1)\left[\frac{(2\Delta+1)!\Gamma(\Phi-\varphi+1)\Gamma(\Phi-\delta+1)}{(\Delta-\delta)!(\Delta+\delta)!}\right.$$

$$\left.\frac{\Gamma(\Phi+\varphi+1)\Gamma(\Phi+\delta+1)(\varphi-\Delta-1)!(\varphi+\delta)(\varphi-\delta)(\varphi+\delta-1-P)!(\varphi-\delta-1-Q)!}{(\varphi+\Delta)!\Gamma(2\Phi+1)P!Q!}\right]^{\frac{1}{2}}. \qquad \text{(III.18)}$$

We now find that the sum over z asymptotically approaches the form of a divergent generalized hypergeometric series of unit argument. The sum over $\ell(\ell')$ is convergent for finite ζ and fixed $\ell'(\ell)$. For infinitesimal ζ the sum over ℓ' is eliminated by Kronecker delta functions of the form $\delta_{\ell',\ell}$, $\delta_{\ell',\ell\pm 1}$. The remaining sum over ℓ is divergent. We may remove the divergences very simply by making the replacements

$$\frac{1}{(\ell-x)!(z-x)!\Gamma(\Phi-\varphi-\ell-z+x+1)} = \frac{(-1)^{\ell}\ell!\,e^{-i\pi(\Phi-\varphi)}}{2\pi i(\ell-\ )!\Gamma(\Phi-\varphi+1)}\int_0^{(1+)} du\, u^{-\Phi+\varphi+\ell-1}(1-u)^{\Phi-\varphi}$$

$$\frac{(-1)^{z-x}e^{-i\pi(\Phi-\varphi-\ell)}}{2\pi i}\int_0^{(1+)} dt\, t^{-\Phi+\varphi+\ell+z-x-1}(1-t)^{\Phi-\varphi-\ell}, \qquad \text{(III.19a)}$$

$$\frac{1}{(\ell'-x')!(\varphi-\varphi'+z-x')!\Gamma(\Phi-\varphi-\ell'-z+x'+1)} = \frac{(1)^{\ell'}\ell'!\,e^{-i\pi(\Phi-\varphi')}}{2\pi i(\ell'-x')!\Gamma(\Phi-\varphi'+1)}$$

$$\int_0^{(1+)} du'\, u'^{-\Phi+\varphi'+\ell'-1}(1-u')^{\Phi-\varphi'}\ \frac{(-1)^{\varphi-\varphi'+z-x'}e^{-i\pi(\Phi-\varphi'-\ell)}}{2\pi i}$$

$$\int_0^{(1+)} dt'\, t'^{-\Phi+\varphi+\ell'+z-x'-1}(1-t')^{\Phi-\varphi'-\ell'}. \qquad \text{(III.19b)}$$

We perform first the sums over x, x', r, and r', which are all finite, then the sum over z, then the contour integrals over t and t', then the sums over ℓ and ℓ', then the contour integrals over u and u'. In (III.19) all the contour integrals are taken as starting at zero, circling unity once in the positive (counter-clockwise) direction, then terminating at zero. Other replacements of factors with Euler-Pochhammer contour integrals which achieve the convergence of all series in (III.18) may easily be found.

We may perform the analytic continuation of (III.18) to the representation functions of the positive discrete series ($\Delta \geqq -\delta \geqq \Phi+1 \geqq 1$ or $\frac{1}{2}$), obtaining in this case

$$\langle \Phi\Delta;\varphi'\delta';P'Q'|e^{iL_{04}\zeta}|\Phi\Delta;\varphi\delta;PQ\rangle = (-1)^{P'+1+\frac{1}{2}(\delta'+\varphi'+\varphi-\delta)-\Phi}$$

$$\left[\frac{(2\Delta+1)!(-\delta'+\Phi)!(\varphi'+\Phi)!}{(\Delta-\delta')!(\Delta+\delta')!}\ \frac{(\varphi'-\Delta-1)!(\varphi'+\delta')(\varphi'\ \delta')(\varphi'+\delta'-1-P')!(\varphi'-\delta'-1-Q')!}{(-\delta'-\Phi-1)!(\varphi'-\Phi-1)!(\varphi'+\Delta)!(2\Phi)!P'!Q'!}\right]^{\frac{1}{2}}$$

$$\sum_{\substack{\ell r\\ \ell' r'}} \frac{1}{2\pi i}\int_0^{(1+)} dt \sum_{zxx'}$$

$$\frac{(-1)^{x'}(\varphi'+\delta'-1-r')!(\Delta-\delta'+r')!(-\delta'-\Phi-1+\ell')!}{x'!(\ell'-x')!(\varphi'-\delta'+\ell')!r'!(\varphi'+\delta'-1-P'-r')!(\Delta-\delta'-Q'+r')!(\varphi'-\Delta-1-r')!}$$

$${}_3F_2(-P',\ -r',\ -\Phi+\varphi+\ell'+z-x'\,|\,\varphi-\varphi'+z-x'+1,\ \varphi'+\delta'-P'-r'\,|\,1)$$

$${}_3F_2(-Q',\ -\varphi'+\Delta+1+r',\ -\ell'+x'\,|\,x'+1,\ \Delta-\delta'-Q'+r'+1\,|\,1)$$

$$\frac{(\Delta-\delta-Q+\ell+r-x)!(\varphi-\varphi'+P'+z-x')!(x+Q)!(\varphi-\varphi'+r'+z-x')!}{(\varphi-\varphi'+z-x')!(-\delta'-\Phi-\varphi'+\varphi+P'+\ell'+r'+z-x')!(Q'+\delta'-Q-\delta-\ell'-r'+x'+\ell+r-x)!}$$

$$\frac{(\varphi-\Delta-1-r+x)!}{(Q+x-Q'-x')!(\varphi-\varphi'-r+x+r'-x')!(\ell-x-\ell'+x')!}\left(\cosh\frac{\zeta}{2}\right)^{2\Phi+\delta'+\delta-\varphi'-\varphi-2\ell'-2\ell}$$

$$\left(\sinh\frac{\zeta}{2}\right)^{\delta'-\varphi'-\delta+\varphi-2\ell'+2\ell}$$

$${}_2F_1(-\Delta+\delta'+Q'-\ell'-r'+x',\ \Phi+\delta'-P'+\varphi'-\varphi-\ell'-r'-z+x'$$
$$|\,\delta'+Q'-\delta-Q-\ell'-r'+x'+\ell+r-x+1\,|\,-\sinh^2\frac{\zeta}{2})$$

$${}_2F_1(-x'-Q',\ \varphi-\varphi'+P'+z-x'+1\,|\,Q-Q'+x-x'+1\,|\,-\sinh^2\frac{\zeta}{2})$$

$${}_2F_1(-\varphi'+\Delta+1+r'-x',\ \varphi-\varphi'+r'+z-x'+1\,|\,\varphi-\varphi'+r'-x'-r+x+1\,|\,-\sinh^2\frac{\zeta}{2})$$

$${}_2F_1(-\ell'+x',\ \Phi-\varphi-\ell'-z+x'+1\,|\,\ell-\ell'-x+x'+1\,|\,-\sinh^2\frac{\zeta}{2})$$

(equation continued)

$$ {}_3F_2(-P,-r,-\Phi+\varphi+\ell+z-x\,|\,z-x+1,\ \varphi+\delta-p-r\,|\,1) $$

$$ {}_3F_2(-Q,-\varphi+\Delta+1+r,\ -\ell+x\,|\,x+1,\ \Delta-\delta-Q+r+1\,|\,1) $$

$$ \frac{(-1)^{x+r+\ell}\, t^{-\Phi+\varphi+\ell+z-x-1}(1-t)^{\Phi-\varphi-\ell}(-\delta-\Phi+\ell-1)!(\varphi+\delta-1-r)!(\Delta-\delta+r)!(-\Phi+\varphi+\ell-1)!}{x!(\varphi-\delta+\ell)!r!(\varphi+\delta-1-P-r)!(\Delta-\delta-Q+r)!(\varphi-\Delta-1-r)!} $$

$$ \left[\frac{(2\Delta+1)!(\Phi-\delta)!(\varphi+\Phi)!(\varphi-\Delta-1)!(\varphi+\delta)(\varphi-\delta)(\varphi+\delta-1-P)!(\varphi-\delta-1-Q)!}{(\Delta-\delta)!(\Delta+\delta)!(\varphi+\Delta)!(\varphi-\Phi-1)!(-\delta-\Phi-1)!(2\Phi)!P!Q!}\right]^{\frac{1}{2}}, \tag{III.20} $$

where the order in which summations and integration are to be performed are explicitly indicated. We may observe that the condition for the convergence of the series

$$ {}_2F_1(-\Phi+\varphi,\ -\Phi-\delta\,|\,\varphi-\delta+1\,|\,1) \tag{III.21} $$

is that Re $(\Phi) > -\frac{1}{2}$; hence the sum over z is the only one which we need to regularize by means of an Euler-Pochhammer contour. We consider the discrete representation $\Phi = -\frac{1}{2}$ to be merely the limit point of the continuous series of representations $\Phi = -\frac{1}{2}+i\rho$, $-\infty < \rho < +\infty$.

The negative discrete series ($\Delta \geqq \delta \geqq \Phi + 1 \geqq 1$ or $\frac{1}{2}$) is most conveniently obtained by use of basis states expressed in terms of the hypergeometric function (III.9) instead of (III.10). We obtain the result in this case:

$$ \langle \Phi\Delta;\varphi'\delta';P'Q'\,|\,e^{iL_{04}\zeta}\,|\,\Phi\Delta;\varphi\delta;PQ\rangle = (-1)^{-\frac{1}{2}(\delta+\delta'+\varphi+\varphi')+Q+\Phi+\Delta} $$

$$ \sum_{\ell'r'\ell r} \frac{1}{2\pi i}\int_0^{(1+)} dt \sum_{zxx'} \left[\frac{(2\Delta+1)!(\delta'+\Phi)!(\varphi'+\Phi)!(\varphi'-\Delta-1)!(\varphi'+\delta')(\varphi'-\delta')}{(\Delta+\delta')!(\Delta-\delta')!(\varphi'+\Delta)!(\delta'-\Phi-1)!}\right. $$

$$ \left.\frac{(\varphi'+\delta'-1-P')!(\varphi'-\delta'-1-Q')!}{(\varphi'-\Phi-1)!(2\Phi)!P'!\,Q'!}\right]^{\frac{1}{2}} \frac{(-1)^{\ell'+r'+x'}(\varphi'+\delta'-1-r')!(\Delta-\delta'+r')!}{x'!(\varphi'+\delta'+\ell')!(\varphi'+\delta'-1-r'-P')!r'!} $$

$$ \frac{(-\Phi+\delta'+\ell'-1)!(-\Phi+\varphi'+\ell'-1)!}{(\Delta-\delta'+r'-Q')!(\varphi'-\Delta-1-r')!}(t)^{-\Phi+\varphi+\ell'+z-x'-1}(1-t)^{\Phi-\varphi'-\ell'} $$

$$ {}_3F_2(-\ell'+x',\ -P',\ -r'\,|\,x'+1,\ \varphi'+\delta'-r'-P'\,|\,1) $$

$$ {}_3F_2(-\Phi+\varphi+\ell'+z-x',\ -Q',\ -\varphi'+\Delta+1+r'\,|\,\varphi-\varphi'+z-x'+1,\ \Delta-\delta'+r'-Q'+1\,|\,1) $$

(equation continued)

$$\frac{(\varphi'+\delta'-P'+\ell'-r'-x'-1)!\,(x'+P')!\,(Q+z-x)!\,(r'+x')!\,(\varphi-\Delta-1-r+z-x)!}{(-\Phi-\Delta+\varphi+\delta+Q+\ell-r+z-x-1)!\,(Q'-Q+\delta-\delta'+\ell'-r'-x'-\ell+r+x)!\,(P'-P+x'-x)!}$$

$$\frac{1}{(r'+x'-r-x)!\,(\ell'-x'-\ell+x)!}\ \left(\cosh\frac{\zeta}{2}\right)^{-2\Phi+\varphi+\varphi'+\delta+\delta'+2\ell+2\ell'}$$

$$\left(\sinh\frac{\zeta}{2}\right)^{\varphi'-\varphi+\delta-\delta'+2\ell'-2\ell}$$

$${}_2F_1\left(-Q'-\varphi+\varphi'-z+x',\ P'+x'+1\,\middle|\,P'-P+x'-x+1\,\middle|\,-\sinh^2\frac{\zeta}{2}\right)$$

$${}_2F_1\left(-\Phi-\Delta+\varphi+\delta'+Q'+\ell'-r'+z-x',\ \varphi'+\delta'-P'+\ell'-r'-x'\right.$$
$$\left.\middle|\,Q'-Q+\delta-\delta'+\ell'-r'-x'-\ell+r+x+1\,\middle|\,-\sinh^2\frac{\zeta}{2}\right)$$

$${}_2F_1\left(-\varphi+\Delta+1+r'-z+x',\ r'+x'+1\,\middle|\,r'+x'-r-x+1\,\middle|\,-\sinh^2\frac{\zeta}{2}\right)$$

$${}_2F_1\left(-\Phi+\varphi+\ell'+z-x',\ \ell'-x'+1\,\middle|\,\ell'-x'-\ell+x+1\,\middle|\,-\sinh^2\frac{\zeta}{2}\right)$$

$${}_3F_2(-\ell+x,\ -P,\ -r\,|\,x+1,\ \varphi+\delta-r-P\,|\,1)$$

$${}_3F_2(-\Phi+\varphi+\ell+z-x,\ -Q,\ -\varphi+\Delta+1+r\,|\,z-x+1,\ \Delta-\delta+r-Q+1\,|\,1)$$

$$\frac{(-1)^x(\varphi+\delta-1-r)!\,(\Delta-\delta+r)!\,(-\Phi+\delta+\ell-1)!}{x!\,(\ell-x)!\,(z-x)!\,(\varphi+\delta+\ell)!\,(\varphi+\delta-1-r-P)!\,r!\,(\Delta-\delta+r-Q)!\,(\varphi-\Delta-1-r)!}$$

$$\left[\frac{(2\Delta+1)!\,(\delta+\Phi)!\,(\varphi+\Phi)!\,(\varphi-\Delta-1)!\,(\varphi+\delta)\,(\varphi-\delta)\,(\varphi+\delta-1-P)!\,(\varphi-\delta-1-Q)!}{(\Delta-\delta)!\,(\Delta+\delta)!\,(\varphi+\Delta)!\,(\delta-\Phi-1)!\,(\varphi-\Phi-1)!\,(2\Phi)!\,P!\,Q!}\right]^{\frac{1}{2}}. \tag{III.22}$$

In both (III.20) and (III.22) we have defined our phases simply by straightforward application of phase conventions in Sp(4), inverting the gamma functions wherever necessary.

We use (III.18), then, for the representation functions of the continuous class, $\Phi = -\frac{1}{2}+i\rho$, $-\infty < \rho < +\infty$ for the principal series, and $\Phi = -\frac{1}{2}+\sigma$, $0 < \sigma < \frac{3}{2}$ for the representations in the supplementary series $\frac{1}{4} > -\Phi(\Phi+1) > -2$. We may then study the forms of this function under the Legendre reflection $\Phi \to -\Phi-1$ by applying the Euler transform to the hypergeometric function in the basis state (III.8):

$$
{}_2F_1\left(-\Phi+\varphi,\ -\Phi-\delta\,\middle|\,\varphi-\delta+1\,\middle|\,-\frac{a_{34}}{a_{12}}\right) = \left(\frac{a_{12}+a_{34}}{a_{12}}\right)^{2\Phi+1}
$$

$$
{}_2F_1\left(\Phi-\delta+1,\ \Phi+\varphi+1\,\middle|\,\varphi-\delta+1\,\middle|\,-\frac{a_{34}}{a_{12}}\right)
$$

$$
= (a_{12})^{-2\Phi-1}\ {}_2F_1\left(\Phi-\delta+1,\ \Phi+\varphi+1\,\middle|\,\varphi-\delta+1\,\middle|\,-\frac{a_{34}}{a_{12}}\right). \tag{III.23}
$$

We may set

$$
(a_{12}+a_{34})^{2\Phi+1} = 1, \tag{III.24}
$$

since this combination of spinors is invariant under all the generators of the group. We may then apply our algorithm for the construction of representation functions from spinor basis states to determine the resulting form of the matrix elements.

References

1. W. J. Holman, III, Jour. Math. Phys. 10, 1710 (1969).
2. W. J. Holman, III, Jour. Math. Phys. 10, 1888 (1969).
3. S. Ström, Arkiv för Fysik 30, 455 (1965).
4. S. Ström, Arkiv för Fysik 40, 1 (1969).
5. R. T. Sharp and H. von Baeyer, Jour. Math. Phys. 7, 1105 (1966).
6. A. Bandzaitis, A. Karosiene, and A. Jucys, Liet. Fiz. Rin. 4, 457 (1964).
7. R. T. Sharp, Nucl. Phys. A95, 222 (1967).
8. J. G. Kuriyan, N. Mukunda, and E. C. G. Sudarshan, Comm. Math. Phys. 8, 204 (1968).

ON THE INDUCED REPRESENTATIONS OF THE (1+4) DE SITTER GROUP AND THEIR REDUCTIONS†

Staffan Ström
Center for Particle Theory
Department of Physics
University of Texas at Austin
Austin, Texas 78712

I. Introduction

The present contribution to this symposium is aimed at giving a survey of those aspects of the representation theory for $\overline{SO_0(1,4)}$‡ which are obtained from the induction procedure. In the course of this survey we will advocate the use of a 2x2 quaternion-matrix description of $\overline{SO_0(1,4)}$. This will allow us to display the induced representations of a group of 2x2 quaternion-matrices, isomorphic to $\overline{SO_0(1,4)}$, in a way which is very similar to the SL(2,C) description of $\overline{SO_0(1,3)}$: A restriction of the quaternions to complex numbers will give us the corresponding SL(2,C) relations. We will construct the unitary representations of $\overline{SO_0(1,4)}$ belonging to the principal continuous series in a form where they are explicitly reduced with respect to a maximal compact subgroup. A more general reduction procedure from the theory of induced representations will then be applied to the problem of reducing the previously constructed representations with respect to representations of a noncompact subgroup, isomorphic to $\overline{SO_0(1,3)}$, a reduction which has obvious applications in the theories which use a global $\overline{SO_0(1,4)}$ space-time group, replacing the Poincaré group.

---

†Presented at the Symposium on De Sitter and Conformal Groups, University of Colorado, Summer 1970.

‡We denote by $\overline{SO_0(1,n)}$ the universal covering group of the group $SO_0(1,n)$ of real linear homogeneous transformations of $(x_0, x_1 \ldots x_n)$ which leave the quadratic form $x_0^2 - x_1^2 - \ldots - x_n^2$ invariant and which are continuously connected to the identity. Similarly $\overline{SO_0(n)}$ denotes the universal covering group of the identity component of the n-dimensional rotation group $SO_0(n)$.

In Section II we briefly review some structural properties (of a sufficiently general class of Lie groups) which will be used in the construction of the induced representations. In Section III we introduce the quaternion matrix group and in Section IV we give some details of the induced representations of $\overline{SO_0(1,4)}$. The alternative way of decomposing the representations is given in Section V together with the transformation connecting the two basis systems involved.

II. Some Structural Properties

In this section we give a brief review of those decompositions of a certain class of groups and their Lie algebras which are of relevance for the induced representations. The structural properties of $\overline{SO_0(1,3)}$ and $\overline{SO_0(1,4)}$ which we will need are actually only those which they have in common with a much larger class of groups. Therefore we let, for the time being, G denote a real, connected semisimple Lie group with a finite center and we introduce the following notations:

$\tilde{g}$: The Lie algebra of G.

K: A maximal compact subgroup of G (all maximal compact subgroups in G are conjugate to each other).

$\tilde{k}$: The Lie algebra of K.

$K(X,Y) \equiv \text{trace}\,\{\text{ad}\,X, \text{ad}\,Y\}$ (the Killing form), where ad X, $X \in \tilde{g}$ is the transformation defined by ad X; $Y \overset{\text{ad}\,X}{\to} [X,Y]$ for all $Y \in \tilde{g}$.

$\tilde{p}$: The set of all $X \in \tilde{g}$ such that $K(X,Y) = 0$ for all $Y \in \tilde{k}$.

We then have the following properties: ($\oplus$ denotes the algebraic direct sum)

$$\tilde{g} = \tilde{k} \oplus \tilde{p} \quad \text{(the Cartan decomposition).} \tag{II.1}$$

$$X \neq 0,\ X \in \tilde{k} \Rightarrow K(X,X) < 0. \tag{II.2}$$

$$x \neq 0,\ X \in \tilde{p} \Rightarrow K(X,X) > 0. \tag{II.3}$$

$$[\tilde{k},\tilde{p}] \subset \tilde{p} \quad \text{(The "k-vector" property of the elements of the set } \tilde{p}.) \tag{II.4}$$

$$[\tilde{p},\tilde{p}] \subset \tilde{k}. \tag{II.5}$$

We further introduce:

$\tilde{a}$: An abelian subalgebra of $\tilde{g}$ which is contained in $\tilde{p}$ and which is maximal among those having these properties.

$\tilde{g}^\alpha$: The set of all elements $X \in \tilde{g}$ such that for all $h \in \tilde{a}$ one has $[h,X] = \alpha(h)X$ (i.e. the "root" $\alpha(h)$ is in the dual space of $\tilde{a}$ and it is independent of the elements in $\tilde{g}^\alpha$).

We then have

$$[\tilde{g}^{\alpha}, \tilde{g}^{\beta}] \subset \tilde{g}^{\alpha+\beta} . \qquad \text{(II.6)}$$

$$\alpha+\beta = 0 \Rightarrow K(\tilde{g}^{\alpha}, \tilde{g}^{\beta}) = 0. \qquad \text{(II.7)}$$

which are the main tools used in proving the various decompositions mentioned below (note that we are here considering real Lie algebras; notable differences between the roots of real and complex Lie algebras are (i) if $\alpha \neq 0$ is a root, 2α may also be a root, and (ii) the multiplicity of a root α (i.e. the dimension of $\tilde{g}^{\alpha}$) can be greater than one). The basis of the decompositions used in the construction of the induced representation is the result

$$\tilde{g} = \sum_{\alpha} \oplus \tilde{g}^{\alpha} \qquad \text{(II.8)}$$

and we therefore consider it in more detail. It is easy to show that

$$\tilde{g}^{o} = \tilde{a} \oplus \tilde{m} \quad \text{where } \tilde{m} = \tilde{g}^{o} \cap \tilde{k}.$$

In order to simplify the following exposition we now restrict ourselves to the case of $SO_o(1,n)$†. Then it is easily seen that $\tilde{a}$ is one-dimensional ($\tilde{p}$ is then spanned by the n "accelerations," any one of which can be chosen as $\tilde{a}$) and consequently the roots are just numbers. We introduce

$$\tilde{n}^{+} = \sum_{\alpha > 0} \tilde{g}^{\alpha}, \qquad \tilde{n}^{-} = \sum_{\alpha < 0} \tilde{g}^{\alpha}$$

i.e. (II.8) now reads

$$\tilde{g} = \tilde{n}^{+} \oplus \tilde{a} \oplus \tilde{m} \oplus \tilde{n}^{-} . \qquad \text{(II.9)}$$

The root property (II.6) furthermore gives

$$[\tilde{m}, \tilde{n}^{\pm}] \subset \tilde{n}^{\pm}, \qquad [\tilde{a}, n^{\pm}] \subset \tilde{n}^{\pm}$$

†We do this only in order to avoid the more involved definition of "positive roots" in the general case. In fact, many of the properties given below for $SO_o(1,n)$ are also valid in the more general case considered up to now.

i.e.

$$\tilde{t}^{\pm} = \tilde{n}^{\pm} \oplus \tilde{a} \oplus \tilde{m} \qquad \text{(II.10)}$$

is a subalgebra of $\tilde{g}$ and $\tilde{n}^{\pm}$ is invariant in $\tilde{t}^{\pm}$ (actually $\tilde{n}^{\pm}$ are nilpotent and $\tilde{a} \oplus \tilde{n}^{\pm}$ are solvable, facts which will not be of immediate use for our present purposes). It is easy to see that the structure of the Lie algebra of $\overline{SO_0(1,n)}$ can be summarized in the following way:

i) $\tilde{a}$ is a one-parameter Lie algebra.
ii) $\tilde{m}$ is isomorphic to the Lie algebra of $SO_0(n-1)$.
iii) $\tilde{n}^{\pm}$ are (n-1) parameter abelian Lie algebras.
iv) The roots are ± 1 with multiplicity (n-1).

The last two properties are clearly demonstrated e.g. by the following choice of $\tilde{a}$ and of bases in $\tilde{n}^{\pm}$:

$$\tilde{a}: \; A_1 = e_{10} + e_{01}, \; \tilde{n}^{\pm}: \; N_i^{\pm} \equiv e_{i0} + e_{0i} \pm (e_{1i} - e_{i1})$$

$$i = 2,3\ldots n$$

where e_{ij} is the matrix $e_{ij} = \{\delta ij\}$.

Besides the above mentioned decompositions of $\tilde{g}$ we will also make use of the Iwasawa decomposition

$$\tilde{g} = \tilde{n}^{+} \oplus \tilde{a} \oplus \tilde{k} \qquad \text{(II.11)}$$

which is easily derived from (II.8). Concerning the corresponding decompositions on the global, group level we will need only the following result: (II.11) has a complete analogy on the group level, i.e. we have ($A, N^{\pm}, M, T^{\pm}$ are the groups which have $\tilde{a}, \tilde{n}^{\pm}, \tilde{m}, \tilde{t}^{\pm}$ as Lie algebras)

$$\overline{SO_0(1,n)} = N^{+}AK \quad (\text{where } K = \overline{SO_0(n)}) \qquad \text{(II.12)}$$

whereas (II.9) can be used to derive that the set of elements

$$g = n^{+}amn^{-}, \quad n^{\pm} \in N^{\pm}, \; m \in M, \; a \in A \qquad \text{(II.13)}$$

form an open dense set in $\overline{SO_0(1,n)}$. In the next section we will give the quaternion matrix form of the decompositions introduced above, for the special case of $\overline{SO_0(1,4)}$.

III. Quaternion Matrix Description of $\overline{SO_0(1,4)}$

The SL(2,C) description of $\overline{SO_0(1,3)}$ is widely used and appreciated as a simplifying device for performing explicit calculations concerning $\overline{SO_0(1,3)}$. The possibility of using an analogous 2x2 quaternion matrix description of $\overline{SO_0(1,4)}$ seems to be less widely known. By giving the relevant formulae in a form which is closely related to the SL(2,C) formalism for $\overline{SO_0(1,3)}$ we hope to convince our audience about the advantages involved.

Let Q and R denote the real quaternions and the real numbers respectively. For an element $x \in Q$ we write

$$x \equiv x_1 + ix_2 + jx_3 + kx_4 \, , \; x_i \in R$$

where $ij = -ji = k$ etc. $i^2 = j^2 = k^2 = -1$. It will be convenient to use the notations

$$\bar{x}_1 \equiv x_1 - ix_2 - jx_3 - kx_4$$

$$\tilde{x} \equiv -j\bar{x}j \equiv x_1 + ix_2 - jx_3 + kx_4$$

$$|x| \equiv (x \cdot \bar{x})^{\frac{1}{2}}$$

and we note that $\overline{(xy)} = \bar{y} \cdot \bar{x}, \widetilde{(xy)} = \tilde{y} \cdot \tilde{x}, |x \cdot y| = |x| \cdot |y|; x,y \in Q$. The set of matrices g of the form

$$g = \begin{pmatrix} \alpha & \beta \\ \gamma & \delta \end{pmatrix}$$

where $\alpha, \beta, \gamma, \delta \in Q$ satisfy

$$\tilde{\alpha}\gamma = \widetilde{(\tilde{\alpha}\gamma)}, \; \tilde{\beta}\delta = \widetilde{(\tilde{\beta}\delta)}, \; \tilde{\alpha}\delta - \tilde{\gamma}\beta = 1 \qquad \text{(III.1)}$$

can be shown to form a group which we will denote $\mathcal{G}$.[1] According to (III.1) we have imposed six real conditions on the original sixteen real parameters in g, i.e. $\mathcal{G}$ is a ten-parameter group. Consider furthermore the real linear homogeneous transformations of the real variables $x_0, x_1 \ldots x_4$ induced by $g \in \mathcal{G}$ according to

$$x \overset{g}{\rightarrow} x' = g x g^{\dagger} \qquad \text{(III.2)}$$

where

$$\tilde{X} \equiv \begin{pmatrix} x_o+x_3 & , & x_1-ix_2-kx_4 \\ x_1+ix_2+kx_4 & , & x_o-x_3 \end{pmatrix} \tag{III.3}$$

and

$$(\mathcal{g}^\dagger)_{rs} = \bar{\mathcal{g}}_{sr}$$

From (III.1) it then follows that this transformation leaves the quadratic form

$$x_o^2 - x_1^2 - \ldots - x_4^2$$

invariant and just as in the case of SL(2,C) and $\overline{SO_o(1,3)}$ one finds that $\mathcal{G}$ is isomorphic to $\overline{SO_o(1,4)}$. By omitting the coordinate x_4 and restricting $\alpha,\beta,\gamma,\delta$ to complex numbers we get back the usual SL(2,C) formalism. We may remark that if

$$\begin{pmatrix} \xi^1 \\ \eta^1 \end{pmatrix} = \mathcal{g}\begin{pmatrix} \xi \\ \eta \end{pmatrix} \qquad \xi,\xi^1,\eta,\eta^1 \in Q, \ \mathcal{g} \in \mathcal{G}$$

it follows that

$$\tilde{\xi}^1\eta^1 - \tilde{\eta}^1\xi^1 = \tilde{\xi}\eta - \tilde{\eta}\xi \ .$$

a relation which is reminiscent of the local isomorphism between $SO_o(1,4)$ and Sp(1,1).[2)]

By considering similarity transformations $C\,\mathcal{G}\,C^{-1}$ of $\mathcal{G}$ with nonsingular quaternion matrices C we can obtain other realizations of $\overline{SO_o(1,4)}$. Here we mention only one such example: the choice

$$C(j) \equiv (\tfrac{1}{2})^{\frac{1}{2}}\begin{pmatrix} 1, & -j \\ -j, & 1 \end{pmatrix}$$

gives the group, denoted G, of matrices

$$g = \begin{pmatrix} a & b \\ c & d \end{pmatrix}$$

where

$$a\bar{b} = \bar{c}d, \quad |a|^2 - |c|^2 = 1, \quad |d|^2 - |b|^2 = 1$$

In this case, one considers the transformations

$$X \xrightarrow{g} X' = gX g^{\dagger}$$

where

$$X = C^{-1}\tilde{X}C = \begin{pmatrix} x_o & , & \bar{x} \\ x & , & x_o \end{pmatrix} x_o \in R, \ x \in R \ .$$

The realization G is particularly convenient if one wants to make use of the SO(3) x SO(3) structure of SO(4).[1)]

The subgroups mentioned in the previous section have the following realizations as subgroups of $\mathcal{G}$: we start by choosing $\mathfrak{a}$ as the group of accelerations in the 3-direction, i.e.

$$\mathfrak{a} \ni a(t) \equiv \begin{pmatrix} e^{t/2} & , & 0 \\ 0 & , & e^{-t/2} \end{pmatrix} \quad t \in R$$

Then

$$\mathfrak{m} \ni m \equiv \begin{pmatrix} \tilde{\bar{u}} & , & 0 \\ 0 & , & u \end{pmatrix} \quad u \in U \quad \text{where U is the set of } x \in Q \text{ with } |x| = 1.$$

$$\mathfrak{n}^{+} \ni n^{+}(u) \equiv \begin{pmatrix} 1 & \mu \\ 0 & 1 \end{pmatrix} \quad \mu \in Z$$

$$\mathfrak{n}^{-} \ni n^{-}(z) \equiv \begin{pmatrix} 1 & 0 \\ z & 1 \end{pmatrix} \quad z \in Z$$

where Z is the set of $x \in Q$ for which $x = \tilde{x}$

$$\mathcal{K} \ni k \equiv \begin{pmatrix} \tilde{\bar{n}} , & -\tilde{\bar{r}} \\ r , & n \end{pmatrix} \text{where } |r|^2 + |n|^2 = 1,\ n\tilde{r} = \widetilde{(n\tilde{r})}$$

The general result that the elements of the form (II.11) form an open dense set in the group can in the present case more specifically be formulated as follows: all $g = \begin{pmatrix} \alpha & \beta \\ \gamma & \delta \end{pmatrix}$ with $\delta \neq 0$ can be written $g = n^{+}(\mu)a(t)mn^{-}(z)$. We note that a first step towards this decomposition is obtained as follows

$$\begin{pmatrix} \alpha , & \beta \\ \gamma , & \delta \end{pmatrix} = \begin{pmatrix} \tilde{\delta}^{-1} , & \beta \\ 0 , & \delta \end{pmatrix} \cdot \begin{pmatrix} 1 , & 0 \\ \delta^{-1}\gamma , & 1 \end{pmatrix} = \tau^{+} n^{-}$$

where $n^{-} \in \mathfrak{n}^{-}$ and $\tau \in \mathfrak{T}^{+} \equiv \mathfrak{n}^{+}\mathfrak{a}\mathfrak{m}$ ($\delta^{-1}\gamma \in Z$ as a consequence of the conditions (III.1)). Again, by restricting the quaternion matrices to complex matrices we get the familiar analogous SL(2,C) decompositions.

IV. Induced Representations of $\mathcal{G}$

We start by giving a short review of the main steps involved in the construction of the induced representations. For further mathematical details we refer to the very readable survey given by W. H. Klink[3] and to the original work of Mackey.[4]

The representation space will be a certain function space. What are the natural candidates? Obviously various spaces of functions defined on the group constitute such candidates since then the operators representing the group elements are naturally defined (by left or right multiplication of the argument). However it is obvious that by taking the space of all square integrable functions f(g):

$$\{f,f\} \equiv \int_G |f(g)|^2 dg < \infty \qquad \text{(IV.1)}$$

and defining the group transformations by multiplication of the argument we would arrive at a representation which is not irreducible (the left or right regular representation). Our aim is to construct unitary irreducible representations (UIR's) and we could now proceed in two different ways: one would be to try to reduce the regular representation into its irreducible parts; another would be to make suitable restrictions on the function space from the beginning, while retaining the desirable feature that the functions are defined on the whole group, and thereby try to obtain irreducible representations. The induced representations may, for our purposes, be looked upon as an elaboration of this second approach (cf. the earlier development of the theory of multiplier representations by Bargmann[5)] and by Gel'fand and Neumark[6)]). We note in passing that the first approach would take us out of the framework of the induced representations: in the reduction of the regular representation into irreducible parts there occurs representations which are not realized as induced representations (the discrete representations).

In the regular representation we have in the scalar product (IV.1) the ordinary product $\bar{f}(g)\cdot f(g)$ of the functions. Consequently this product varies over the whole group. We could then say that our problem is to find a product other than $\bar{f}\cdot f$, which is a function over some set which is smaller than the whole group while the functions are still defined on the whole group. One way of achieving this is the following: consider a (closed) subgroup H of G and a UIR $D^{(\alpha)}(h)$ of H ($h \in H$) and consider functions f(g) which satisfy

$$f(hg) = D^{(\alpha)}(h)\, f(g), \quad h \in H, \quad g \in G \tag{IV.2}$$

(sometimes called the covariance condition), i.e. we then assume that the f(g)'s are vector functions in a representation space $H^{(\alpha)}$ where $D^{(\alpha)}(h)$ is realized, i.e. in more detail (IV.2) reads

$$f_{(m)}(hg) = \sum_{(n)} D^{(\alpha)}_{(m)(n)}(h)\, f_{(n)}(g) \tag{IV.3}$$

where (m) ((n)) is the set of indices which characterizes a basis vector for the UIR $D^{(\alpha)}(h)$, characterized by the set of indices (α). $D^{(\alpha)}_{(m)(n)}(h)$ are the matrix elements of $D^{(\alpha)}(h)$ in this basis. The scalar product in $H^{(\alpha)}$ is

$$(f^1(g), f^2(g))_{(\alpha)} \equiv \sum_{(m)} \overline{f^1_{(m)}(g)}\, f^2_{(m)}(g) . \tag{IV.4}$$

From the unitarity of $D^{(\alpha)}(h)$ it follows that $(f^1(g), f^2(g))_{(\alpha)}$ depends only on the right cosets in G with respect to H:

$$(f^1(hg), f^2(hg))_{(\alpha)} = (f^1(g), f^2(g))_{(\alpha)} \quad .$$

In the above $D^{(\alpha)}$ may be finite- or infinite-dimensional. However, in our application we will only consider the case of a finite-dimensional $D^{(\alpha)}$. We then define a Hilbert space $\mathcal{H}^{(\alpha)}$, later to become our representation space, by completion with respect to the scalar product

$$(f^1, f^2) \equiv \int_{H \backslash G} (f^1(g), f^2(g))_{(\alpha)} \, d\mu(z) \qquad \text{(IV.5)}$$

where $d\mu(z)$ is the quasi-invariant measure (i.e. it sends sets of measure zero into sets of measure zero; it will be enough for us to find one such measure cf. e.g. Ref. 4) on the set of cosets. We can now define a unitary representation $\mathcal{D}^{(\alpha)}(g)$ of G on $\mathcal{H}^{(\alpha)}$ in the following way:

$$\mathcal{D}^{(\alpha)}(g) \, f(g_o) = (\sigma(z,g))^{\frac{1}{2}} f(g_o g) \qquad \text{(IV.6)}$$

where z and g_o belong to the same coset and where $\sigma(z,g)$ is the function which appears in the transformation of the quasi-invariant measure $d\mu(z)$: multiplication from the right with the element $g \in G$ induces a transformation

$$z \overset{g}{\rightarrow} z'(g)$$

in the set of cosets and then

$$d\mu(z) \rightarrow d\mu(z'(g)) \equiv \sigma(z,g) \, d\mu(z).$$

It follows that $\mathcal{D}^{(\alpha)}(g)$ in (IV.6) is unitary. The group property

$$\mathcal{D}^{(\alpha)}(g_1) \, \mathcal{D}^{(\alpha)}(g_2) = \mathcal{D}^{(\alpha)}(g_1 g_2)$$

requires that $\sigma(z,g)$ satisfies (the "multiplier condition")

$$\sigma(z, g_1 g_2) = \sigma(z, g_1) \, \sigma(z'(g_1), g_2) \qquad \text{(IV.7)}$$

a relation which follows from the very definition of the cosets of G with respect to H. The representation $\mathcal{D}^{(\alpha)}(g)$ obtained in this way is called an induced representation and it is said to be induced from the representation $D^{(\alpha)}(h)$ of the subgroup H. We note that in the above procedure the choice of H and $d\mu(z)$ fixed the general framework but it does not provide us with numbers which characterize the representation. These numbers, i.e. (α), come exclusively from the covariance condition, i.e. the choice of $D^{(\alpha)}(h)$. By choosing the subgroup H "sufficiently large" we may hope to obtain a representation $\mathscr{D}^{(\alpha)}(g)$ characterized by "sufficiently many" indices (α) so as to ensure its irreducibility.

We now consider the class of groups $\overline{SO_0(1,n)}$ treated in Section II and choose $H = T^+ \equiv N^+AM$. We define the principal continuous series of representations of $\overline{SO_0(1,n)}$ as those which are obtained by inducing from T^+ and by choosing $D^{(\alpha)}(t^+) \equiv D^{(\alpha_1)}(a)\, D^{(\alpha_2)}(m)$ where $D^{(\alpha_1)}(a)$ and $D^{(\alpha_2)}(m)$ are arbitrary UIR's of A and M respectively (we recall that N^+ is invariant in T^+, i.e. $D^{(\alpha_1)}(a)\, D^{(\alpha_2)}(m)$ is a UIR of T^+). We note that $D^{(\alpha_1)}(a)D^{(\alpha_2)}(m)$ is finite-dimensional and the index (α_1) characterizing the UIR's of the abelian group A is "continuous," hence the name.

The coset space of interest for the principal continuous series of representations is thus $T^+\backslash\overline{SO_0(1,n)}$. From (II.10) and (II.11) it follows that $T^+\backslash\overline{SO_0(1,n)}$ is isomorphic to $\mathfrak{M}\backslash\mathcal{K}$, i.e. it is compact. Furthermore, according to (II.13) there is a one-to-one correspondence between "almost all" elements of $T^+\backslash\overline{SO_0(1,n)}$ and the elements of N^-, i.e. we expect a compactification of N^- to be isomorphic to $T^+\backslash\overline{SO_0(1,n)}$ (cf. below). From now on we restrict ourselves to the case of $\overline{SO_0(1,4)}$ and we use the notations of Section III. There is then a one-to-one correspondence between the elements of $N^{\pm}$ and Z and we simplify the notations slightly and use μ and z to denote both elements in Z and the corresponding elements in $\mathfrak{n}^+$ and $\mathfrak{n}^-$ respectively. The observation that we can describe the coset space of interest as $\mathfrak{M}\backslash\mathcal{K}$ is important since in physical applications we are interested in decomposing the representations with respect to UIR's of $\mathcal{K}$. The realization of $\mathfrak{T}^+\backslash\mathcal{G}$ in terms of Z is, however, most convenient in some contexts and therefore we will develop both to a certain extent. We therefore now consider in some detail the various steps involved, for the case of the group $\mathcal{G}$, in the construction of the principal continuous series of representations (further details are found in Refs. 1 and 7).

An explicit formula for the transformation of the cosets under multiplication from the right is easily obtained by considering the elements $z \in \mathfrak{n}^-$ as representatives of the cosets. If

$$g = \begin{pmatrix} \alpha & \beta \\ \gamma & \delta \end{pmatrix} \quad \text{with } \delta \neq 0$$

and if we introduce the notation

$$z_g = \tau(z.g) \text{ where } z, z.g \in \mathfrak{n}^-$$

it follows that

$$z.g = (z\beta + \delta)^{-1}(z\alpha + \gamma) \qquad \text{(IV.8)}$$

which should be compared with the corresponding fractional transformation obtained in the case of SL(2,C). Note, however, that the order of the quaternion factors in (IV.8) is important.

Consider next the description of $\mathfrak{J}^+\backslash\mathfrak{G}$ in terms of $\mathfrak{m}\backslash\mathfrak{G}$. This is obtained by considering elements $k \in \mathfrak{K}$ and $z \in \mathfrak{n}^-$ which belong to the same coset with respect to $\mathfrak{J}^+$, i.e. k and z for which

$$k = \tau\, z \quad . \qquad \text{(IV.9)}$$

Obviously all other elements in $\mathfrak{K}$ which differ from k in (IV.9) by a left factor $m \in \mathfrak{m}$ also belong to the same coset as z. The relation (IV.9) can be used to obtain the desired relations, for any suitable parameters in $\mathfrak{m}\backslash\mathfrak{K}$, which correspond to (IV.8) (cf. Ref. 7). We denote by $d\mu(k)$ and $d\mu(m)$ the normalized invariant measures on $\mathfrak{K}$ and $\mathfrak{m}$ respectively. Then $d\mu(k)/d\mu(m)$ is a suitable quasivariant measure on $\mathfrak{m}\backslash\mathfrak{K}$ and by direct calculations one finds

$$d\mu(k) = d\mu(z)\, d\mu(m)$$

where

$$d\mu(z) = \frac{4}{\pi^2} \cdot \frac{dz}{(1+|z|^2)^3} \qquad \text{(IV.10)}$$

(i.e. $\int_{\mathfrak{n}^-} d\mu(z) = 1$; actually we compactify Z by the addition of a point z_∞ corresponding to g:s with $\delta = 0$. Since

$$\begin{pmatrix} \alpha & \beta \\ \gamma\ , & 0 \end{pmatrix} \cdot \begin{pmatrix} 0 & 1 \\ -1\ , & 0 \end{pmatrix} \in \mathfrak{J}^+$$

these elements constitute only one coset). Since $d\mu(m)$ is normalized we can therefore write

$$\int_{\mathfrak{n}^-} \varphi(g)\, d\mu(z) = \int_{\mathfrak{K}} \varphi(g)\, d\mu(k) \qquad \text{(IV.11)}$$

if $\varphi(g)$ is a class function with respect to $\mathfrak{J}^+$: $\varphi(\tau g) = \varphi(g)$, $\tau \in \mathfrak{J}^+$.

We also need the transformation properties of the measures. They are most easily expressed in the Z-formalism: from (IV.8) and (IV.10) we get immediately

$$d\mu(z.g) = \sigma(z, g)\, d\mu(z)$$

where

$$\sigma(z.g) = \left[\frac{1+|z|^2}{|z\alpha+\gamma|^2+|z\beta+\delta|^2}\right]^3 \tag{IV.12}$$

Our representation space is to be constructed from functions in a representation space for a UIR of $\mathfrak{m}\cdot\mathfrak{a}$. Since A is one-dimensional and $\mathfrak{m}$ is isomorphic to SU(2) the UIR's of $\mathfrak{a}\cdot\mathfrak{m}$ are characterized by $\alpha_1 \equiv \rho$ and $\alpha_2 \equiv \ell$ where ρ is an arbitrary real number and ℓ is a non-negative integer or half-integer. Thus we shall consider vector functions f^ℓ with components f^ℓ_m, $m=-\ell, -\ell+1 \ldots \ell-1, \ell$, which satisfy

$$f^\ell_m(\tau g) = e^{i\rho t} \sum_{n=-\ell}^{\ell} D^\ell_{mn}(m)\, f^\ell_n(g) \tag{IV.13}$$

where $\tau = \mu a(t) m$ and $D^\ell_{mn}(m)$ are Wigner matrices, and we write

$$(f^\ell, f'^\ell)_{[\ell]} \equiv \sum_{m=-\ell}^{\ell} \cdot \overline{f^\ell_m} \cdot f'^\ell_m$$

The representation space for the representation (ρ,ℓ) is denoted $\mathcal{H}^{(\rho,\ell)}$ and it is obtained from the scalar product

$$(f^\ell, f'^\ell) \equiv \int_Z (f^\ell(g), f'^\ell(g))_{[\ell]}\, d\mu(z)$$

From (IV.11) it is clear that $\mathcal{H}^{(\rho,\ell)}$ can equivalently be characterized as the Hilbert space which has the scalar product

$$(f^\ell, f'^\ell) \equiv \int_K (f^\ell(g), f'^\ell(g))_{[\ell]}\, d\mu(k) \tag{IV.14}$$

A unitary representation (ρ,ℓ) of $\mathcal{G}$ is then defined in $\mathcal{H}^{(\rho,\ell)}$ by

$$\mathcal{D}^{(\rho,\ell)}(g)\, f^\ell_m(g_0) = (\sigma(z,g))^{\frac{1}{2}} \cdot f^\ell_m(g_0 g) \tag{IV.15}$$

It can be shown that for $\rho \neq 0$ and ℓ a non-negative half-integer and for ρ arbitrary real and ℓ non-negative integer, (ρ,ℓ) is also irreducible.[1] The representations (ρ,ℓ) and $(-\rho,\ell)$ are equivalent.

In physical applications the decomposition of (ρ,ℓ) with respect to UIR's of K is often of immediate interest. We consider this decomposition in some more detail. It will then be convenient to rewrite (IV.15) in a form where only functions on $\mathcal{K}$ appear. We write $\delta(g)$ to denote that δ is the (22) element of g and we use $k(g)$ to denote an element in $\mathcal{K}$ which belongs to the same right coset of $\mathcal{G}$ with respect to $\mathcal{J}^+$ as g i.e. if $g = \tau z$ then $k(g) = \tau' z$. Using

$$|\delta(g k^{-1}(g))|^2 = |\gamma|^2 + |\delta|^2, \quad g = \begin{pmatrix} \alpha & \beta \\ \gamma & \delta \end{pmatrix}$$

and (IV.12) the defining relation (IV.15) can be rewritten

$$\mathcal{D}^{(\rho,\ell)}(g)\, f^{\ell}_{m}(g_o) = \left|\frac{\delta(g_o k^{-1}(g_o))}{\delta(g_o\, k^{-1}(g_o g))}\right|^3 f^{\ell}_{m}(g_o g) \qquad \text{(IV.16)}$$

$$(g_o = \tau_o\, z)$$

Next we choose a representative $g_o = k_o \in \mathcal{K}$ and in analogy with the notation $z.g$ introduced above we write $k(k_o g) \equiv k_o . g$ i.e. $k_o g = \tau(k_o . g)$ and $k_o . g$ is defined only up to a left factor $m \in \mathfrak{m}$. Furthermore if $\tau = \mu a(t) m$ we write $t = t(\tau)$, $m = m(\tau)$ and also in general $t = t(g)$ if $g = \mu a(t) k$, $k \in \mathcal{K}$. Then

$$|\delta(\tau)| = |\delta(m a(t))| = |\delta(m)| \cdot |\delta a(t)| = |\delta(a(t))| = e^{-\frac{t(\tau)}{2}}$$

With these notations and using (IV.13) we get the following final expression from (IV.16)

$$\mathcal{D}^{(\rho,\ell)}(g)\, f^{\ell}_{m}(k_o) = |\delta(k_o g (k_o . g)^{-1})|^{-3-2i\rho} \cdot$$
$$\sum_{n=-\ell}^{\ell} D^{\ell}_{mn}(m(k_o g (k_o . g)^{-1}))\, f^{\ell}_{n}(k_o . g) \qquad \text{(IV.17)}$$

which explicitly defines the representation (ρ,ℓ) on functions defined on $\mathcal{K}$. The indefiniteness in the element $k_o . g$ was a left factor $m \in \mathfrak{m}$. Equation (IV.17) still defines (ρ,ℓ) uniquely: from (IV.13), the representation property of $D^{\ell}_{mn}(m)$ and the fact that $|\delta(m)| = 1$ it is easily seen that the r.h.s. of (IV.17) is independent of the choice of $k_o . g$.

The UIR's of $\mathcal{K}$ can be characterized by two numbers p and q which are both integral or half-integral and where $p \geq |q|$. The reduction of the representations (ρ,ℓ) with respect to UIR's of $\mathcal{K}$ is well known.[8] According to Ref. 8 the representation space $\mathcal{H}^{(\rho,\ell)}$ can be constructed as an infinite direct sum of representation spaces $\mathcal{H}^{(p,q)}$

for the UIR's (p,q) of $\mathcal{K}$, where p takes the values ℓ, $\ell+1$, $\ell+2$,... and q takes the values $-\ell$, $-\ell+1$,...$\ell-1$, ℓ for each value of p and each representation (p,q) occurs with multiplicity one.

In our realization of $\mathcal{H}^{(p,\ell)}$ the spaces $\mathcal{H}^{(p,q)}$ are spaces of functions defined on $\mathcal{K}$ and, according to (IV.13), satisfying

$$f^{\ell}_{m}(m k) = \sum_{\ell=-n}^{n} D^{\ell}_{mn}(m)\, f^{\ell}_{n}(k) \tag{IV.18}$$

Once we have chosen bases in the spaces $\mathcal{H}^{(p,q)}$ and thus in $\mathcal{H}^{(p,\ell)}$, the definitions (IV.17) and (IV.14) of the operator $\mathcal{D}^{(p,\ell)}(g)$ and the scalar product immediately give an integral formula for the matrix elements of an arbitrary finite transformation. A convenient choice is the following: we choose an angular momentum basis for the UIR's (j) of $\mathfrak{m}$ and construct $\mathcal{H}^{(p,q)}$ as a direct sum of representation spaces $\mathcal{H}^{(j)}$ of the UIR's (j). Then the values $j = |q|, |q|+1 \ldots p-1, p$ all occur once. The matrix elements in this basis of a transformation $k \in \mathcal{K}$ are denoted $R^{jj'}_{mm'}(k;\, p,q)$. If we parameterize k according to

$$k = m_1 k_{34} m_2 \tag{IV.19}$$

where $m_1, m_2 \in \mathfrak{m}$ and $k_{34} \equiv k_{34}(\psi)$ is a rotation of an angle ψ, $0 \le \psi \le \pi$ in the (34)-plane, $R^{jj'}_{mm'}(k;\, p,q)$ decomposes as follows

$$R^{jj'}_{mm'}(k;\, p,q) = \sum_{r=-\min(j,j')}^{r=\min(j,j')} D^{j}_{m,r}(m_1) R^{jj'}_{r}(\psi;pq)\, D^{j'}_{rm'}(m_2) \tag{IV.20}$$

where $R^{jj'}_{r}(\psi;p,q)$ are well known functions. Any fixed row in this general matrix is again an angular momentum basis now realized in terms of functions defined on $\mathcal{K}$. By considering the row ℓ, m we obtain a set of functions which furthermore satisfy the covariance condition (IV.18):

$$R^{\ell,j'}_{m,m'}(m \cdot k;\, p,q) = \sum_{s=-\ell}^{s=\ell} D^{\ell}_{m,s}(m)\, R^{\ell j'}_{sm'}(k;\, pq) \tag{IV.21}$$

The dimension of a UIR (p,q) is $(p+1)^2 - q^2$, i.e. the $R^{jj'}_{mm'}$ satisfy

$$\int d\mu(k)\, \overline{R^{j_1 j_2}_{m_1 m_2}(k;\, p,q)}\, R^{j_3 j_4}_{m_3 m_4}(k;\, p',\, q') =$$

$$((p+1)^2 - q^2)^{-1} \cdot \delta_{m_1 m_3}\, \delta_{m_2 m_4}\, \delta_{j_1 j_3}\, \delta_{j_2 j_4}\, \delta_{p\, p'}\, \delta_{q\, q'}$$

and consequently the functions

$$N(p,q;\, \rho,\ell)((p+1)^2 - q^2)^{\frac{1}{2}} \cdot R^{\ell j'}_{mm'}(k;\, pq), \tag{IV.22}$$

where $N(p,q;\, \rho,\ell)$ is a phase factor and where p,q,j',m' take the above-mentioned values, constitute an orthonormal basis in $\mathcal{H}^{(\rho,\ell)}$. We consider briefly the general matrix elements in this basis. They are denoted $D^{pqp'q'}_{jmj'm'}(g;\rho,\ell)$. In order to obtain the simplest possible form for these elements we use the decomposition

$$g = k_1 a_3(t) k_2,\quad k_1,\, k_2 \in K,\ t \geq 0$$

(valid for all $g \in G$). The only new functions to be determined are then the matrix elements of the "boost" $a_3(t)$, which we denote $A^{pqp'q'}_j(t;\, \rho,\ell)$ (they are diagonal in the j and m indices and independent of m). Thus the general matrix elements are

$$D^{pqp'q'}_{jmj'm'}(g;\rho,\ell) = \sum_{j''=\max(|q|,|q'|)}^{j''=\min(p,p')}\ \sum_{m''=-j''}^{m''=j''} R^{jj''}_{mm''}(k_1;\, pq)$$

$$A^{pqp'q'}_{j''}(t;\, \rho,\ell)\, R^{j''j'}_{m''m'}(k_2;\, p',\, q') \tag{IV.23}$$

and the integral formula for $A^{pqp'q'}_j(t;\, \rho,\ell)$ which one obtains from (IV.14) is

$$A^{pqp'q'}_j(t;\, \rho,\ell) = \overline{N(p,q,\rho,\ell)}\, N(p',q';\, \rho,\ell)((p+1)^2-q^2)((p'+1)^2-q'^2))^{\frac{1}{2}}$$

$$\cdot (2j+1)^{-\frac{1}{2}} \cdot \sum_k \int_0^{\pi} \frac{2}{\pi} \sin^2\psi\, d\psi \cdot R^{\ell,j}_k(\psi;pq)(\mathrm{Ch}t - \cos\psi\, \mathrm{Sh}t)^{-3/2-i\rho} R^{\ell j}_k(\psi';p'q')$$

where

$$\cos \psi' = \frac{\cos \psi - \text{Tght}}{1 - \cos \psi \, \text{Tght}}$$

The differential properties of $A_j^{pqp'q'}(t;\, \rho, \ell)$ are obtained from a consideration of the action of the generator of the boost a_3 (t) on the general matrix element (IV.23). We refer to Ref. 7 for more details. Other properties of these functions may be studied by using analytic continuation from corresponding matrix elements in a UIR of SO(5).[9] The advantage of using bases and matrix elements in which the covariance condition takes a simple form will be seen again in the next section where we consider a different decomposition of the representations (ρ, ℓ).

V. Decomposition of the Representations (ρ, ℓ) with Respect to UIR's of a Noncompact Subgroup

In the present section we illustrate further the power of the theory of induced representation by applying it to the problem of decomposing the representations (ρ, ℓ) with respect to UIR's of a noncompact subgroup which is isomorphic to $\overline{SO_0(1,3)}$. As is well illustrated by contributions to this symposium, there is presently a vivid interest in exploring the assumption that the apparent Poincaré invariance is in fact only approximate and should be replaced by an exact $\overline{SO_0(1,4)}$ invariance, i.e. we should use a $\overline{SO_0(1,4)}$ invariant particle classification and $\overline{SO_0(1,4)}$ covariant field equations, etc. The decomposition we are going to consider here will in such a framework replace the decomposition of the UIR's (m,s) of the Poincaré group corresponding to a positive m^2 and spin s with respect to UIR's of the Lorentz group.[10] Particular cases may also be of interest in special dynamical models where SO(1,4) appears as the dynamical group (or a subgroup of it).

The decomposition treated in Section IV was relatively simple inasmuch as the representation space $\mathcal{H}^{(\rho,\ell)}$ could be constructed as a discrete direct sum of representation spaces $\mathcal{H}^{(p,q)}$. When we consider a decomposition with respect to UIR's of a noncompact subgroup we expect that $\mathcal{H}^{(\rho,\ell)}$ has to be constructed as a direct integral of representation spaces. Before entering upon the details of the problem at hand we give first a very brief review of some relevant results from the theory of induced representation. The use of these results will enable us to give the solution of our problem in global form.

The problem we are considering is thus of the following kind: suppose we have two subgroups H_1 and H_2 of a group G; suppose further that $L_1(H_1)$ is a UIR of H_1 and that a representation $\mathcal{D}^{L_1(H_1)}$ of G has been construced by induction, how is then $\mathcal{D}^{L_1(H_1)}$ decomposed with respect to UIR's of H_2? The key to the resolution of this problem is provided (under some weak conditions on H_1 and H_2, see Ref. 4) by the double cosets of G with respect to H_1 and H_2, i.e. the sets S_γ of the form

$$S_\gamma \equiv H_1 g_\gamma H_2, \quad g_\gamma \text{ fix}, \; g_\gamma \in G.$$

By considering sufficiently many g_γ we get all elements in G, i.e. there is a minimal set of indices I, $\gamma \in I$ such that

$$G = \bigcup_{\gamma \in I} H_1 g_\gamma H_2$$

Correspondingly we split up a function f(g) defined on G into components $f_\gamma(g)$ according to

$$f_\gamma(g) \equiv f(g) \qquad \text{for } g \in S_\gamma \text{ of}$$

$$f_\gamma(g) = 0 \qquad \text{otherwise}$$

i.e. the representation space $\mathcal{H}^{L_1(H_1)}$ for the representation $\mathcal{D}^{L_1(H_1)}$ is decomposed into the direct sum (in a generalized sense, depending on the structure of I) of subspaces $\mathcal{H}_\gamma^{L_1(H_1)}$ formed by the functions $f_\gamma(g)$.

If $g_\gamma \in S_\gamma$ then obviously $g_\gamma h_2 \in S_\gamma$ for $h_2 \in H_2$, i.e. $\mathcal{H}_\gamma^{L_1(H_1)}$ is invariant under right translation by an element in H_2, i.e. $\mathcal{H}_\gamma^{L_1(H_1)}$ is the representation space for a (usually very general) representation of H_2. Thus once the set I has been determined the remaining task is to decompose the representation of H_2 acting in $\mathcal{H}_\gamma^{L_1(H_1)}$, and denoted $D_\gamma^{(H_1)}$, into irreducible parts. Now it can be shown (Mackey) that the $D_\gamma^{(H_1)}$'s can be constructed as induced representations, induced from particular subgroups H_γ of H_2 defined by

$$H_\gamma = g_\gamma^{-1} H_1 g_\gamma \cap H_2 \tag{V.1}$$

In order to make this plausible we define

$$f^\gamma(h_2) \equiv f_\gamma(g_\gamma h_2), \qquad h_2 \in H_2$$

Let $h_\gamma \in H_\gamma$. Then, with $h_\gamma = g_\gamma^{-1} h_1 g_\gamma$, $h_1 \in H_1$,

$$f^{\gamma}(h_{\gamma} h_2) \equiv = f^{\gamma}(g_{\gamma}^{-1} h_1 g_{\gamma} h_2) = f_{\gamma}(h_1 g_{\gamma} h_2) = L_1(h_1) f_{\gamma}(g_{\gamma} h_2) =$$

$$= L_1(g_{\gamma} h_{\gamma} g_{\gamma}^{-1}) f^{\gamma}(h_2)$$

i.e. the $f^{\gamma}(h_2)$'s satisfy a modified covariance condition:

$$f^{\gamma}(h_{\gamma} h_2) = L_1^{\gamma}(h_{\gamma}) f^{\gamma}(h_2) \tag{V.2}$$

where L_1 completely determines L_1^{γ}:

$$L_1^{\gamma}(h_{\gamma}) \equiv L_1(g_{\gamma} h_{\gamma} g_{\gamma}^{-1})$$

The norm in $\mathcal{H}_{\gamma}^{L_1(H_1)}$ is (cf. Section IV)

$$\int d\mu(z) || f^{\gamma}(g) ||_{L_1}$$

where

$$|| f^{\gamma}(g) ||_{L_1} = || f^{\gamma}(h_1 g_{\gamma} h_2) ||_{L_1} = || f(g_{\gamma} h_2) ||_{L_1} = || f^{\gamma}(h_2) ||$$

$$\equiv || f^{\gamma}(h_2) ||_{L_1^{\gamma}}$$

which according to the above is a class function with respect to H_{γ}, i.e. we may consider it as a function defined on $H_{\gamma} \backslash H_2$. Thus, after a corresponding factorization of the measure, the representation spaces $\mathcal{H}_{\gamma}^{L_1(H_1)}$ are seen to carry an induced representation of H_2, induced from H_{γ} and the norm in $\mathcal{H}_{\gamma}^{L_1(H_1)}$ is given by

$$||f||_{\gamma}^2 \equiv \int_{H_{\gamma} \backslash H_2} d\mu'(\xi) || f^{\gamma}(h_2) ||_{L_1^{\gamma}} \tag{V.3}$$

In our application we shall consider the subgroup $\mathfrak{B}$ of $\mathfrak{G}$ which is isomorphic to $SO_0(1,3)$ and which acts in the (0,1,2,4) space. Its maximal compact subgroup is then $\mathfrak{m}$ and all elements $\mathcal{b} \in \mathfrak{B}$ can be written

$$\mathcal{b} = m_1 a_4(t) m_2 \tag{V.4}$$

The representations (ρ, ℓ) of $\mathfrak{G}$ were constructed by induction from the subgroup $\mathfrak{J}^+$, i.e. we need the double cosets of $\mathfrak{G}$ with respect to $\mathfrak{J}^+$ and $\mathfrak{B}$. Using e.g. the parameters $g = n^+(\mu)\, a(t)\, k$ with k parametrized according to (V.19), one finds that any element of $\mathfrak{G}$ can be written in one of the following ways

$$\text{(i) } g = \tau e^+ b, \quad \text{(ii) } g = \tau e^- b, \quad \text{(iii) } g = \tau e^0 b$$

where $\tau \in \mathfrak{J}^+$, $b \in \mathfrak{B}$ and where $e^+ = k_{34}(0)$ (i.e. the unit element) $e^- = k_{34}(\pi)$, $e^0 = k_{34}(\pi/2)$.

The three classes can also be characterized as follows

(i) The g's with $|\delta| > |\gamma|$
(ii) The g's with $|\delta| < |\gamma|$
(iii) The g's with $|\delta| = |\gamma|$

From this it is clear that class (iii) consists of elements characterized by one parameter less than the elements in (i) and (ii). Therefore this class will not contribute to the integral formulae below and we will omit it completely in the following, i.e. the indices (γ) are taken as (+) and (-) and we write

$$f_m^{\ell\pm}(b) \equiv f_m^{\ell}(e^{\pm} b)$$

Both the groups H_γ, i.e. H_+ and H_- are isomorphic to $\mathfrak{m}$ and therefore we may write the modified covariance condition as

$$f_m^{\ell\pm}(mb) = \sum_{n=-\ell}^{\ell} D_{mn}^{\ell}(m^{\pm})\, f_n^{\ell\pm}(b) \qquad \text{(V.5)}$$

where $m \in \mathfrak{m}$, $b \in \mathfrak{B}$ and $m^{\pm} \equiv e^{\pm} m (e^{\pm})^{-1}$. It remains to express the measures in variables relating to the new subgroups. We recall that the formula

$$||f^{\ell}||^2 \equiv (f^{\ell}, f^{\ell}) = \int_{\mathfrak{K}} \sum_{n=-\ell}^{\ell} |f_n^{\ell}(k)|^2\, d\mu(k)$$

for the norm in $\mathcal{H}^{(\rho,\ell)}$ was obtained from the fact that $\mathfrak{J}^+\backslash\mathfrak{G}$ and $\mathfrak{m}\backslash\mathfrak{K}$ were isomorphic and since $\mathfrak{m}$ is compact we could write the norm as an integral over $\mathfrak{K}$. The norm of the functions $f_n^{\ell\pm}(b)$ is now given by an integral over $\mathfrak{m}\backslash\mathfrak{B}$ (cf. (V.3)), i.e. it can be written as an integral over $\mathfrak{B}$. The explicit relation between the parameters and measures is obtained e.g. from a comparison between the parametrization $g = \tau k$

and cases (i) and (ii) with $\mathscr{b}$ parametrized according to (V.4). One finds:

$$\text{i)}\quad 0 \le \psi < \frac{\pi}{2},\ k_{34}(\psi) = \tau\, a_4(t),\ \tau \in \mathfrak{J}^+,\ \cos\psi = (\mathrm{Ch}\,t)^{-1}$$

$$\text{ii)}\quad \frac{\pi}{2} > \psi \ge \pi,\ k_{34}(\psi) = \tau\, a_4(t),\ \tau \in \mathfrak{J}^+,\ \cos\psi = (\mathrm{Ch}\,t)^{-1}$$

After introducing the new variable t into $d\mu(k)$ and denoting by $d\mu(\mathscr{b})$ an invariant measure on $\mathfrak{B}$ (i.e. with a suitably chosen constant) we get

$$||f^{\ell}||^2 = \sum_{\gamma:+,-}\ \sum_{n=-\ell}^{\ell} \int_{\mathfrak{B}} (\mathrm{Ch}\ t(\mathscr{b}))^{-3}\,|f_n^{\ell\gamma}(\mathscr{b})|^2\, d\mu(\mathscr{b}) \tag{V.6}$$

With

$$h_n^{\ell\pm}(\mathscr{b}) \equiv (\mathrm{Ch}\ t(\mathscr{b}))^{-3/2}\, f_n^{\ell\pm}(\mathscr{b}),\quad ||h^{\ell\pm}(\mathscr{b})||^2_{[\ell]} \equiv (h^{\ell\pm}(\mathscr{b}), h^{\ell\pm}(\mathscr{b}))_{[\ell]}$$

equation (V.6) reads

$$||f^{\ell}||^2 = \sum_{\gamma} \int_{\mathfrak{B}} ||h^{\ell\gamma}(\mathscr{b})||^2_{[\ell]}\, d\mu(\mathscr{b}) \tag{V.7}$$

Next we determine the action of the representation $\mathfrak{D}^{(\rho,\ell)}$ on the functions $h_n^{\ell\pm}(\mathscr{b})$. From the general formula (IV.16) it follows that

$$\mathfrak{D}^{(\rho,\ell)}(\mathscr{b}) f_m^{\ell\pm}(\mathscr{b}_o) = \left| \frac{\delta(e^{\pm}\mathscr{b}_o(k.e^{\pm}\mathscr{b}_o)^{-1})}{\delta(e^{\pm}\mathscr{b}_o\mathscr{b}(k.e^{\pm}\mathscr{b}_o\mathscr{b})^{-1})} \right|^3 f_m^{\ell\pm}(\mathscr{b}_o\mathscr{b}) \tag{V.8}$$

One finds that $|\delta(e^{\pm}\mathscr{b}(k.e^{\pm}\mathscr{b})^{-1}| = (\mathrm{Ch}\ t(\mathscr{b}))^{\frac{1}{2}}$, i.e. (V.8) can be written

$$\mathfrak{D}^{(\rho,\ell)}(\mathscr{b})\, h_m^{\ell\pm}(\mathscr{b}_o) = h_m^{\ell\pm}(\mathscr{b}_o\mathscr{b}) \tag{V.9}$$

We denote by $\mathfrak{H}^{(\rho,\ell)\pm}$ the Hilbert spaces of functions $h_n^{\ell\pm}(\mathscr{b})$ defined as above and with the norms

$$||h^{\ell\pm}||^2 \equiv \int_{\mathfrak{B}} d\mu(\mathscr{b})\ ||h^{\ell\pm}(\mathscr{b})||_{[\ell]} \tag{V.10}$$

Thus we have

$$\mathcal{H}^{(\rho,\ell)} = \mathcal{H}^{(\rho,\ell)+} \oplus \mathcal{H}^{(\rho,\ell)-} \tag{V.11}$$

We recall that the general theory gave us information about the existence of induced representations of $\mathcal{B}$ on $\mathcal{H}^{(\rho,\ell)\pm}$ (and how these could be constructed) but it did not give the decomposition of these representations into irreducible parts. Equation (V.9) shows that $\mathcal{D}^{(\rho,\ell)}(b)$ is closely related to the regular representation in both $\mathcal{H}^{(\rho,\rho)+}$ and $\mathcal{H}^{(\rho,\rho)-}$. It is not the whole regular representation since the functions $h_m^{\ell\pm}(b)$ are not general square integrable functions on $\mathcal{B}$; they must further satisfy (V.5). However, the relation to the regular representation will be important for our purposes since its decomposition into irreducible parts is known[6] and the restriction given by the condition (V.5) is furthermore easily introduced into that decomposition. As is well known the regular representation of $\mathcal{B}$ can be decomposed into a (generalized) direct sum of UIR's of $\mathcal{B}$ characterized by two real numbers (ν,ℓ_o) where $\nu \geq 0$ and where ℓ_o is an integer or half-half-integer (the representations (ν,ℓ_o) of $\mathcal{B}$ all belong to the principal continuous series). This decomposition is most conveniently expressed in terms of the generalized Fourier coefficients with respect to the UIR's of $\mathcal{B}$ of the functions $h_n^{\ell\pm}(b)$ (i.e. rather than just writing down the relevant relations in operator form, we give explicitly the expansion and transformation coefficients). In order to define these Fourier coefficients we need the matrix elements of a general element $b \in \mathcal{B}$ in a UIR (ν,ℓ_o). We choose to consider the matrix elements in a standard angular momentum basis and they will be denoted $D^{jj'}_{mm'}(b;\ell_o,\nu)$ (the properties of these functions are well known). The Fourier coefficients of $h_n^{\ell\pm}(b)$ are thus defined by

$$h^{jj'\ell\pm}_{mm'n}(\ell_o,\nu) \equiv \int_{\mathcal{B}} D^{jj'}_{mm'}(b^{-1};\ell_o,\nu)\, h_n^{\ell\pm}(b)\, d\mu(b) \tag{V.12}$$

An important consequence of (V.12) is now that many of these coefficients are zero. With

$$D^{jj'}_{mm'}(m_1 a_4(t) m_2) = \sum_{k'=-\min(j,j')}^{k'=\min(j,j')} D^{j}_{mk'}(m_1)\, A^{jj'}_{k'}(t,\ell_o,\nu)\, D^{j'}_{k'm'}(m_2)$$

and using (V.5) it follows that the r.h.s. of (V.12) will contain

$$\int D^{j'}_{k'm'}(m_1^{-1})\, D^{\ell}_{nk}(m_1^{\pm})\, d\mu(m_1) = \frac{\delta_{j'\ell}\,\delta_{m'n}\,\delta_{kk'}}{2j+1}$$

and thus we may write

$$\int_{\mathcal{B}} D^{jj'}_{mm'}(b^{-1};\ell_o,\nu)\, h^{\ell\pm}_{n}(b)\, d\mu(b) \equiv \delta_{j'\ell}\,\delta_{m'n}\, h^{j\ell\pm}_{mn}(\ell_o,\nu) \qquad \text{(V.13)}$$

This result provides the restriction on the possible values of ℓ_o that can occur in the decomposition of (ρ,ℓ). The possible values of j' are $|\ell_o|$, $|\ell_o|+1, \ldots$, i.e. the Fourier coefficients are nonzero only if $\ell \geq |\ell_o|$, i.e. for fixed ℓ only a finite number of ℓ_o-values contribute. A transformation $b \to b\, b_1$ in $h^{\ell\pm}_n(b)$ induces a transformation

$$\mathcal{h}^{j\ell\pm}_{mn}(\ell_o,\nu) \to \mathcal{h}'^{j\ell\pm}_{mn}(\ell_o,\nu) = \int_{\mathcal{B}} d\mu(b)\, D^{j\ell}_{mn}(b^{-1},\ell_o,\nu)\, h^{\ell\pm}_{n}(b\, b_1)$$

$$= \sum_{j'=|\ell_o|,|\ell_o|+1,\ldots} \; \sum_{m'=-j'}^{j'} D^{jj'}_{mm'}(b_1;\ell_o,\nu)\, \mathcal{h}^{j'\ell\pm}_{m'n}(\ell_o,\nu) \qquad \text{(V.14)}$$

We introduce the notation $\mathcal{H}^{\ell\pm}(\ell_o,\nu)$ for the Hilbert space of the coefficients

$$\left\{\mathcal{h}^{j\ell\pm}_{mn}(\ell_o,\nu)\right\} \quad \begin{array}{l} |n| \leq \ell \\ |m| \leq j,\; j \geq |\ell_o| \end{array}$$

with the norm

$$||\mathcal{H}^{\ell\pm}(\ell_o,\nu)||^2 \equiv \sum_{j\geq|\ell_o|} \; \sum_{|m|<j} \; \sum_{|n|\leq\ell} |\mathcal{h}^{j\ell\pm}_{mn}(\ell_o,\nu)|^2$$

The Plancherel formula for the UIR's of $\mathcal{B}$ then states that

$$||h^{\ell\pm}||^2 = (2\pi^4)^{-2} \sum_{\ell_o} \int_0^{\infty} (\ell_o^{\,2}+\nu^2)\, d\nu\, ||\mathcal{H}^{\ell\pm}(\ell_o,\nu)||^2 \qquad \text{(V.15)}$$

where the sum goes over the values $\pm\ell$, $\pm(\ell-1), \ldots \pm\frac{1}{2}$ or 0 for ℓ_o. This formula together with (V.11) expresses the desired result. In analogy with (V.15) we may formally write

$$\mathcal{H}^{(\rho,\ell)\pm} = (2\pi^4)^{-2} \sum_{\ell_o} \int_0^\infty (\ell_o^{\,2}+\nu^2)\, d\nu \mathcal{H}^{\ell\pm}(\ell_o,\nu)$$

where the range of summation is the same as in (V.15).

Above we have shown how functions $f_n^\ell(k)$, square integrable over K, are associated with functions $h_n^{\ell\pm}(b)$, square-integrable over B. In physical applications more explicit results are needed such as transformation formulae referring to specific bases. We therefore derive the relevant results in the angular momentum basis for the UIR's of K used above, i.e. we shall consider the Fourier coefficients of those particular functions $h_n^{\ell\pm}(b)$ which are associated with the orthonormal basis (IV.21). In general we have

$$h_m^{\ell\pm}(a_4(t)) = (\mathrm{Ch}\ t)^{-3/2} f_m^\ell(\tau^{-1} k_{34}(\psi^\pm))$$

where we have put $\psi \equiv \psi^+$ for $0 \le \psi < \frac{\pi}{2}$ and $\psi \equiv \psi^-$ for $\frac{\pi}{2} > \psi \ge \pi$. $\psi^\pm$ and t are related by $\cos\psi^\pm = (\mathrm{Ch}\ t)^{-1}$. From (IV.13) it follows that

$$h_m^{\ell\pm}(a_4(t)) = (\mathrm{Ch}\ t)^{-3/2-i\rho} f_m^\ell(k_{34}(\psi^\pm))$$

and in general we get

$$h_m^{\ell\pm}(m_1\, a_4(t) m_2) = (\mathrm{Ch}\ t)^{-3/2-i\rho} f_m^\ell(m_1^{\pm} k_{34}(\psi^\pm) m_2) \qquad \text{(V.16)}$$

Thus we choose to consider the basis functions

$$f_m^\ell(m_1^\pm k_{34}(\psi^\pm) m_2) \equiv [(p+1)^2 - q^2]^{\frac{1}{2}} N(p,q,\ell,\rho)\, R_{mm'}^{\ell j'}(m_1^\pm k_{34}(\psi^\pm) m_2)$$

The corresponding $h_m^{\ell\pm}$-functions are denoted $r_{mm'}^{\ell j'\pm}(b;p,q;\rho)$ and according to (V.16) we have

$$r_{mm'}^{\ell j'\pm}(b;p,q;\rho) = (\mathrm{Ch}\ t)^{-3/2-i\rho}[(p+1)^2 - q^2]^{\frac{1}{2}} N(p,q;\ell;\rho)\cdot$$

$$\sum_k D_{m,k}^\ell(m_1^\pm)\, R_k^{\ell j'}(\psi^\pm;p,q)\, D_{km'}^{j'}(m_2)$$

For the Fourier coefficients we write

$$\int_{\mathcal{B}} d\mu(\mathscr{b})\, D^{j''j'''}_{m''m'''}(\mathscr{b}^{-1};\ell_o,\nu)\, r^{\ell j'\pm}_{mm'}(\mathscr{b};pq;\rho) \equiv$$

$$\delta_{\ell,j'''}\cdot\delta_{m,m'''}\, \mathfrak{R}^{j''\ell j'\pm}_{m''mm'}(\ell_o,\nu;p,q;\rho)$$

Integration of the compact variables yields

$$\mathfrak{R}^{j''\ell j'\pm}_{m''mm'}(\ell_o,\nu;p,q;\rho) = \delta_{m,m''}\,\delta_{j',j''}\,\mathfrak{R}^{\ell j'\pm}_{mm'}(\ell_o,\nu;p,q;\rho)$$

where

$$\mathfrak{R}^{\ell j'\pm}_{mm'}(\ell_o,\nu;p,q;\rho) = N(pq;\ell\rho)\,\frac{(-1)^{\ell-j'}\;2(p+1)^2-q^2\,^{\frac{1}{2}}}{\pi(2\ell+1)(2j'+1)}$$

$$\sum_k (\pm1)^{m-k}\int_0^\infty (\mathrm{Ch}\,t)^{-3/2-i\rho}\,A^{j'\ell}_{-k}(t;\ell_o,\nu)\,R^{\ell j'}_k(\psi^\pm;p,q)(\mathrm{Ch}\,t)^2\,dt$$

where the integrand contains only well known functions. The $\mathfrak{R}^{\ell j'\pm}_{mm'}(\ell_o,\nu;p,q;\rho)$'s may formally be looked upon as the matrix elements in the unitary transformation which connects the "$\mathcal{K}$-basis" $|jmpq\rangle$ and the "$\mathcal{B}$ basis" $|jm\ell_o,\nu\rangle$.

The decomposition formula relating the $D^{pqp'q'}_{jmj'm'}(\mathscr{b};\ell,\rho)$'s and the $D^{jj'}_{mm'}(\mathscr{b};\ell_o,\nu)$'s is

$$D^{pqp'q'}_{jmj'm'}(\mathscr{b};\ell,\rho) = (2\pi^4)^{-2}\sum_\gamma\sum_{\ell_o}\int(\ell_o^2+\gamma^2)\,d\nu\;\cdot$$

$$D^{jj'}_{mm'}(\mathscr{b};\ell_o,\nu)\cdot\sum_n \overline{\mathfrak{R}^{\ell j\gamma}_{nm}(\ell_o,\nu;pq;\rho)}\,\mathfrak{R}^{\ell j'\gamma}_{nm'}(\ell_o,\nu;p',q';\rho) \tag{V.17}$$

It is obtained as follows: a general scalar product in $\mathcal{H}^{(\rho,\ell)}$ can be written

$$(f'^{\ell},f''^{\ell}) = \sum_\gamma\int_{\mathcal{B}} d\mu(\mathscr{b})\,(h'^{\ell\gamma}(\mathscr{b}),h''^{\ell\gamma}(\mathscr{b}))_{[\ell]} =$$

$$(2\pi^4)^{-2}\sum_\gamma\sum_{\ell_o}\int_0^\infty(\ell_o^2+\nu^2)\,d\nu\sum_j\sum_m\sum_n \overline{\mathcal{H}'^{j\ell\gamma}_{mn}(\ell_o,\nu)}\;\mathcal{H}^{j\ell\gamma}_{mn}(\ell_o,\nu)$$

(V.17) is now obtained by substituting

$$f'^{\ell}_{m} \to N(p,q,\ell,\rho)[(\rho+1)^2-q^2]^{\frac{1}{2}} \mathfrak{R}^{\ell j \ell}_{nm}(k;\, p,q)$$

$$f''^{\ell}_{m} \to \mathfrak{D}^{\ell,\rho}(b)\; N(p',q',\ell,\rho)[(p'+1)^2-q'^2]^{\frac{1}{2}} \mathfrak{R}^{\ell j'}_{nm'}(k;\, p'q')$$

where the latter element according to (V.14) has the Fourier coefficients

$$\sum_{j''} \sum_{m''} D^{jj''}_{mm''}(b;\, \ell_o,\nu)\, \mathfrak{R}^{j''\ell j'\pm}_{m''mm'}(\ell_o,\nu;\, p',q';\, \rho)\ .$$

Conclusion

We have given a survey of a class of induced representations of $\overline{SO_0(1,4)}$ and their decompositions. The theory of induced representations has been shown to provide a suitable framework for the introduction of and detailed investigation of various bases of particular interest from a physical point of view.

References

1. R. Takahashi, Bull. Soc. Math. de France 91, 289 (1963).
2. S. Helgason, Differential Geometry and Symmetric Spaces (Academic Press, New York, 1962).
3. W. Klink, "Induced Representation Theory and the Poincaré Group," in Lectures in Theoretical Physics, Vol. XI-D, ed. by K. T. Mahanthappa and W. E. Brittin (Gordon and Breach, New York, 1968), p. 43.
4. See e.g. G. Mackey, Induced Representations of Groups and Quantum Mechanics (W. A. Benjamin, New York, 1967) for a survey and G. Mackey, Bull. Ann. Math. Soc. 69, 628 (1963), for a guide to the literature.
5. V. Bargmann, Ann. Math. 48, 568 (1967).
6. M. Neumark, Linear Representations of the Lorentz Group (Pergamon Press, Oxford, 1964).
7. S. Ström, Arkiv Fysik 40, 1 (1969).
8. See e.g. J. Dixmier, Bull. Soc. Math. de France 89, 9 (1961).
9. Cf. the contribution to this symposium by W. J. Holman.
10. H. Joos, Fortschr. Phys. 10, 65 (1962).

Section C: O(4,2) and O(3,2)

GENERALIZED DIRAC AND MAJORANA REPRESENTATIONS OF $\overline{SO}(3,2)$†

Leon Jaffe
Center for Particle Theory
Department of Physics
University of Texas
Austin, Texas 78712

Abstract

We consider the use of the Gel'fand-Zetlin formulae for $\overline{SO}(5)$ to obtain those unitary irreducible representations of $\overline{SO}(3,2)$ which have a discrete singleton reduction with respect to $\overline{SO}(3,1)$ [the "discrete Lorentz" representations]. These discrete Lorentz representations can be regarded as generalizations of the Majorana and Dirac representations in certain cases. The method used also generalizes easily to discuss an analogous class of representations of $\overline{SO}(n,2)$.

Introduction

The unitary irreducible representations (UIR) of $S = \overline{SO}(3,2)$ [$\overline{SO}(p,q)$ denotes the universal covering group of $SO(p,q)$] have been considered in Ref. 1 for those UIR which have a singleton reduction with respect to its maximal pseudo-compact subgroup $K = \overline{SO}(3) \otimes \overline{SO}(2)$. [A singleton reduction of a representation of group with respect to a subgroup means that each irreducible representation of the subgroup occurs at most once in the reduction.] These will simply be referred to as "singleton" UIR of S.

However, for many physical applications we are interested in S, for example, because it contains L_+, the covering group of the proper Lorentz group ($L_+ \cong \overline{SO}(3,1)$) as a subgroup. It therefore becomes useful to know the reduction of representations of S with respect to L_+.

We consider those UIR of $\overline{SO}(3,2)$ which have a discrete singleton reduction with respect to $\overline{SO}(3,1)$. [A discrete reduction of a representation of a group with respect to a subgroup means that when restricting the representation of the group to a representation of the subgroup we obtain a direct discrete sum of irreducible representations of the subgroup.] These will be referred to as "discrete Lorentz" UIR of S.

†Presented at the Symposium on De Sitter and Conformal Groups, University of Colorado, Summer 1970.

We arrive at the remarkable result that all the "discrete Lorentz" UIR are nothing but "singleton" UIR of S.

We thus independently arrive at many of the singleton UIR obtained by Ehrman. In fact, we obtain all those singleton UIR in which for each value of the angular momentum the eigenvalues of the generator of SO(2), A_{54}, are bounded.

Section 1

The real Lie algebra of S = $\overline{SO}(3,2)$ is given by the following commutation results [CR]:

$$[A_{\alpha\beta}, A_{\gamma\delta}] = g_{\alpha\delta} A_{\beta\gamma} + g_{\beta\gamma} A_{\alpha\delta} - g_{\alpha\gamma} A_{\beta\delta} - g_{\beta\delta} A_{\alpha\gamma} \quad (1)$$

where the metric $g_{\alpha\beta}$ is given by:

$$g_{\alpha\beta} = +1 \quad \text{for } \alpha = \beta = 1,2,3$$
$$= -1 \quad \text{for } \alpha = \beta = 4 \text{ or } 5$$
$$= 0 \quad \text{otherwise.}$$

$A_{\alpha\beta} = -A_{\beta\alpha}$ are the basis elements of the Lie algebra, with $A^{+}_{\alpha\beta} = -A_{\alpha\beta}$ for a UIR.

Section 2

A UIR for $\overline{SO}(3,1)$ is specified by:[2)] [In the notation of Ref. 3 $m_{21} = m$, $m_{31} = \ell$, $|m_{41}| = \ell_0$, $(m_{42}+1)^2 = \ell_1^2$]

$$A_{21}|m_{41}\ m_{42}\ m_{31}\ m_{21}\rangle = i\, m_{21}|m_{41}\ m_{31}\ m_{21}\rangle \quad (2)$$

$$A_{32}|m_{41}\ m_{42}\ m_{31}\ m_{21}\rangle = A(m_{21})|m_{41}\ m_{42}\ m_{31}\ m_{21}+1\rangle - A(m_{21}-1)|m_{41}\ m_{42}\ m_{31}\ m_{21}-1\rangle \quad (3)$$

$$-iA_{43}|m_{41}\ m_{42}\ m_{31}\ m_{21}\rangle = B(m_{31})|m_{41}\ m_{42}\ m_{31}+1\ m_{21}\rangle - B(m_{31}-1)|m_{41}\ m_{42}\ m_{31}-1\ m_{21}\rangle + iC_2|m_{41}\ m_{42}\ m_{31}\ m_{21}\rangle \quad (4)$$

where:

$$A(m_{21}) = \tfrac{1}{2}[(m_{31}+\tfrac{1}{2})^2 - (m_{21}+\tfrac{1}{2})^2]^{\frac{1}{2}} \quad (5)$$

$$B(m_{31}) = \tfrac{1}{2}[m_{21}{}^2 - (m_{31}+1)^2]^{\frac{1}{2}} \left\{ \frac{[m_{41}{}^2-(m_{31}+1)^2][(m_{42}+1)^2-(m_{31}+1)^2]}{(m_{31}+1)^2[(m_{31}+1)^2-\frac{1}{4}]} \right\}^{\frac{1}{2}} \quad (6)$$

$$C_2 = \frac{m_{21}\, m_{41}(m_{42}+1)}{(m_{31}+1)\, m_{31}} \quad (7)$$

The actions of the remaining $A_{\alpha\beta}$ are specified by the CR.

The above basis is analogous to that of $\overline{SO}(4)$ where the ranges of m_{21}, m_{31} are specified by $|m_{21}| \leq m_{31}$; $|m_{41}| \leq m_{31} \leq m_{42}$. Finite dimensional representations of $\overline{SO}(3,1)$ have the same range, but for unitary representations m_{42} is no longer the maximum value of m_{31} and to satisfy unitarity m_{42} has the following ranges:

A. $m_{41} = 0$; $|m_{42}+1| = 1$
[The identity representation]
B. $m_{41} = 0$; $0 < |m_{42}+1| < 1$
C. $2|m_{41}|$ integral or zero; $m_{42}+1 = 0$
D. $2|m_{41}|$ integral or zero; $m_{42}+1 = iy$, y real and $\neq 0$.

N.B. $[m_{41}, (m_{42}+1)]$ specify the same UIR as $[-m_{41}, -(m_{42}+1)]$. The ranges of m_{21}, m_{31} remain: $|m_{21}| \leq m_{31}$, $|m_{41}| \leq m_{31}$, except for the identity representation where $m_{21} = m_{31} = 0$.

Section 3

Since we consider UIR of S with a discrete singleton reduction with respect to $\overline{SO}(3,1)$ we may write

$$A_{54}|m_{41}\, m_{42}\, m_{31}\, m_{21}\rangle = \sum_{\substack{m_{41}',\, m_{42}' \\ m_{31}',\, m_{21}'}} |m_{41}'\, m_{42}'\, m_{31}'\, m_{21}'\rangle\, C^{m_{41}' m_{42}' m_{31}' m_{21}'}_{m_{41} m_{42} m_{31} m_{21}} \quad (8)$$

In determining these UIR of S it is sufficient to determine the action of A_{54} on a $\overline{SO}(3,1)$ basis as the action of A_{5i} [i=1,2,3] follow from the CR.

The matrix elements $C^{m_{41}' m_{42}' m_{31}' m_{21}'}_{m_{41} m_{42} m_{31} m_{21}}$ are determined from the CR, the pertinent ones being:

$$[A_{21}, A_{54}] = [A_{32}, A_{54}] = 0 \quad (9)$$

$$[A_{43}, [A_{54}, A_{43}]] = -A_{54} \tag{10}$$

$$[A_{54}, [A_{54}, A_{43}]] = -A_{43} \tag{11}$$

the remaining CR following from the Jacobi identity.

There are several ways to proceed now:

(a) The above commutation relations can be used to obtain recurrence relations for the matrix elements which then have to be solved. The solution of the recurrence relations following from (9) and (10) has been given by Gel'fand et al.[3] and hence it remains to solve those following from (11).

(b) The above approach is tedious and can be somewhat simplified by using the CR of A_{54} with the Casimir operators of $\overline{SO}(3,1)$, obtaining the same recurrence relations. This approach was used in Ref. 4.

(c) The solution of the CR for $\overline{SO}(5)$ obtained by Gel'fand and Zetlin[5] can be used to obtain the matrix elements for $\overline{SO}(3,2)$. This approach lacks rigour but as the solution has been obtained for $\overline{SO}(n+2)$ in general it has the advantage that it can be generalized to discuss $\overline{SO}(n,2)$.

Section 4

If we put $A_{\alpha\beta} = [g_{\alpha\alpha}\, g_{\beta\beta}]^{\frac{1}{2}} B_{\alpha\beta}$ we see that $B_{\alpha\beta}$ satisfy the CR of the Lie algebra of $\overline{SO}(5)$. We can thus write down the matrix elements for $B_{\alpha\beta}$ from Ref. 5 and obtain in addition to Eqs. (2) to (7) the following:

$$B_{54}|m_{41}\ m_{42}\rangle = A(m_{41})|m_{41}+1\ m_{42}\rangle - A(m_{41}-1)|m_{41}-1\ m_{42}\rangle + A(m_{42})|m_{41}\ m_{42}+1\rangle - A(m_{42}-1)|m_{41}\ m_{42}-1\rangle \tag{12}$$

where labels which do not change have been omitted and:

$$A(m_{41}) = \tfrac{1}{2}[(m_{31}+\tfrac{1}{2})^2-(m_{41}+\tfrac{1}{2})^2]^{\frac{1}{2}}\left[\frac{[(m_{51}+\frac{1}{2})^2-(m_{41}+\frac{1}{2})^2][(m_{52}+\frac{3}{2})^2-(m_{41}+\frac{1}{2})^2]}{[(m_{42}+1)^2-m_{41}^2][(m_{42}+1)^2-(m_{41}+1)^2]}\right]^{\frac{1}{2}} \tag{13}$$

$$A(m_{42}) = \tfrac{1}{2}[(m_{31}+\tfrac{1}{2})^2-(m_{42}+\tfrac{3}{2})^2]^{\frac{1}{2}}\left[\frac{[(m_{51}+\frac{1}{2})^2-(m_{42}+\frac{3}{2})^2][(m_{52}+\frac{3}{2})^2-(m_{42}+\frac{3}{2})^2]}{[m_{41}^2-(m_{42}+1)^2][m_{41}^2-(m_{42}+2)^2]}\right]^{\frac{1}{2}} \tag{14}$$

For $\overline{SO}(5)$ the Gel'fand-Zetlin representation is obtained via the following reduction procedure:

$$\overline{SO}(5) \supset \overline{SO}(4) \supset \overline{SO}(3)$$

where each reduction is a discrete singleton reduction. The use of their results for our class of UIR of $\overline{SO}(3,2)$ remains valid since these UIR are obtained by the following reduction procedure:

$$\overline{SO}(3,2) \supset \overline{SO}(3,1) \supset \overline{SO}(3)$$

each reduction being again a discrete singleton reduction.

For $\overline{SO}(5)$ we have:

$$|m_{41}| \leq m_{31} \leq m_{42}$$

$$|m_{41}| \leq m_{51} \leq m_{42} \leq m_{52}$$

and

$$m_{51} = \max |m_{41}| = \min m_{42}$$

$$m_{52} = \max m_{42}$$

However for $\overline{SO}(3,2)$ m_{42} and hence m_{51}, m_{52} have completely different ranges for a UIR.

Now the use of the Gel'fand-Zetlin formula poses some problems:

(1) When a zero occurs in denominator of either of expressions (13), (14), then for $\overline{SO}(5)$ there are at least two zero factors in the numerator and the entire expression is zero. For $\overline{SO}(3,2)$ this does not hold true, and we have the following set of rules:

(a) If a zero occurs in the denominator, then the part of the numerator independent of m_{31} must be zero.

(b) There must be at least as many zero factors in the numerator as there are in the denominator.

(c) It is sufficient (though not necessary) for a matrix element to be zero if there are more zero factors in the numerator than in the denominator.

(2) When $m_{41}'(m_{42}'+1) = 0$ then $m_{41} = m_{41}'$, $m_{42}+1 = m_{42}'+1$ specifies the same representation as $m_{41} = -m_{41}'$, $m_{42}+1 = -(m_{42}'+1)$ and this gives rise to the following problems:

(a) If $m_{41} = (m_{42}+1) = 0$ then $|m_{41}+1, m_{42}\rangle$, $|m_{41}\ m_{42}+1\rangle$ specify the same states as $|m_{41}-1, m_{42}\rangle$ and $|m_{41}, m_{42}-1\rangle$ respectively and Eq. (12) is no longer meaningful. In this case it should read:

$$B_{54}|m_{41}\ m_{42}\rangle = \alpha A(m_{41})|m_{41}+1,\ m_{42}\rangle + \beta A(m_{42})|m_{41}\ m_{42}+1\rangle \quad (15)$$

where

$$A(m_{41}) = \tfrac{1}{2}[(m_{31}+\tfrac{1}{2})^2 - (m_{41}+\tfrac{1}{2})^2]^{\frac{1}{2}} \quad (16)$$

$$A(m_{42}) = \tfrac{1}{2}[(m_{31}+\tfrac{1}{2})^2 - (m_{42}+\tfrac{3}{2})^2]^{\frac{1}{2}} \quad (17)$$

(b) If $m_{41} = m_{41}' = \pm\frac{1}{2}$, $m_{42}+1 = 0$ then $m_{41} = m_{41}' \mp 1$ respectively specifies the same UIR as $m_{41} = m_{41}'$. Taking $m_{41} = \frac{1}{2}$ Eq. (12) remains valid if we put $|m_{41}-1, m_{42}\rangle = \gamma|m_{41}, m_{42}\rangle$.

(c) If $m_{41} = 0$, $m_{42}'+1 = \pm\frac{1}{2}$ then $m_{42} = m_{42}' \mp 1$ respectively specifies the same UIR as $m_{42} = m_{42}'$. Taking $m_{42}+1 = \frac{1}{2}$ Eq. (12) remains valid with $|m_{41}\ m_{42}-1\rangle = \delta|m_{41}\ m_{42}\rangle$.

α, β, γ, δ have to be determined from the CR, which is nevertheless a simpler task for these special cases than for the general case.

These rules can be justified by examining the recurrence relations from which (13) and (14) are derived.[6]

Section 5

Since $A_{54} = -B_{54}$ and $A_{54}^+ = -A_{54}$ for a UIR we have the condition that $A(m_{41})$, $A(m_{41}-1)$, $A(m_{42})$, $A(m_{42}-1)$ are real.

A. Suppose $m_{41} = 0$; $|m_{42}+1| = 1$. We may take $m_{42} = 0$ as $m_{42}+1 = \pm 1$ specify exactly the same UIR. Since $m_{42} = 0$, $m_{41} = \pm 1$ specifies a non-unitary representation we must have $A(m_{41}) = A(m_{41}-1) = 0$, and we must have, for example, $m_{51} = 0$. [This uses rule 1.] Also $m_{41} = 0$, $m_{42} = 1$ is non-unitary and hence $A(m_{42}) = 0$, giving $m_{52} = 0$. $A(m_{42}-1)$ is now undetermined by (14), but it must be zero for otherwise we have the UIR $(m_{41}, m_{42}+1) = (0,0)$ and $A(m_{42}-\frac{1}{2}\pm\frac{1}{2})$ are singular ($m_{31} \neq 0$).

So the identity representation of $\overline{SO}(3,1)$ can only occur in the reduction of a UIR of $\overline{SO}(3,2)$ if the latter is the identity representation of $\overline{SO}(3,2)$.

So here we have:

$$\underline{m_{41} = m_{42} = 0;\ m_{51} = m_{52} = 0}$$

B. Suppose $m_{41} = 0$; $m_{42} = m_{42}'$, $0 < |m_{42}'+1| < 1$. We may take $0 < m_{42}'+1 < 1$. With this range for m_{42}, $m_{41} = \pm 1$ is non-unitary and hence $A(m_{41}) = A(m_{41}-1) = 0$, giving $m_{51} = 0$. $m_{41} = 0$, $m_{42} = m_{42}'+1$ is non-unitary and hence $A(m_{42}') = 0$ giving $m_{52} = m_{42}'$. Hence $A(m_{42}'-1) \neq 0$ and it is real. Now $m_{41} = 0$, $m_{42} = m_{42}'-2$ is non-unitary and hence $A(m_{42}'-2) = 0$ giving $(m_{52}+\frac{3}{2})^2 = (m_{42}'-\frac{1}{2})^2$. Hence $m_{42}'+\frac{3}{2} = \frac{1}{2} - m_{42}'$ and $m_{52} = m_{42}' = -\frac{1}{2}$.

So finally we have that $m_{41} = 0$, $m_{42}+1 = \frac{1}{2}$ and:

$$B_{54}|m_{41}\ m_{42}\rangle = \delta A(m_{42}-1)|m_{41}\ m_{42}\rangle \tag{18}$$

From the CR we obtain $\delta^2 = -1$ and $\delta = \pm i$ specify inequivalent UIR.

So here we have [for both these UIR]:

$$\underline{m_{41} = 0,\ m_{42}+1 = \tfrac{1}{2};\ m_{51} = 0,\ m_{52}+\tfrac{3}{2} = 1.}$$

C. $2|m_{41}|$ integral or zero; $m_{42} = m_{42}'$, $m_{42}'+1 = 0$. Here $m_{42} = m_{42}'\pm 1$ is non-unitary if $m_{41} \neq 0$. If $m_{41} = 0$ then $m_{42} = m_{42}'\pm 1$ is the identity representation which does not occur in any SO(3,2) UIR apart from the identity UIR. So in both cases $A(m_{42}) = A(m_{42}-1) = 0$, giving $m_{51} = 0$.

Let $\underline{m}_{41}$ be the minimum value of $|m_{41}|$ occurring in the reduction.

(a) $\underline{m}_{41} \geq 1$. When $|m_{41}| = \underline{m}_{41}$ we may take $m_{41} = \underline{m}_{41}$. Clearly $A(m_{41}-1) = 0$ for $\underline{m}_{41}$ to be the minimum of $|m_{41}|$ in the reduction. Hence $m_{52} + \frac{3}{2} = \underline{m}_{41} - \frac{1}{2}$. [If $\underline{m}_{41} = 1$ we have in the denominator of $A(m_{41}-1)$ the factor $(m_{42}+1)^2 - (\underline{m}_{41}-1)^2 = 0^2$, requiring two zero factors in the numerator.] The reality condition is satisfied.

So the range of m_{41}, m_{42} and the values of m_{51}, m_{52} are given by:

$$\underline{m_{52}+2 \leq m_{41},\ m_{42}+1 = 0;\ m_{51} = 0,\ \tfrac{1}{2} \leq m_{52} + \tfrac{3}{2}}$$

(b) $\underline{m}_{41} = \frac{1}{2}$. We can take $m_{41} = \underline{m}_{41} = \frac{1}{2}$ and then:

$$B_{54}|\underline{m}_{41}\ m_{42}\rangle = A(\underline{m}_{41})|\underline{m}_{41}+1\ m_{42}\rangle + \gamma A(\underline{m}_{41}\ \ 1)|\underline{m}_{41}\ m_{42}\rangle \tag{19}$$

The CR give $\gamma^2 = -1$ and $\gamma = \pm i$ specify inequivalent UIR. The reality of $A(m_{41})$, $A(m_{41}-1)$ gives $0 \leq (m_{52}+\frac{3}{2})^2 \leq 1$. So we have the following:

1. $\underline{m_{41} = \tfrac{1}{2},\ m_{42}+1 = 0;\ m_{51} = 0,\ m_{52}+\tfrac{3}{2} = 1}$

2. $\underline{\tfrac{1}{2} \leq m_{41},\ m_{42}+1 = 0;\ m_{51} = 0,\ 0 \leq m_{52} + \tfrac{3}{2} < 1}$

remembering that each case specifies two inequivalent UIR.

(c) $\underline{m}_{41} = 0$. In this case we have Eqs. (15), (16), (17). Clearly $B = 0$ for $A(m_{42}) = A(m_{42}-1) = 0$. From the CR we see that $\alpha = \sqrt{2}$. So we have:

$$\underline{0 \leq m_{41},\ m_{42}+1 = 0;\ m_{51} = 0,\ m_{52} + \tfrac{3}{2} = \tfrac{1}{2}.}$$

D. Suppose $m_{42} = m_{42}'$, $m_{42}'+1 = iy$, $y \neq 0$. Then $m_{42} = m_{42}' \pm 1$ is non-unitary and hence $A(m_{42}) = A(m_{42}'-1) = 0$, giving: $m_{51} = m_{42}' = m_{52}$ and

$$A(m_{41}) = \tfrac{1}{2}[(m_{\ 1}+\tfrac{1}{2})^2 - (m_{41}+\tfrac{1}{2})^2]^{\frac{1}{2}} \tag{20}$$

This gives the following UIR:

1. $0 \le |m_{41}|$, $m_{42}+1 = iy$, $y > 0$; $m_{51} = m_{52} = m_{42}$
2. $\tfrac{1}{2} \le |m_{41}|$, $m_{42}+1 = iy$, $y > 0$; $m_{51} = m_{52} = m_{42}$

We thus obtain the UIR indicated diagrammatically in the column under the L_+ reduction in the appendix. (α,β) denotes an L_+ UIR with $m_{41} = \alpha$, $m_{42}+1 = \beta$. $(\alpha,\beta) - (\alpha',\beta')$ indicates that there are non-zero matrix elements of B_{54} mapping states in (α,β) into states in (α',β'). $(\overline{\alpha,\beta})$ indicates that there are non-zero matrix elements of B_{54} mapping (α,β) into itself.

We obtain a very restricted class of UIR of $\overline{SO}(3,2)$, the confinement to a discrete reduction with respect to $\overline{SO}(3,1)$ being the main restrictive factor. For example none of these discrete Lorentz UIR extend to a UIR of $\overline{SO}(4,2)$. For $\overline{SO}(n,1)$ one obtains all the UIR using this method[2)] as all the UIR of $\overline{SO}(n,1)$ have a discrete singleton $\overline{SO}(n)$ reduction. But all the UIR of $\overline{SO}(n,2)$ do not have a discrete singleton $\overline{SO}(n,1)$ reduction.

Section 6

The reduction of these "discrete Lorentz" UIR has been done in Ref. 4 and the results are included here in diagrammatic form in the appendix.

The following is the connection between the notation used above and that in Ref. 4:

$$\Gamma_o = iB_{54}$$

$$m = m_{21}$$

$$j = m_{31}$$

$$k = m_{41}$$

$$c = m_{42}+1$$

$$|D\ kc\ jm\rangle = (i)^k |m_{41}\ m_{42}\ m_{31}\ m_{21}\rangle$$

The relationship between D_1, D_2 and m_{51}, m_{52} is given by:

$$D_1 = \tfrac{5}{2} - [m_{51}+\tfrac{1}{2}]^2 - [m_{52}+\tfrac{3}{2}]^2$$

$$D_2 = -\{[m_{51}+\tfrac{1}{2}]^2 - \tfrac{1}{4}\}\{(m_{52}+\tfrac{3}{2})^2 - \tfrac{1}{4}\} \ .$$

$[\alpha,\beta]$ specifies a UIR of $\overline{SO}(3) \otimes \overline{SO}(2)$ with $m_{31}=\alpha$ and β being the eigenvalue of Γ_0. Solid lines indicate non-zero matrix elements for B_{43}.

The method used to obtain the $\overline{SO}(3) \otimes \overline{SO}(2)$ reduction in Ref. 4 is roughly as follows:

The subspace of an $\overline{SO}(3,2)$ UIR space with m_{31} fixed is finite dimensional and we can effect a similarity transformation in this subspace which takes the basis $|m_{41}\ m_{42}\ m_{31}\ m_{21}\rangle$ into a basis $|\alpha\ \mu\ m_{31}\ m_{21}\rangle$ which is diagonal with respect to B_{54}, and

$$B_{54}|\alpha\ \mu\ m_{31}\ m_{21}\rangle = i\,\mu|\alpha\ \mu\ m_{31}\ m_{21}\rangle$$

$|\alpha\ \mu\ m_{31}\ m_{21}\rangle$ then specifies the $\overline{SO}(3) \otimes \overline{SO}(2)$ basis. We then show that the $\overline{SO}(3) \otimes \overline{SO}(2)$ reduction is a singleton reduction, the label α being redundant. Putting:

$$|\mu\ m_{31}\ m_{21}\rangle = \sum_{m_{41},\, m_{42}} |m_{41}\ m_{42}\ m_{31}\ m_{21}\rangle C^{\mu}_{m_{41}\ m_{42}}$$

and

$$A_{43}|\mu\ m_{31}\ m_{21}\rangle = \sum_{\mu',\ m_{31}'} |\mu'\ m_{31}'\ m_{21}\rangle\ C^{\mu'\ m_{31}'}_{\mu\ m_{31}}$$

we obtain recurrence relations for the expansion coefficients $C^{\mu}_{m_{41}m_{42}}$. When these are solved the matrix elements of A_{43}, $C^{\mu'\ m_{31}'}_{\mu\ m_{31}}$ can easily be determined.

This method has the advantage over merely comparing the Casimir operators with the singleton UIR Ehrman obtain in that it can be generalized to discuss the $\overline{SO}(n) \otimes \overline{SO}(2)$ basis of the UIR of $\overline{SO}(n,2)$ analogous to those obtained above for $\overline{SO}(3,2)$.

Conclusion

The discrete Lorentz UIR include the four Majorana UIR. Class C(a), C(b)2, C(c) can be considered as generalizations of the Majorana UIR given by C(b)1.

Class D1, D2 UIR provide a natural generalization of the finite non-unitary Dirac representation to a unitary representation for integral and half-integral spin respectively. To see this note that the Dirac representation has the following reductions with respect to $\overline{SO}(3,1)$ and $\overline{SO}(3) \otimes \overline{SO}(2)$ respectively:

$$(-\tfrac{1}{2}, \tfrac{3}{2}) - (\tfrac{1}{2}, \tfrac{3}{2})$$

$$[-\tfrac{1}{2}, \tfrac{1}{2}] - [\tfrac{1}{2}, \tfrac{1}{2}]$$

Clearly class D2 UIR is a unitary generalization of this and class D1 UIR is the integral counterpart of D2. It may be useful to investigate these UIR as representing an infinite tower of particles with the same internal quantum numbers, but different spin.[7] The expansion coefficients connecting the L_+ basis to the compact basis are explicitly determined in Ref. 4 for these generalized Dirac UIR, and also the matrix elements of A_{42} [which are omitted in Ref. 1].

Mathematically the above is interesting as a prelude to discussing $\overline{SO}(n,2)$ in order to see what happens to the ambiguities in using the Gel'fand-Zetlin formulae in a simple case.

Acknowledgments

The author acknowledges useful discussions with A. Böhm and F. Schwarz.

Appendix

	$m_{51}+\frac{1}{2}$	$m_{52}+\frac{3}{2}$	L_+ Reduction	K Reduction
B	$\frac{1}{2}$	1	$(0, \frac{1}{2})$	(i) $[\frac{1}{2},0]$ — $[\frac{3}{2},1]$ — $[\frac{5}{2},2]$ — etc. (ii) $[-\frac{1}{2},0]$ — $[-\frac{3}{2},1]$ — $[-\frac{5}{2},2]$ — etc.
C (a)	$\frac{1}{2}$	$\ell-\frac{1}{2}$ $[2\ell$ integral $\ell\geq 1]$	$(\ell,0)$ $(\ell+1,0)$ $(\ell+2,0)$ etc.	$[0,\ell]$ $[-1,\ell+1]$ $[1,\ell+1]$ $[-2,\ell+2]$ $[0,\ell+2]$ $[2,\ell+2]$ etc.
C(b) 1	$\frac{1}{2}$	1	$(\frac{1}{2},0)$	(i) $[1,\frac{1}{2}]$ — $[2,\frac{3}{2}]$ — $[3,\frac{5}{2}]$ — etc. (ii) $[-1,\frac{1}{2}]$ — $[-2,\frac{3}{2}]$ — $[-3,\frac{5}{2}]$ — etc.
C(b) 2	$\frac{1}{2}$	(i) $0<m_{52}+\frac{3}{2}<1$ (ii) $0=m_{52}+\frac{3}{2}$. $[M=\pm(m_{52}+\frac{3}{2})]$	(i) $(\frac{1}{2},0)$, $(\frac{3}{2},0)$, $(\frac{5}{2},0)$, etc. (ii) $(\frac{1}{2},0)$, $(\frac{3}{2},0)$, $(\frac{5}{2},0)$, etc.	$[M,\frac{1}{2}]$ $[M-1,\frac{3}{2}]$ $[M+1,\frac{3}{2}]$ $[M-2,\frac{5}{2}]$ $[M,\frac{5}{2}]$ $[M+2,\frac{5}{2}]$ etc.

	$m_{51}+\frac{1}{2}$	$m_{52}+\frac{3}{2}$	L_+ Reduction	K Reduction
C(c)	$\frac{1}{2}$	$\frac{1}{2}$	(0, 0) (1, 0) (2, 0) etc.	[0,0] [-1,1] [1,1] [-2,2] [2,0] [2,2] etc.
D1	$c-\frac{1}{2}$	$c+\frac{1}{2}$	(0, c) (-1, c) (1, c) (-2, c) (2, c) etc. etc.	[0,0] [-1,1] [0,1] [1,1] [-2,2] [-1,2] [0,2] [1,2] [2,2] etc.
D2	$c-\frac{1}{2}$	$c+\frac{1}{2}$	($-\frac{1}{2}$, c) — ($\frac{1}{2}$, c) ($-\frac{3}{2}$, c) ($\frac{3}{2}$, c) ($-\frac{5}{2}$, c) ($\frac{5}{2}$, c) etc. etc.	[$-\frac{1}{2}$, $\frac{1}{2}$] [$\frac{1}{2}$, $\frac{1}{2}$] [$-\frac{3}{2}$, $\frac{3}{2}$] [$-\frac{1}{2}$, $\frac{3}{2}$] [$\frac{1}{2}$, $\frac{3}{2}$] [$\frac{3}{2}$, $\frac{3}{2}$] [$-\frac{5}{2}$, $\frac{5}{2}$] [$-\frac{3}{2}$, $\frac{5}{2}$] [$-\frac{1}{2}$, $\frac{5}{2}$] [$\frac{1}{2}$, $\frac{5}{2}$] [$\frac{3}{2}$, $\frac{5}{2}$] [$\frac{5}{2}$, $\frac{5}{2}$] etc.

References

1. J. B. Ehrman, Proc. Cambridge Phil. Soc. 53 (1957); Thesis, Princeton, 1954.
2. F. Schwarz, University of Texas preprint CPT-59 (1970).
3. Gel'fand, Minlos and Shapiro, Representations of the Rotation and Lorentz Groups (Pergamon Press, New York, 1963).
4. L. Jaffe, University of Texas preprint CPT-31 (1969). [To be published in Jour. Math. Phys.]
5. I. M. Gel'fand and M. L. Zetlin, DAN SSR 71, 1017-1020 (1950). For notational changes see Ref. 2.
6. These recurrence relations are given in L. Jaffe, University of Texas preprint, CPT-28.
7. A. Böhm, these Proceedings.

SOME CLASSES OF UNITARY IRREDUCIBLE REPRESENTATIONS OF THE GROUP $SO_o(4,2)$†

Fritz Schwarz
Center for Particle Theory
Department of Physics
The University of Texas
Austin, Texas 78712

Abstract

The identity component $SO_o(4,2)$ of the group $O(4,2)$ is usually called the conformal group. We determine the following unitary irreducible representations (UIR's) of the group $SO_o(4,2)$:

a. The UIR's which contain only the discrete series of $SO_o(4,1)$.
b. The UIR's which change the discrete label of the $SO_o(4,1)$ representations only.
c. The UIR's which remain irreducible under restriction to $SO_o(4,1)$.

We begin always with the Lie algebra of the conformal group and construct an irreducible representation of antihermitian operators of it. All representations obtained in this way can be extended to a UIR of the conformal group.

I. Introduction

We denote by $O(4,2)$ the group of linear homogeneous transformations of the 6-dimensional real vector space, which leave the quadratic form $x_1^2 + x_2^2 + x_3^2 + x_4^2 - x_5^2 - x_6^2$ invariant. The identity component $SO_o(4,2)$ of this group is usually called the conformal group. By $so(4,2)$ we denote the Lie algebra of $SO_o(4,2)$.

The conformal group has been of considerable interest in theoretical physics, see for example Ref. 1, and has been studied therefore by a number of authors.[2)] In these references the unitary irreducible representations (UIR's) obtained are either reduced with respect

†Presented at the Symposium on De Sitter and Conformal Groups, University of Colorado, Summer 1970.

to the maximal compact subgroup, or a basis in a Cartan-subalgebra is diagonalized. The representations given in this paper are reduced in the chain $SO(4) \subset SO_0(4,1) \subset SO_0(4,2)$ and we derive the following classes of UIR's:

a. The UIR's which contain only representations of the discrete series of $SO_0(4,1)$.
b. The UIR's which change the discrete label of $SO_0(4,1)$ only.
c. The UIR's which remain irreducible under restriction to $SO_0(4,1)$.

The group $SO_0(4,2)$ contains 7 subgroups of the type

$$g_{ij}(t) = \begin{bmatrix} 1 & & & & \\ & \ddots & & & \\ & & \cos t \cdots \sin t & & \\ & & \vdots \qquad \vdots & & \\ & & -\sin t \cdots \cos t & & \\ & & & \ddots & \\ & & & & 1 \end{bmatrix} \qquad \begin{matrix} 1 \leq i < j \leq 4 \\ \\ 5 \leq i < j \leq 6 \end{matrix} \tag{1}$$

and 8 subgroups of the type

$$g_{ij}(t) = \begin{bmatrix} 1 & & & & \\ & \ddots & & & \\ & & \cosh t \cdots \sinh t & & \\ & & \vdots \qquad \vdots & & \\ & & -\sinh t \cdots \cosh t & & \\ & & & \ddots & \\ & & & & 1 \end{bmatrix} \qquad \begin{matrix} 1 \leq i \leq 4 \\ \\ 5 \leq j \leq 6 \end{matrix} \tag{2}$$

The matrix $g_{ij}(t)$ corresponds to a rotation or pseudorotation in the x_i-x_j-plane respectively. The basis elements A_{ij} of the Lie algebra so(4,2) are defined by

$$A_{ij} = \frac{d}{dt} g_{ij}(t)\Big|_{t=0} \tag{3}$$

and obey the commutation relations

$$[A_{ij}, A_{kl}] = g_{jk}A_{il} + g_{il}A_{jk} - g_{ik}A_{jl} - g_{jl}A_{ik}$$

with $g_{ii} = +1$ for $1 \leq i \leq 4$, $g_{ii} = -1$ for $i = 5,6$ and $g_{jk} = 0$ for $j \neq k$. From (4) it follows that a representation of the Lie algebra is completely determined, if the operators $A_{i,i+1}$ with $1 \leq i \leq 5$ are known, because the other operators A_{ij} can be expressed through them with the relations (4). We denote an operator of a representation, corresponding to the generator A_{ij}, by the same symbol. In an UIR of the group $SO_0(4,2)$ the operators A_{ij} are antihermitian, e.g.

$$A_{ij} = - A^{+}_{ij} \tag{5}$$

All representations of the Lie algebra so(4,2) derived in this paper can be extended to a UIR of the conformal group. This follows immediately from Ref. 3, see especially the end of page 572. Further we assume that each $SO_0(4,1)$ representation occurs at most with multiplicity one. Actually, we believe that this is not an additional restriction.

We want to use as far as possible the results of Ref. 4 for the special case SO(6), which are given in the appendix. For that reason it is convenient to define new generators B_{ij} by $A_{ij} = \sqrt{g_{ii}}\sqrt{g_{jj}}\, B_{ij}$, see Ref. 5. A simple calculation shows that if the A_{ij} obey the relations (4), the B_{ij} obey the commutation relations (A.1) of the Lie algebra so(6). However, the condition (5) for the unitarity of the group representation has to be replaced by

$$B_{ij} = - \epsilon_{ij} B^{+}_{ij} \tag{6}$$

where

$$\epsilon_{ij} = \begin{cases} +1 & \text{for } 1 \leqq i, j \leqq 4 \text{ or } 5 \leqq i, j \leqq 6 \\ -1 & \text{for } 1 \leqq i \leqq 4 \text{ and } 5 \leqq j \leqq 6 \end{cases}$$

This means we can take from the appendix all those results for the derivation of which only the commutation relations are used, and then we impose the new condition (6). In Section II we derive the UIR's of the $SO_0(4,1)$ De Sitter group, instead of changing the notation of Ref. 6 to ours, because this is a good example of how easy is to derive also a wide class of UIR's of non-compact pseudorotation groups by suitable use of the results of Ref. 4. In Sections III, IV and V we come to our main task, the determination of the classes of UIR's of $SO_0(4,2)$, specified under point a, b, and c respectively. All results are collected in some tables at the end of this paper.

II. The UIR's of the De Sitter Group $SO_0(4,1)$

For the generator B_{45} we have $A_{45} = iB_{45}$ and $B^{+}_{45} = B_{45}$. From the appendix we get the following expression for B_{45} as a result of the commutation relations

$$B_{45}|m_{41}\ m_{42}\rangle = A(m_{41})|m_{41}+1,\ m_{42}\rangle + A(m_{42})|m_{41},\ m_{42}+1\rangle$$
$$- A(m_{41}-1)|m_{41}-1,\ m_{42}\rangle - A(m_{42}-1)|m_{41},m_{42}-1\rangle \tag{7}$$

with the matrix elements

$$A(m_{41}) = \tfrac{1}{2}\sqrt{(m_{31}+\tfrac{1}{2})^2-(m_{41}+\tfrac{1}{2})^2}$$
$$\sqrt{\frac{[(z_{51}+\frac{1}{2})^2-(m_{41}+\frac{1}{2})^2][(z_{52}+\frac{3}{2})^2-(m_{41}+\frac{1}{2})^2]}{[(m_{42}+1)^2-m_{41}^2][(m_{42}+1)^2-(m_{41}+1)^2]}} \tag{8a}$$

$$A(m_{42}) = \tfrac{1}{2}\sqrt{(m_{31}+\tfrac{1}{2})^2-(m_{42}+\tfrac{3}{2})^2}$$
$$\sqrt{\frac{[(z_{51}+\frac{1}{2})^2-(m_{42}+\frac{3}{2})^2][(z_{52}+\frac{3}{2})^2-(m_{42}+\frac{3}{2})^2]}{[m_{41}^2-(m_{42}+1)^2][m_{41}^2-(m_{42}+2)^2]}} \tag{8b}$$

The labels m_{21}, m_{31}, m_{41} and m_{42} specify an irreducible representation of SO(2), SO(3) and SO(4) respectively, and z_{51} and z_{52} are complex numbers $z_{ij} = x_{ij} + iy_{ij}$. They have to be chosen in such a way that the SO(4)-labels m_{41} and m_{42} have the region defined by (A.4) and that B_{45} is hermitian.

Equation (A.4) requires that $|m_{41}| \leqq m_{41}^{max} = m_{42}^{min} \leqq m_{42}$. For this to be true we must have $A(m_{41}^{max}) = A(m_{42}^{min}-1) = 0$, and we get a condition for one of the two constants z_{5i}, and we choose z_{51}:

$$(z_{51}+\tfrac{1}{2})^2 = (m_{41}^{max}+\tfrac{1}{2})^2 = (m_{42}^{min}+\tfrac{1}{2})^2 \tag{9}$$

From (9) it follows that $z_{51} = m_{51} = m_{41}^{max} = m_{42}^{min}$, so that we have $|m_{41}| \leqq m_{51} \leqq m_{42}$.

Now we have to determine the restrictions for z_{52} which result from the hermiticity of B_{45}. The hermiticity means that the square roots in (8) are purely imaginary, i.e., the expressions under the square roots have to be real and negative. The requirement of reality restricts z_{52} to one of the following two possibilities:

$$z_{52} + \tfrac{3}{2} = iy_{52} \tag{10a}$$

$$z_{52} = x_{52} \tag{10b}$$

The fact that the expressions under the square roots in (8) have to be negative gives the additional requirement:

$$(z_{52} + \tfrac{3}{2})^2 \leqq (m_{41} + \tfrac{1}{2})^2 \tag{11}$$

We discuss now the condition (11) for the two possibilities (10) separately.

In the case (10a), inequality (11) is evidently satisfied for arbitrary real y_{42}. However, to avoid having the same representation occur more than once, we make the further restriction $0 < y_{52}$ and call these representations $D(m_{51}, iy_{52})$. The equality $y_{52} = 0$ will be included in a later case. The SO(4) content is $|m_{41}| \leqq m_{51} \leqq m_{42}$. For (10b) there are different possibilities. If $x_{52} = m_{52}$, integer or half integer together with the SO(4)-labels, it may happen that m_{41} is further restricted, i.e., $m_{41}^{min} \leqq m_{41}$ for m_{41} positive and $m_{41} \leqq m_{41}^{max}$ for m_{41} negative, because for

$$(m_{52} + \tfrac{3}{2})^2 = (m_{41}^{min} - \tfrac{1}{2})^2 = (m_{41}^{max} + \tfrac{1}{2})^2 \tag{12}$$

one has

$$A(m_{41}^{min} - 1) = A(m_{41}^{max}) = 0 \tag{13}$$

A solution of (12) is $m_{41}^{min} = -m_{41}^{max} = m_{52} + 2$ and we have two cases $m_{52} + 2 \leqq \pm m_{41} \leqq m_{51} \leqq m_{42}$ for m_{41} positive or negative respectively, with m_{51} and $m_{52} + 2 = \frac{1}{2}, 1, \frac{3}{2}, \ldots$. We call these representations $D^{\pm}(m_{51}, m_{52})$ corresponding to the sign of m_{41}.

Let us now assume $m_{51} \neq 0$ and integer. Then the smallest value at the right hand side of (11) is evidently $\frac{1}{4}$, so that x_{52} is restricted by

$$0 \leqq (x_{52} + \tfrac{3}{2})^2 \leqq \tfrac{1}{4} \tag{14}$$

Again, we replace (14) by the stronger inequality

$$0 \leqq x_{52} + \tfrac{3}{2} \leqq \tfrac{1}{2} \tag{15}$$

because we want every representation to occur only once. If the $<$-sign on the right side of (14) is valid, the SO(4)-content is $|m_{41}| \leqq m_{51} \leqq m_{42}$. These representations are called $D^{o}(m_{51}, x_{52})$. In either case the range for the discrete $SO_0(4,1)$ label is $m_{51} = 1,2,3\ldots$. If $m_{51} = 0$, the right hand side of (14) can be replaced by $\frac{9}{4}$, and by the same arguments following equation (15) we get

$$0 \leqq x_{52} + \tfrac{3}{2} \leqq \tfrac{3}{2} \quad . \tag{16}$$

If the <-sign at the right side is valid, we have $m_{41} = m_{51} = 0 \leqq m_{42}$ with the class $D^1(x_{52})$; the = -sign leads to the trivial case $A(m_{41}) = A(m_{42}) \equiv 0$. The results of this section are collected in Table 1.

III. The UIR's of $SO_0(4,2)$ Which Contain Only The Discrete Series of $SO_0(4,1)$

In this section we want to determine those UIR's of the group $SO_0(4,2)$ which contain only representations of the discrete series $D^{\pm}(m_{51}, m_{52})$ of $SO_0(4,1)$. The generator A_{56} is connected with B_{56} through $A_{56} = -B_{56}$, and (6) gives in this case $B^{+}_{56} = -B_{56}$; i.e., B_{56} is antihermitian. As a result of the commutation relations, we get from the appendix the following expression for B_{56}

$$B_{56}|m_{51}, m_{52}\rangle = B(m_{51})|m_{51}+1, m_{52}\rangle + B(m_{52})|m_{51}, m_{52}+1\rangle$$
$$+ i\,C_4|m_{51}, m_{52}\rangle - B(m_{51}-1)|m_{51}-1, m_{52}\rangle - B(m_{52}-1)|m_{51}, m_{52}-1\rangle \tag{17}$$

with the matrix elements

$$B(m_{51}) = \sqrt{[m_{41}{}^2 - (m_{51}+1)^2][(m_{42}+1)^2 - (m_{51}+1)^2]} \cdot \sqrt{\frac{[z_{61}{}^2-(m_{51}+1)^2][(z_{62}+1)^2-(m_{51}+1)^2][(z_{63}+2)^2-(m_{51}+1)^2]}{(m_{51}+1)^2[4(m_{51}+1)^2-1][(m_{52}+2)^2-(m_{51}+1)^2][(m_{52}+1)^2-(m_{51}+1)^2]}} \tag{18a}$$

$$B(m_{52}) = \sqrt{[m_{41}{}^2 - (m_{52}+1)^2][(m_{42}+1)^2-(m_{52}+2)^2]} \cdot \sqrt{\frac{[z_{61}{}^2-(m_{52}+2)^2][(z_{62}+1)^2-(m_{52}+2)^2][(z_{63}+2)^2-(m_{52}+2)^2]}{(m_{52}+2)^2[4(m_{52}+2)^2-1][(m_{51}+1)^2-(m_{52}+2)^2][m_{51}{}^2-(m_{52}+2)^2]}} \tag{18b}$$

$$C_4 = \frac{m_{41}(m_{42}+1)z_{61}(z_{62}+1)(z_{63}+2)}{m_{51}(m_{51}+1)(m_{52}+1)(m_{52}+2)} \tag{19}$$

The labels m_{51} and m_{52} specify a representation of the discrete series $D^{\pm}(m_{51}, m_{52})$ of $SO_0(4,1)$ and have the range given in Table 1. Resulting from the commutations alone, the labels z_{61}, z_{62} and z_{63} are complex numbers. They are strongly restricted by the requirements that the $SO_0(4,1)$ labels are limited to the right range, that B_{56} is antihermitian, and that the matrix elements contain no divergences. From Table 1 it is clear that we must have $m_{52} \leqq m_{52}{}^{max} \leqq m_{51}{}^{min} \leqq m_{51}$. For this to be true we need $B(m_{52}{}^{max}) = B(m_{51}{}^{min}-1) = 0$, and we get a condition for one of the constants z_{6j}; we choose z_{61}:

$$z_{61}{}^2 = (m_{52}{}^{max} + 2)^2 = (m_{51}{}^{min})^2 \tag{20}$$

That means, $z_{61} = m_{61}$, $|m_{61}| = \frac{1}{2}, 1, \frac{3}{2}, \ldots$, integer or half integer together with the $SO_0(4,1)$ labels, and we have $m_{52} + 2 \leqq |m_{61}| \leqq m_{51}$. Evidently different representations belong to $\pm m_{61}$ because m_{61} occurs also in C_4. The antihermiticity of B_{56} requires that the matrix elements (18) and (19) are real. The reality of C_4 restricts the remaining labels z_{62} and z_{63} to one of the following two possibilities:

$$z_{62}+1 = x_{62}+1 \text{ and } z_{63}+2 = x_{63}+2 \tag{21a}$$

$$z_{62}+1 = iy_{62} \text{ and } z_{63}+2 = iy_{63} \tag{21b}$$

$$z_{62}+1 = 0, \ z_{63}+2 = x_{63}+2 \text{ or } iy_{63} \tag{21c}$$

The reality of the matrix elements (18) requires in addition

$$[(z_{62}+1)^2-(m_{51}+1)^2][(z_{63}+2)^2-(m_{51}+1)^2] \geqq 0 \tag{22a}$$

$$[(z_{62}+1)^2-(m_{52}+2)^2][(z_{63}+2)^2-(m_{52}+2)^2] \geqq 0 \tag{22b}$$

Now it is easy to see that the matrix elements (18) and (19) contain divergencies if z_{62} and z_{63} are arbitrary, and m_{52} is not suitably bounded from below. A simple consideration shows that these divergencies can only be avoided in one of the three cases which are possible according to (22), i.e. if

$$(z_{62}+1)^2 \leqq (m_{52}+2)^2 \tag{23a}$$

$$(z_{63}+2)^2 \leqq (m_{52}+2)^2 \tag{23b}$$

Eqs. (23) show that m_{52} has to be bounded from below, and this excludes also (21b). To limit the range of m_{52}, we choose $z_{62} = m_{62}$, integer or half integer together with m_{61}. We have $B(m_{52}-1) = 0$ for $m_{62} = m_{52}$, i.e.,

$$m_{62}+2 \leqq m_{52}+2 \leqq m_{61} \leqq m_{51} \tag{24}$$

It remains to determine the range of m_{62} and the conditions for z_{63}. Let us assume at first that $m_{62}+1 = 1, \frac{3}{2}, 2\ldots$. Then from (23b) and (24) it follows that x_{63} is restricted by

$$0 \leqq (x_{63}+2)^2 \leqq (m_{62}+2)^2 \tag{25}$$

Again, we want to avoid having the same representation occur more than once and replace (25) therefore by the stronger inequality

$$0 \leqq x_{63}+2 \leqq m_{62}+2 \tag{26}$$

If the $<$-sign at the right side of (26) is valid, there are no further restrictions on m_{52}. However, if $x_{63}+2 = m_{62}+2$, it is easy to see from (18b) that $m_{52}+2$ is fixed to the single value $m_{52}+2 = m_{62}+2$.

Now we assume $m_{62}+1 = \frac{1}{2}$. In the matrix element (18a) the factor $[(m_{62}+1)^2-(m_{52}+1)^2]$ cancels with the same factor in the denominator, and similarly in the matrix element (18b). This means $m_{52}+2$ always has the range $m_{52}+2 = -m_{61}\ldots+m_{61}$ with m_{61} half-integer. From (8) we see that m_{52} occurs in the $SO_0(4,1)$ representations only in the form $(m_{52}+\frac{3}{2})^2$, and a simple consideration shows that the $SO_0(4,1)$ representations occurring in this case are actually labeled by $m_{52}+2 = \frac{1}{2}, \frac{3}{2}, \ldots m_{61}$. From (23b) we get the condition

$$0 \leqq x_{63}+2 \leqq \tfrac{1}{2} \tag{27}$$

for x_{63}. If the $<$-sign is valid at the right side, there are no further restrictions for the other labels. If $x_{63}+2 = \frac{1}{2}$, either $m_{52}+2$ is fixed to the single value $\frac{1}{2}$ and $|m_{61}| = \frac{1}{2}, \frac{3}{2}, \ldots$, or $\frac{3}{2} \leqq m_{52}+2 \leqq |m_{61}|$ with $|m_{61}| \leqq \frac{3}{2}, \frac{5}{2}, \ldots$.

The last possibility we have to consider is $m_{62}+1 = 0$. The whole discussion is similar to the case $m_{62}+1 = \frac{1}{2}$ and we don't repeat it here completely. However, this case is different from all the other cases in that m_{61} is always positive, because $C_4 = 0$. The second possibility given in (21c) has to be excluded, because in this case the matrix element (18b) is not always real.

IV. The UIR's of $SO_o(4,2)$ Which Change The Discrete $SO_o(4,1)$ Label Only

Now we determine the UIR's of $SO_o(4,2)$ which contain only those $SO_o(4,1)$ representations which differ by the discrete label m_{51}. This is evidently the fact in the following two cases:

$$z_{52} + \tfrac{3}{2} = \text{constant} \tag{28a}$$

$$z_{52} + \tfrac{3}{2} = \pm \tfrac{1}{2} \tag{28b}$$

To determine the generator B_{56} we start again with the matrix elements (18) and (19). However, the label z_{52} is not necessarily discrete now and is, therefore, replaced by z_{52}. Clearly z_{52} is restricted from the beginning to one of the values in Table 1. In the case (28a) we need evidently

$$B(z_{52}) = B(z_{52} - 1) = 0 \tag{29}$$

This requires $(z_{62}+1)^2 = (z_{52}+1)^2$ and $(z_{63}+2)^2 = (z_{52}+2)^2$ with $z_{52}+1$, $z_{52}+2 \neq 0$. With these restrictions the matrix elements (18a) and (19) simplify to

$$B(m_{51}) = \sqrt{[m_{41}{}^2 - (m_{51}+1)^2][(m_{42}+1)^2 - (m_{51}+1)^2]}$$

$$\sqrt{\frac{z_{61}{}^2 - (m_{51}+1)^2}{(m_{51}+1)^2[4(m_{51}+1)^2 - 1]}} \tag{30}$$

$$C_4 = \pm \frac{m_{41}(m_{42}+1)z_{61}}{m_{51}(m_{51}+1)} \quad . \tag{31}$$

In the case (28b) the conditions for the matrix elements (18b) are

$$\begin{aligned} B(z_{52}) &= 0 \qquad \text{for} \qquad z_{52} + 2 = 1 \\ B(z_{52}-1) &= 0 \qquad \text{for} \qquad z_{52} + 2 = 0 \end{aligned} \tag{32}$$

To cancel the divergencies in (18) and (19) we must have $z_{62}+1 = 0$, and the conditions (32) give immediately $(z_{63}+2)^2 = 1$. With these special values for z_{62} and z_{63} the matrix element (19a) simplifies to (30), from (18b) we get

$$B(z_{52}) = \sqrt{[m_{41}{}^2 - (z_{52}+2)^2][(m_{42}+1)^2 - (z_{52}+2)^2]} \cdot$$

$$\sqrt{-\frac{[z_{61}{}^2 - (z_{52}+2)^2][1 - (z_{52}+2)^2]}{[4(z_{52}+2)^2 - 1][(m_{51}+1)^2 - (z_{52}+2)^2][m_{51}{}^2 - (z_{52}+2)^2]}}$$

and $C_4 = 0$. If the operator B_{56} with the matrix elements (30), (33) and $C_4 = 0$ is applied to a vector with $z_{52}+2 = +1$, the result is a new vector with $z_{52}+2 = 0$, multiplied by a factor which is given by the right hand side of (31), the value of $B(z_{52}-1)$ for $z_{52}+2 = 1$. A vector with $z_{52}+2 = 0$ is transformed into a new one with $z_{52}+2 = 1$, multiplied again by the factor (31). Now, $SO_0(4,1)$ representations which differ only in the label z_{52} with $z_{52}+2 = 0$ or 1 are identical, and a simple consideration shows that the case (28b) is included in (28a) if we admit the values $z_{52}+2 = 0$ and 1.

Considerations similar to those in the last section restrict z_{61} to m_{61} with the condition $|m_{61}| = m_{51}$, integer or half integer at the same time as m_{51}.

From Table 1 we get the ranges for m_{61}, z_{62} and z_{63}. For $z_{52}+\frac{3}{2} = iy_{52}$ we have $|m_{61}| = 0, \frac{1}{2}, 1 \ldots$, for $z_{52} = x_{52}$, $0 \leqq x_{52} + \frac{3}{2} < \frac{1}{2}$ the range of m_{61} is $|m_{61}| = 1,2,3\ldots$ with the SO(4) content $m_{41} = 0$, $m_{51} \leqq m_{42}$. If z_{52} is fixed to a value that specifies one of the discrete series $D^{\pm}(m_{51}, m_{52})$, the resulting representations are already contained in Section III. The results of this section and Section III are collected in Table 2.

V. The UIR's of $SO_0(4,2)$ Which Contain Only A Single $SO_0(4,1)$ Representation

In this last section we determine those UIR's of $SO_0(4,2)$ which remain irreducible when restricted to $SO_0(4,1)$. This is the fact if, in addition to (28), one of the following two conditions is fulfilled:

$$m_{51} + \tfrac{1}{2} = \text{constant} \tag{34a}$$

$$m_{51} + \tfrac{1}{2} = \pm\tfrac{1}{2} \tag{34b}$$

For the possibility (34a) we have

$$B(m_{51}) = B(m_{51} - 1) = 0 \tag{35}$$

If $m_{61}{}^2 = (m_{51}+1)^2$ and $m_{51}+1 \neq 0$, the first matrix element is zero, and the second simplifes to

$$B(m_{51}-1) = \sqrt{(m_{41}^2 - m_{51}^2)[(m_{42}^2+1)^2 - m_{51}^2]}\sqrt{\frac{1}{m_{51}^2(2m_{51}-1)}} \quad (36)$$

The only possibility that (36) is zero is $m_{41}^2 = m_{51}^2$ with $m_{51}^2 \neq 0$. The second case (34b) has to be treated in the same way as (28b). The discussion is exactly the same with the result that this possibility is included if we admit in the first case the values $m_{51}+1$ and m_{51} equal to zero, and in addition $z_{63}+2 = 0$. From Table 1 it is easy to see that the requirements we need can be fulfilled only if either the $SO_o(4,1)$ representation belongs to one of the discrete series $D^{\pm}(m_{51}, m_{52})$ with $m_{52}+2 = |m_{41}| = m_{51}$, or if it belongs to the series $D^1(x_{52})$ with $x_{52}+\frac{3}{2} = \frac{1}{2}$.

The operators B_{45} and B_{56} are in this last case considerably simpler than the general expressions (7) and (17), and we give them for that reason in the simplest form:

$$\begin{aligned} B_{45}|m_{41}, m_{42}\rangle &= \tfrac{1}{2}\sqrt{(m_{31}+\tfrac{1}{2})^2 - (m_{42}+\tfrac{3}{2})^2}\;|m_{41}, m_{42}+1\rangle \\ &- \tfrac{1}{2}\sqrt{(m_{31}+\tfrac{1}{2})^2 - (m_{42}+\tfrac{1}{2})^2}\;|m_{41}, m_{42}-1\rangle \end{aligned} \quad (37)$$

$$B_{56}|m_{41}, m_{42}\rangle = i(m_{41}+1)|m_{41}, m_{42}\rangle \quad (38)$$

The results of this section are given in Table 2.

Acknowledgments

The author thanks Dr. A. Böhm, Dr. S. Ström and L. Jaffe for interesting discussions and careful reading of the manuscript.

Appendix

At first we give in this appendix the results of Gelfand and Zetlin[4)] for the rotation group SO(6). The notation has been changed slightly in agreement with Ref. 7. The infinitesimal generators A_{ij} $(1 \leq i < j \leq 6)$ obey the commutation relations

$$[A_{ij}, A_{kl}] = \delta_{jk}A_{il} + \delta_{il}A_{jk} - \delta_{ik}A_{jl} - \delta_{jl}A_{ik} \quad (A.1)$$

and they are antihermitian

$$A_{ij} = -A_{ij}^{+} \quad (A.2)$$

An irreducible representation of SO(6) is determined by three numbers, m_{61}, m_{62} and m_{63}. We denote a vector in a representation space by $|m_{ij}\rangle$, where m_{ij} is an abbreviation for a complete set of labels, which determine an irreducible representation and specify each vector in the representation space uniquely. The complete scheme is

$$|m_{ij}\rangle = \left| \begin{matrix} m_{61} & m_{62} & m_{63} \\ m_{51} & m_{52} & \\ m_{41} & m_{42} & \\ m_{31} & & \\ m_{21} & & \end{matrix} \right| \tag{A.3}$$

All labels m_{ij} are integer or half-integer at the same time and obey the conditions

$$\begin{aligned} |m_{61}| &\leqq m_{51} \leqq m_{62} \leqq m_{52} \leqq m_{63} \\ |m_{41}| &\leqq m_{51} \leqq m_{42} \leqq m_{52} \\ |m_{41}| &\leqq m_{31} \leqq m_{42} \\ |m_{21}| &\leqq m_{31} \end{aligned} \tag{A.4}$$

We denote the operators in an irreducible representation, which correspond to the infinitesimal generators A_{ij}, by the same symbols. They are given by

$$A_{12}|m_{21}\rangle = i\, m_{21}|m_{21}\rangle \tag{A.5}$$

$$A_{23}|m_{21}\rangle = A(m_{21})|m_{21}+1\rangle - A(m_{21}-1)|m_{21}-1\rangle \tag{A.6}$$

$$A_{34}|m_{31}\rangle = B(m_{31})|m_{31}+1\rangle + iC_2|m_{31}\rangle - B(m_{31}-1)|m_{31}-1\rangle \tag{A.7}$$

$$\begin{aligned} A_{45}|m_{41}, m_{42}\rangle = {} & A(m_{41})|m_{41}+1, m_{42}\rangle + A(m_{42})|m_{41}, m_{42}+1\rangle \\ & - A(m_{41}-1)|m_{41}-1, m_{42}\rangle - A(m_{42}-1)|m_{41}, m_{42}-1\rangle \end{aligned} \tag{A.8}$$

$$A_{56}\,|m_{51}, m_{52}\rangle = B(m_{51})|m_{51}+1, m_{52}\rangle + B(m_{52})|m_{51}, m_{52}+1\rangle + $$
$$+ iC_4\,|m_{51}, m_{52}\rangle - B(m_{51}-1)|m_{51}-1, m_{52}\rangle$$
$$- B(m_{52}-1)|m_{51}, m_{52}-1\rangle \tag{A.9}$$

The labels in the state vectors which do not change have always been omitted. The matrix elements are

$$A(m_{21}) = \tfrac{1}{2}\sqrt{(m_{31}+\tfrac{1}{2})^2 - (m_{21}+\tfrac{1}{2})^2} \tag{A.10}$$

$$A(m_{41}) = \tfrac{1}{2}\sqrt{(m_{31}+\tfrac{1}{2})^2 - (m_{41}+\tfrac{1}{2})^2}\;\cdot$$
$$\cdot\sqrt{\frac{[(m_{51}+\tfrac{1}{2})^2 - (m_{41}+\tfrac{1}{2})^2][(m_{52}+\tfrac{3}{2})^2 - (m_{41}+\tfrac{1}{2})^2]}{[(m_{42}+1)^2 - m_{41}{}^2][(m_{42}+1)^2 - (m_{41}+1)^2]}} \tag{A.11}$$

$$A(m_{42}) = \tfrac{1}{2}\sqrt{(m_{31}+\tfrac{1}{2})^2 - (m_{42}+\tfrac{1}{2})^2}\;\cdot$$
$$\cdot\sqrt{\frac{[(m_{51}+\tfrac{1}{2})^2 - (m_{42}+\tfrac{3}{2})^2][(m_{52}+\tfrac{3}{2})^2 - (m_{42}+\tfrac{3}{2})^2]}{[(m_{41}{}^2 - (m_{42}+1)^2][m_{41}{}^2 - (m_{42}+2)^2]}} \tag{A.12}$$

$$B(m_{31}) = \sqrt{m_{21}{}^2 - (m_{31}+1)^2}\sqrt{\frac{[m_{41}{}^2 - (m_{31}+1)^2][(m_{42}+1)^2 - (m_{31}+1)^2]}{(m_{31}+1)^2\,[4(m_{31}+1)^2 - 1]}} \tag{A.13}$$

$$B(m_{51}) = \sqrt{[m_{41}{}^2 - (m_{51}+1)^2][(m_{42}+1)^2 - (m_{51}+1)^2]}$$
$$\sqrt{\frac{[m_{61}{}^2 - (m_{51}+1)^2][(m_{62}+1)^2 - (m_{51}+1)^2][(m_{63}+2)^2 - (m_{51}+1)^2]}{(m_{51}+1)^2[4(m_{51}+1)^2-1][(m_{52}+2)^2-(m_{51}+1)^2][(m_{52}+1)^2-(m_{51}+1)^2]}} \tag{A.14}$$

$$B(m_{52}) = \sqrt{[m_{41}^2 - (m_{52}+2)^2][(m_{42}+1)^2 - (m_{52}+2)^2]}$$

$$\sqrt{\frac{[m_{61}^2-(m_{52}+2)^2][(m_{62}+1)^2-(m_{52}+2)^2][(m_{63}+2)^2-(m_{52}+2)^2]}{(m_{52}+2)^2[4(m_{52}+2)^2-1][(m_{51}+1)^2-(m_{52}+2)^2][m_{51}^2-(m_{52}+2)^2]}} \tag{A.15}$$

$$C_2 = \frac{m_{21}\ m_{41}\ (m_{42}+1)}{m_{31}\ (m_{31}+1)} \tag{A.16}$$

$$C_4 = \frac{m_{41}\ (m_{42}+1)\ m_{61}\ (m_{62}+1)(m_{63}+2)}{m_{51}\ (m_{51}+1)(m_{52}+1)(m_{52}+2)} \tag{A.17}$$

Let us now describe by an example how one derives these results for SO(n+1) if they are already known for SO(n). We choose n = 4, the general case is a straightforward generalization. Evidently the problem reduces immediately to determining an antihermitian operator A_{45} which obeys the commutation relations (A.1) with $1 \leqq i,j,k,l \leqq 5$. Now it can easily be shown that the commutation relations for SO(5) are equivalent to those of SO(4), which are given by (A.1) with $1 \leqq i,j,k,l \leqq 4$, and the following ones:

$$[A_{i,i+1}, A_{45}] = 0 \qquad \text{for } 1 \leqq i \leqq 3 \tag{A.18}$$

$$[A_{34}, [A_{45}, A_{34}]] = A_{45} \tag{A.19}$$

$$[A_{45}, [A_{45}, A_{34}]] = -A_{34} \tag{A.20}$$

From these commutation relations one gets the expression (A.8) for the operator A_{45}, and a set of recurrence relations for the matrix elements. A solution of these recurrence relations is given by (A.11) and (A.12), with the labels m_{51} and m_{52} replaced by complex numbers z_{51} and z_{52}. The requirement of antihermiticity (A.2) restricts them to m_{51} and m_{52} with the range (A.4).

Table 1

Representations	Conditions for m_{51} & z_{52}	SO(4)-Content
$D(m_{51}, iy_{52})$	$m_{51} = 0, \frac{1}{2}, 1 \ldots$ $z_{52} + \frac{3}{2} = iy_{52}, 0 < y_{52}$	$\lvert m_{41}\rvert \leq m_{51} \leq m_{42}$ 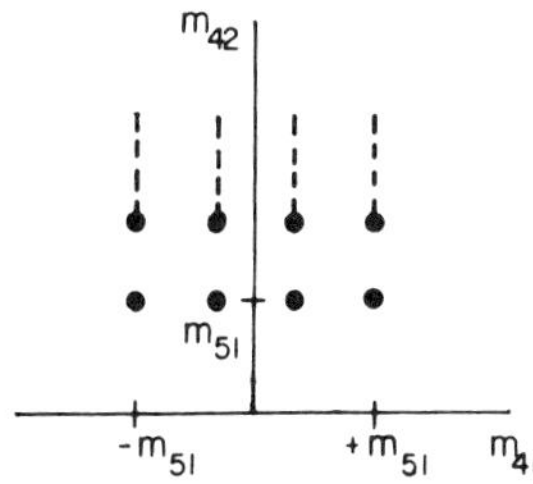
$D^0(m_{51}, x_{52})$	$m_{51} = 1, 2, 3 \ldots$ $z_{52} = x_{52}, 0 \leq x_{52}+\frac{3}{2} < \frac{1}{2}$	
$D^1(x_{52})$	$m_{51} = 0$ $z_{52}=x_{52}; 0 \leq x_{52}+\frac{3}{2} < \frac{3}{2}$	
$D^0(m_{51}m_{52})$	$m_{51} = 1, 2, 3 \ldots$ $z_{52}=m_{52}, m_{52} + 2 = 1$	$m_{41} = 0, m_{51} \leq m_{42}$ 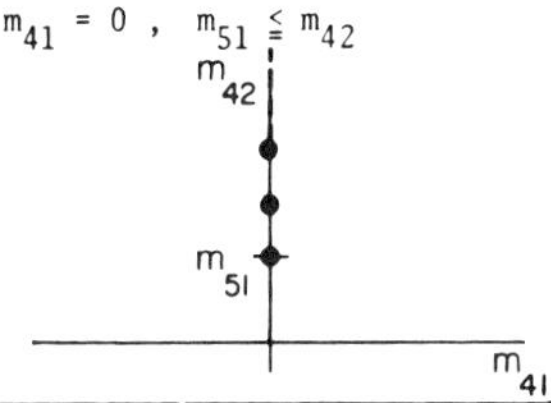
$D^+(m_{51}, m_{52})$	$m_{51} = \frac{1}{2}, 1, \frac{3}{2} \ldots$ $z_{52}=m_{52}, m_{52}+2=\frac{1}{2}, 1, \frac{3}{2} \ldots$ $1 \leq m_{52}+2 \leq m_{51}$ or $\frac{1}{2} \leq m_{52}+2 \leq m_{51}$ for m_{51}, m_{52} integer or half-integer respectively	$m_{52} + 2 \leq m_{41} \leq m_{51} \leq m_{42}$ 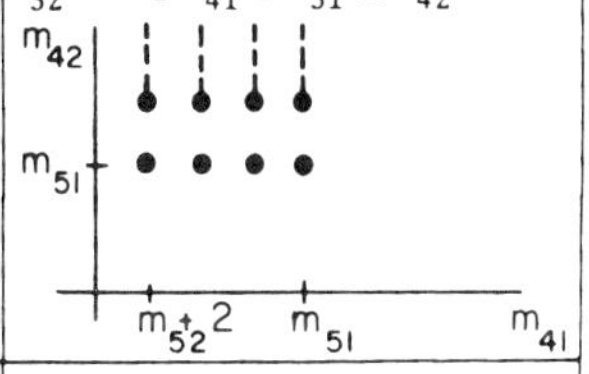
$D^-(m_{51},m_{52})$	The same as for $D^+(m_{51}, m_{52})$	$m_{52} + 2 \leq -m_{41} \leq m_{51} \leq m_{42}$ 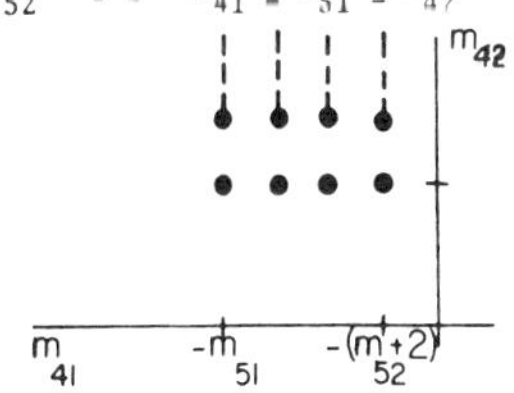

Table 1. Classification of the UIR's of $SO_0(4,1)$. In the diagrams the coordinates of each dot, m_{41} and m_{42}, specify a SO(4) representation. The operator B_{45} is given by (7) and (8).

Table 2

Representations	Conditions for the $SO_0(4,2)$ and $SO_0(4,1)$ Labels	SO(4) Content
$D^+(m_{51},m_{52})$	$m_{61}^2 = (m_{51}+1)^2$ $(z_{62}+1)^2 = (z_{52}+1)^2$ $(z_{63}+2)^2 = (z_{52}+2)^2$ $z_{52} = m_{52}$	$m_{52}+2=m_{41}=m_{51}\leq m_{42}$ 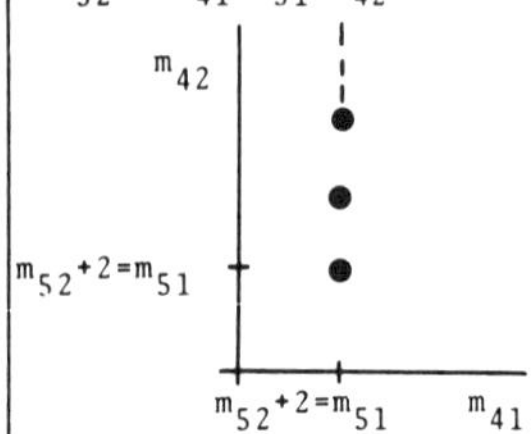
$D^-(m_{51},m_{52})$	The same as for the case $D^+(m_{51},m_{52})$	$m_{52}+2=-m_{41}=m_{51}\leq m_{42}$ 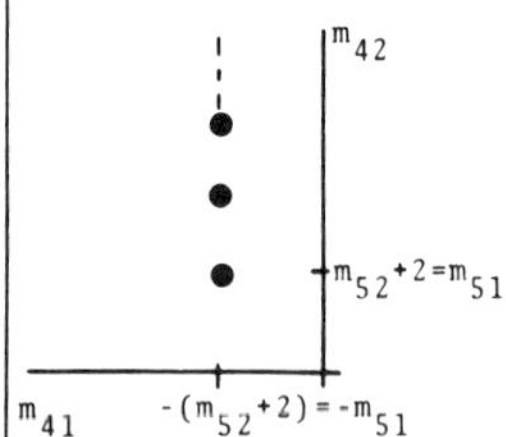
$D^1(x_{52})$	$m_{61}^2 = (m_{51}+1)^2$ $(z_{62}+1)^2 = (z_{52}+1)^2$ $(z_{63}+2)^2 = (z_{52}+2)^2$ $m_{51} = 0$ $z_{52}=x_{52},\ x_{52}+1=0$	$m_{41}=0\leq m_{42}$ 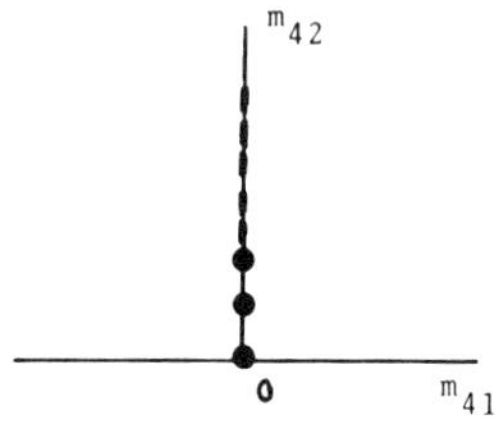

Table 2. The UIR's of $SO_0(4,1)$ which can be extended to a UIR of $SO_0(4,2)$. Each dot in the diagrams determines a SO(4) representation. The operators B_{45} and B_{56} are given by (37) and (38) respectively.

Table 3

Conditions for the $SO_o(4,2)$ Labels $z_{61}=m_{61}$, z_{62} and z_{63}	$SO_o(4,1)$ Content
$z_{62}=m_{62}$; $m_{62}+1=1,\frac{3}{2},2,\ldots;m_{62}+2\leq\lvert m_{61}\rvert$ $z_{63}=x_{63}$ a.) $0\leq x_{63}+2<m_{62}+2$ b.) $x_{63}+2=m_{62}+2$ m_{61} is always positive for $x_{63}+2=0$	$m_{62}+2\leq m_{52}+2\leq\lvert m_{61}\rvert\leq m_{51}$ $m_{62}+2=m_{52}+2\leq\lvert m_{61}\rvert\leq m_{51}$
$z_{62}=m_{62}$; $m_{62}+1=\frac{1}{2}$; $\frac{1}{2}\leq\lvert m_{61}\rvert$ $z_{63}=x_{63}$ a.) $0\leq x_{63}+2<\frac{1}{2}$ b.) $x_{63}+2=\frac{1}{2}$ m_{61} is always positive for $x_{63}+2=0$	$\frac{1}{2}\leq m_{52}+2\leq\lvert m_{61}\rvert\leq m_{51}$ $\frac{1}{2}=m_{52}+2\leq\lvert m_{61}\rvert\leq m_{51}$ or $\frac{3}{2}\leq m_{52}+2\leq m_{61}\leq m_{51}$
$z_{62}=m_{62}$; $m_{62}+1=0$; $1\leq m_{61}$ $z_{63}=m_{63}$ a.) $0\leq x_{63}+2<1$ b.) $x_{63}+2=1$	$1\leq m_{52}+2\leq m_{61}\leq m_{51}$ $1=m_{52}+2\leq m_{61}\leq m_{51}$
$z_{62}=z_{63}=z_{52}$ a.) $\lvert m_{61}\rvert=0,\frac{1}{2},1,\ldots$ b.) $\lvert m_{61}\rvert=0,\frac{1}{2},1,\ldots$ c.) $m_{61}=1,2,3,\ldots$	$z_{52}+\frac{1}{2}=iy_{52}$; $0<y_{52}$; $\lvert m_{61}\rvert\leq m_{51}$ $z_{52}=x_{52}$; $0\leq x_{52}+\frac{3}{2}<\frac{1}{2}$ $z_{52}=x_{52}$; $m_{52}+2=1$, $m_{61}=m_{51}$ and $m_{41}=0$

Table 3. The representations of $SO_o(4,2)$ which are determined in Sections III and IV. The operator B_{56} is given by (17), the matrix elements by (18), (19) or (30) and (31).

References

1. E. Cunningham, Proc. London Math. Soc. 8, 77 (1909).
   H. Bateman, loc. cit. 8, 223 (1910).
   L. Page, Phys. Rev. 49, 254, 466, 946 (1936).
   P. A. M. Dirac, Proc. Roy. Soc. A177, 148 (1938).
   H. A. Kastrup, Phys. Rev. 142, 1060 (1966); 143, 1041 (1966); 150, 1183 (1966); Ann. Phys. 9, 388 (1962); Nucl. Phys. 50, 561 (1966).
   A. O. Barut, Lectures in Theoretical Physics, Vol. XB (Gordon and Breach, New York, 1968), p. 377 and references therein.
   H. Kleinert, Fort. Phys. 16, 1 (1968).
2. Y. Murai, Progr. Theoret. Phys. 9, 147 (1953).
   M. I. Graev, DAN SSSR 98, 517 (1954).
   A. Esteve and P. G. Sona, Nuovo Cimento 32, 473 (1964).
   A. Kihlberg, V. F. Muller and F. Halbwachs, Comm. Math. Phys. 3, 194 (1966).
   Tsu Yao, Jour. Math. Phys. 8, 1931 (1967); 9, 1615 (1968).
   A. O. Barut and A. Böhm, preprint CPT-44, Univ. of Texas at Austin (1970) ; J. Math. Phys. 11, 2938 (1970).
3. E. M. Stein, IAEA Vienna 1965, p. 563.
4. I. M. Gelfand and M. L. Zetlin, DAN SSSR 71, 1017 (1950).
5. A. V. Nikolov, Funkt. Anal. Appl. 2, 94 (1969); preprint Dubna P5-3140 (1967).
6. L. H. Thomas, Ann. Math. 42, 113 (1941).
   T. D. Newton, loc. cit. 51, 730 (1950).
   J. Dixmier, Bull. Soc. Math. France 89, 9 (1961).
   S. Ström, Arkiv Fysik 30, 455 (1965).
   A. Kihlberg and S. Ström, loc. cit. 31, 491 (1966).
7. F. Schwarz, preprint CPT-59, Univ. of Texas at Austin (1970), to appear in Jour. Math. Phys.

UNITARY IRREDUCIBLE REPRESENTATIONS OF SU(2,2), REDUCTION WITH RESPECT TO AN ISO-POINCARÉ SUBGROUP†

Tsu Yao
Department of Physics
University of Pittsburgh
Pittsburgh, Pennsylvania 15213

The noncompact group SU(2,2), which is the covering group of the conformal group, or O(4,2), has attracted some attention in recent years among particle and relativity physicists.[1] In this talk we shall restrict ourselves to the study of the unitary irreducible representations of SU(2,2), and their reduction with respect to a subgroup E(3,1) which is isomorphic to the Poincaré group.[2]

I. Commutation Relations and Subgroups of SU(2,2)

The generators of O(4,2), L_{ab}, obey the following commutation relations,

$$[L_{ab}, L_{cd}] = i\{g_{ac}L_{bd} - g_{ad}L_{bc} - g_{bc}L_{ad} + g_{bd}L_{ac}\} , \qquad (I.1)$$

$$a,b,c,d = 0,1,2,3,5,6$$

with the metric g_{ab} chosen to be

$$g_{11} = g_{22} = g_{33} = g_{55} = -g_{00} = -g_{66} = 1 .$$

Let us concentrate on the 1,2,5,6 space (iso-Minkowski space) and define

†Presented at the Symposium on De Sitter and Conformal Groups, University of Colorado, Summer 1970.

$$\left.\begin{array}{l} \mathcal{L}_{\alpha\beta} \equiv L_{\alpha\beta} \\ \mathcal{P}_{\alpha} = L_{0\alpha} + L_{3\alpha} \end{array}\right\} \quad \text{(Poincaré transformations)}$$

$$\mathcal{R}_{\alpha} \equiv L_{0\alpha} - L_{3\alpha} \qquad \text{(special conformal transformations)}$$

$$L_{03} \qquad \text{(scale transformation)}$$

$$\alpha, \beta = 1,2,5,6 \tag{I.2}$$

Equation (I.1) now becomes

$$[\mathcal{P}_{\alpha}, \mathcal{P}_{\beta}] = 0$$

$$[\mathcal{L}_{\mu\nu}, \mathcal{P}_{\alpha}] = i\{g_{\mu\alpha}\mathcal{P}_{\nu} - g_{\nu\alpha}\mathcal{P}_{\mu}\}$$

$$[\mathcal{L}_{\mu\nu}, \mathcal{L}_{\alpha\beta}] = i\{g_{\mu\alpha}\mathcal{L}_{\nu\beta} - g_{\mu\beta}\mathcal{L}_{\nu\alpha} - g_{\nu\alpha}\mathcal{L}_{\mu\beta} + g_{\nu\beta}\mathcal{L}_{\mu\alpha}\}$$

$$[\mathcal{L}_{\mu\nu}, \mathcal{R}_{\alpha}] = i\{g_{\mu\alpha}\mathcal{R}_{\nu} - g_{\nu\alpha}\mathcal{R}_{\mu}\}$$

$$[\mathcal{R}_{\alpha}, \mathcal{R}_{\beta}] = 0$$

$$[\mathcal{P}_{\alpha}, \mathcal{R}_{\beta}] = -2i\{\mathcal{L}_{\alpha\beta} + g_{\alpha\beta}L_{03}\}$$

$$[L_{03}, \mathcal{L}_{\alpha\beta}] = 0$$

$$[L_{03}, \mathcal{P}_{\alpha}] = -i\mathcal{P}_{\alpha}$$

$$[L_{03}, \mathcal{R}_{\alpha}] = i\mathcal{R}_{\alpha} \tag{I.3}$$

Next, we introduce several noncompact subgroups of SU(2,2). We begin with

(1) E(2) subgroup[3)]

The generators of E(2) are $\mathcal{P}_{\pm}$, and $\mathcal{L}_5$, with

$$\mathcal{P}_{\pm} \equiv \mathcal{P}_1 \pm i\mathcal{P}_2$$

$$\mathcal{L}_5 \equiv \mathcal{L}_{12} \ .$$

The commutation relations are

$$[\mathcal{P}_+, \mathcal{P}_-] = 0$$

$$[\mathcal{L}_5, \mathcal{P}_{\pm}] = \pm\mathcal{P}_{\pm}$$

and the Casimir operator of E(2) is $\mathcal{P}_+\mathcal{P}_-$. We now introduce the basis vector $|\epsilon, m\rangle$, and

$$\mathcal{P}_+\mathcal{P}_-|\epsilon, m\rangle = \epsilon^2|\epsilon, m\rangle$$

$$\mathcal{L}_5|\epsilon, m\rangle = m|\epsilon, m\rangle$$

$$\mathcal{P}_{\pm}|\epsilon, m\rangle = \epsilon\,|\epsilon, m \pm 1\rangle \qquad \text{(I.4)}$$

with $\epsilon^2 > 0, m = 0, \pm\frac{1}{2}, \pm 1 \ldots$

(2) E(3) subgroup

The generators of E(3) are $\vec{\mathcal{P}}$, and $\vec{\mathcal{L}}$, with

$$\vec{\mathcal{P}} = (\mathcal{P}_1, \mathcal{P}_2, \mathcal{P}_5)$$

$$\vec{\mathcal{L}} = (\mathcal{L}_{12}, \mathcal{L}_{25}, \mathcal{L}_{51})$$

The Casimir operators are $\vec{\mathcal{P}}^2$ and $\vec{\mathcal{L}}\cdot\vec{\mathcal{P}}$. Let the basis vector be $|\xi, \epsilon; t, m\rangle$, and $\mathcal{L}_{\pm} = \mathcal{L}_{25} \pm i\mathcal{L}_{51}$. We now have

$$\vec{\mathcal{P}}^2|\xi, \epsilon; t, m\rangle = \xi^2|\xi, \epsilon; t, m\rangle$$

$$\vec{\mathcal{L}}\cdot\vec{\mathcal{P}}|\xi, \epsilon; t, m\rangle = t\xi|\xi, \epsilon; t, m\rangle$$

$$\mathcal{P}_5|\xi, \epsilon; t, m\rangle = (\xi^2 - \epsilon^2)^{\frac{1}{2}}|\xi, \epsilon; t, m\rangle$$

$$\mathcal{L}_{\pm}|\xi, \epsilon; t, m\rangle = \left\{\mp(\xi^2 - \epsilon^2)^{\frac{1}{2}}\frac{\partial}{\partial\epsilon} \pm \tfrac{1}{2}\epsilon(\xi^2 - \epsilon^2)^{-\frac{1}{2}} + t\frac{\xi}{\epsilon} - (m \pm 1)\frac{1}{\epsilon} \times (\xi^2 - \epsilon^2)^{\frac{1}{2}}\right\}|\xi, \epsilon; t, m \pm 1\rangle$$

where the unitary condition has been implemented, and $\xi^2 > 0$, $t = 0, \pm\frac{1}{2}, \pm 1, \ldots$.

(3) E(3,1) subgroup

The generators of E(3,1) are P_μ, and $\mathcal{L}_{\alpha\beta}$, $\mu,\alpha,\beta = 1,2,5,6$. The Casimir operators are $P_\mu P^\mu$ and $W_\mu W^\mu$, where $W_\mu = \frac{1}{2}\epsilon_{\mu\nu\alpha\beta}P^\nu \mathcal{L}^{\alpha\beta}$. Let the basis vector be $|\eta,\xi,\epsilon;\ s,t,m\rangle$. For "time-like" unitary irreducible representations we have[4] $\eta^2 > 0$, $s = 0,\frac{1}{2},1,3/2,\ldots,$ $-s \le t \le s$.

$$P_\mu P^\mu |\eta,\xi,\epsilon;\ s,t,m\rangle = -\eta^2 |\eta,\xi,\epsilon;\ s,t,m\rangle$$

$$W_\mu W^\mu |\eta,\xi,\epsilon;\ s,t,m\rangle = \eta^2 s(s+1)|\eta,\xi,\epsilon;\ s,t,m\rangle$$

$$P_6 |\eta,\xi,\epsilon;\ s,t,m\rangle = (\xi^2+\eta^2)^{\frac{1}{2}} |\eta,\xi,\epsilon;\ s,t,m\rangle$$

$$\begin{aligned}\mathcal{L}_{56}|\eta,\xi,\epsilon;\ s,t,m\rangle = &-i\{(\xi^2+\eta^2)^{\frac{1}{2}}(\xi^2-\epsilon^2)^{\frac{1}{2}}\frac{1}{\xi}\frac{\partial}{\partial\xi} \\ &+\tfrac{1}{2}(\xi^2-\epsilon^2)^{\frac{1}{2}}(\xi^2+\eta^2)^{-\frac{1}{2}} \\ &+\tfrac{1}{2}(\xi^2-\epsilon^2)^{-\frac{1}{2}}(\xi^2+\eta^2)^{\frac{1}{2}}\}|\eta,\xi,\epsilon;\ s,t,m\rangle \\ &+\tfrac{1}{2}i\frac{\epsilon\eta}{\xi^2}\{(s-t)(s+t+1)\}^{\frac{1}{2}}|\eta,\xi,\epsilon;\ s,t+1,m\rangle \\ &-\tfrac{1}{2}i\frac{\epsilon\eta}{\xi^2}\{(s+t)(s-t+1)\}^{\frac{1}{2}}|\eta,\xi,\epsilon;\ s,t-1,m\rangle\end{aligned}$$

$$\begin{aligned}\mathcal{L}_{\pm6}|\eta,\xi,\epsilon;\ s,t,m\rangle = &-i(\xi^2+\eta^2)^{\frac{1}{2}}\{\frac{\epsilon}{\xi}\frac{\partial}{\partial\xi}+\frac{\partial}{\partial\epsilon}+\tfrac{1}{2}\epsilon(\xi^2+\eta^2)^{-1} \\ &\pm(m\pm1)\frac{1}{\epsilon}\mp t\frac{1}{\xi\epsilon}(\xi^2-\epsilon^2)^{\frac{1}{2}}|\eta,\xi,\epsilon;\ s,t,m\pm1\rangle \\ &\mp\tfrac{1}{2}i\frac{\eta}{\xi^2}\{\xi\pm(\xi^2-\epsilon^2)^{\frac{1}{2}}\}\{(s-t)(s+t+1)\}^{\frac{1}{2}} \\ &\times|\eta,\xi,\epsilon;\ s,t+1,m\pm1\rangle\mp\tfrac{1}{2}i\frac{\eta}{\xi^2}\{\xi\mp(\xi^2-\epsilon^2)^{\frac{1}{2}}\} \\ &\times\{(s+t)(s-t+1)\}^{\frac{1}{2}}|\eta,\xi,\epsilon;\ s,t-1,m\pm1\rangle\end{aligned} \tag{I.6}$$

where we have normalized our states to be

$$\langle\eta',\xi',\epsilon';\ s',t',m'|\eta,\xi,\epsilon;\ s,t,m\rangle = \delta(\eta'^2-\eta^2)\,\delta(\xi'^2-\xi^2) \times\delta(\epsilon'^2-\epsilon^2)\,\delta_{s's}\delta_{t't}\delta_{m'm}$$

So far our approach has been simple and standard. All six parameters have simple "physical" interpretation. η is the "mass," ξ is the magnitude of "momentum," ϵ is the projection of "momentum" in the 1-2 plane, s is the "spin," t is the "helicity," and m is the component of "angular momentum" along the 5-axis. We remind the reader that all quantities are referred to in the 1,2,5,6 space.

II. Reduction of Unitary Irreducible Representations of SU(2,2) with Respect to "Time-Like" Representations of E(3,1)

The effect of a finite scale transformation is simple. We have

$$e^{i\zeta L_{03}}|\eta,\xi,\epsilon;\, s,t,m\rangle = e^{-3\zeta}|\eta e^{-\zeta},\, \xi e^{-\zeta},\, \epsilon e^{-\zeta};\, s,t,m\rangle$$

and

$$L_{03}|\eta,\xi,\epsilon,;\, s,t,m\rangle = i\{\eta\frac{\partial}{\partial\eta} + \xi\frac{\partial}{\partial\xi} + \epsilon\frac{\partial}{\partial\epsilon} + 3\}|\eta,\xi,\epsilon;\, s,t,m\rangle \qquad \text{(II.1)}$$

Next, and finally, we come to the lengthy determination of $\mathfrak{R}_\mu$.

$$\mathfrak{R}_5|\eta,\xi,\epsilon;\, s,t,m\rangle = \{(\xi^2-\epsilon^2)^{\frac{1}{2}}[-\frac{\partial^2}{\partial\eta^2} + \frac{\partial^2}{\partial\xi^2} - \frac{\partial^2}{\partial\epsilon^2} + 2\frac{\eta}{\xi}\frac{\partial}{\partial\eta}\frac{\partial}{\partial\xi}]$$

$$+ [\eta(\xi^2-\epsilon^2)^{-\frac{1}{2}} - \frac{1}{\eta}(\xi^2-\epsilon^2)^{\frac{1}{2}}]\frac{\partial}{\partial\eta} + \frac{1}{\xi}[4(\xi^2-\epsilon^2)^{\frac{1}{2}} + \epsilon^2(\xi^2-\epsilon^2)^{-\frac{1}{2}}]\frac{\partial}{\partial\xi}$$

$$+ [\epsilon(\xi^2-\epsilon^2)^{-\frac{1}{2}} - \frac{1}{\epsilon}(\xi^2-\epsilon^2)^{\frac{1}{2}}]\frac{\partial}{\partial\epsilon} + m^2\frac{1}{\epsilon^2}(\xi^2-\epsilon^2)^{\frac{1}{2}} - 2mt\frac{\xi}{\epsilon^2}$$

$$+ t^2(\xi^2-\epsilon^2)^{\frac{1}{2}}[\frac{1}{\eta^2} + 2\frac{1}{\xi^2} + \frac{1}{\epsilon^2}] - s(s+1)\frac{1}{\xi^2}(\xi^2-\epsilon^2)^{\frac{1}{2}}$$

$$-2t\frac{1}{\eta^2\xi}(\xi^2-\epsilon^2)^{\frac{1}{2}}(\xi^2+\eta^2)^{\frac{1}{2}}\beta(s) - \frac{1}{4}(\xi^2-\epsilon^2)^{\frac{1}{2}}(\xi^2+\eta^2)^{-1} + \frac{9}{4}(\xi^2-\epsilon^2)^{-\frac{1}{2}}$$

$$+ \frac{1}{\eta^2}(\xi^2-\epsilon^2)^{\frac{1}{2}}\alpha(s)\}|\eta,\xi,\epsilon;\, s,t,m\rangle$$

$$+ \{(s-t)(s+t+1)\}^{\frac{1}{2}}\frac{\epsilon}{\xi}\{-\frac{1}{\xi}(\xi^2+\eta^2)^{\frac{1}{2}}[\frac{\partial}{\partial\eta} + \frac{1}{\eta}(t+1) + \frac{1}{2}\eta(\xi^2+\eta^2)^{-1}]$$

$$+ \frac{1}{\eta}\beta(s)\}|\eta,\xi,\epsilon;\, s,t+1,m\rangle$$

$$+ \{(s+t)(s-t+1)\}^{\frac{1}{2}}\frac{\epsilon}{\xi}\{\frac{1}{\xi}(\xi^2+\eta^2)^{\frac{1}{2}}[\frac{\partial}{\partial\eta} - \frac{1}{\eta}(t-1) + \frac{1}{2}\eta(\xi^2+\eta^2)^{-1}]$$

$$+ \frac{1}{\eta}\beta(s)\}|\eta,\xi,\epsilon;\, s,t-1,m\rangle$$

(equation continued)

$$+\{(s+t+1)(s+t+2)\}^{\frac{1}{2}} \frac{\epsilon}{\xi\eta} \gamma(s)|\eta,\xi,\epsilon;\ s+1,t+1,m\rangle$$
$$+\{(s+t+1)(s-t+1)\}^{\frac{1}{2}}\ 2\frac{1}{\xi\eta^2}(\xi^2-\epsilon^2)^{\frac{1}{2}}(\xi^2+\eta^2)^{\frac{1}{2}} \gamma(s)|\eta,\xi,\epsilon;\ s+1,t,m\rangle$$
$$-\{(s-t+1)(s-t+2)\}^{\frac{1}{2}} \frac{\epsilon}{\xi\eta} \gamma(s)|\eta,\xi,\epsilon;\ s+1,t-1,m\rangle$$
$$-\{(s-t-1)(s-t)\}^{\frac{1}{2}} \frac{\epsilon}{\xi\eta} \gamma(s-1)|\eta,\xi,\epsilon;\ s-1,t+1,m\rangle$$
$$+\{(s+t)(s-t)\}^{\frac{1}{2}}\ 2\frac{1}{\xi\eta^2}(\xi^2-\epsilon^2)^{\frac{1}{2}}(\xi^2+\eta^2)^{\frac{1}{2}} \gamma(s-1)|\eta,\xi,\epsilon;\ s-1,t,m\rangle$$
$$+\{(s+t-1)(s+t)\}^{\frac{1}{2}} \frac{\epsilon}{\xi\eta} \gamma(s-1)|\eta,\xi,\epsilon;\ s-1,t-1,m\rangle \qquad \text{(II.2)}$$

where $\alpha(s)$, $\beta(s)$, and $\gamma(s)$ are functions of s only, and depend on the eigenvalues of the Casimir operators C_2, C_3, and C_4 of SU(2,2).

The expressions for $\mathfrak{R}_+$, $\mathfrak{R}_-$, and $\mathfrak{R}_6$ can be similarly written down.

$$\mathfrak{R}_+|\eta,\xi,\epsilon;\ s,t,m\rangle$$
$$=\{\epsilon[-\frac{\partial^2}{\partial\eta^2}+\frac{\partial^2}{\partial\xi^2}+\frac{\partial^2}{\partial\epsilon^2}+2\frac{\eta}{\xi}\frac{\partial}{\partial\eta}\frac{\partial}{\partial\xi}+2\frac{\xi}{\epsilon}\frac{\partial}{\partial\epsilon}\frac{\partial}{\partial\xi}+2\frac{\eta}{\epsilon}\frac{\partial}{\partial\epsilon}\frac{\partial}{\partial\eta}]$$
$$+\frac{\eta}{\epsilon}[2(m+1)-\frac{\epsilon^2}{\eta^2}-2t\frac{1}{\xi}(\xi^2-\epsilon^2)^{\frac{1}{2}}]\frac{\partial}{\partial\eta}$$
$$+\frac{\xi}{\epsilon}[2(m+1)+3\frac{\epsilon^2}{\xi^2}-2t\frac{1}{\xi}(\xi^2-\epsilon^2)^{\frac{1}{2}}]\frac{\partial}{\partial\xi}+(2m+7)\frac{\partial}{\partial\epsilon}$$
$$+m^2\frac{1}{\epsilon}+6m\frac{1}{\epsilon}+t^2\epsilon[\frac{1}{\eta^2}+2\frac{1}{\xi^2}-\frac{1}{\epsilon^2}]-s(s+1)\frac{\epsilon}{\xi^2}-t\frac{\epsilon}{\xi}(\xi^2-\epsilon^2)^{-\frac{1}{2}}$$
$$-4t\frac{1}{\xi\epsilon}(\xi^2-\epsilon^2)^{\frac{1}{2}}-2t\frac{\epsilon}{\xi\eta^2}(\xi^2+\eta^2)^{\frac{1}{2}}\beta(s)+\frac{5}{2}\frac{1}{\epsilon}+\frac{5}{2}\frac{\xi^2}{\epsilon}(\xi^2-\epsilon^2)^{-1}$$
$$-\tfrac{1}{4}\epsilon(\xi^2+\eta^2)^{-1}-\frac{9}{4}\epsilon(\xi^2-\epsilon^2)^{-1}+\frac{\epsilon}{\eta^2}\alpha(s)\}|\eta,\xi,\epsilon;\ s,t,m+1\rangle$$
$$+\{(s-t)(s+t+1)\}^{\frac{1}{2}}\frac{1}{\xi}\{\xi+(\xi^2-\epsilon^2)^{\frac{1}{2}}\}\{\frac{1}{\xi}(\xi^2+\eta^2)^{\frac{1}{2}}[\frac{\partial}{\partial\eta}+\frac{1}{\eta}(t+1)$$
$$+\tfrac{1}{2}\eta(\xi^2+\eta^2)^{-1}]-\frac{1}{\eta}\beta(s)\}|\eta,\xi,\epsilon;\ s,t+1,m+1\rangle$$
$$+\{(s+t)(s-t+1)\}^{\frac{1}{2}}\frac{1}{\xi}\{\xi-(\xi^2-\epsilon^2)^{\frac{1}{2}}\}\{\frac{1}{\xi}(\xi^2+\eta^2)^{\frac{1}{2}}[\frac{\partial}{\partial\eta}-\frac{1}{\eta}(t-1)$$
$$+\tfrac{1}{2}\eta(\xi^2+\eta^2)^{-1}]+\frac{1}{\eta}\beta(s)\}|\eta,\xi,\epsilon;\ s,t-1,m+1\rangle$$
$$-\{(s+t+1)(s+t+2)\}^{\frac{1}{2}}\frac{1}{\xi\eta}\{\xi+(\xi^2-\epsilon^2)^{\frac{1}{2}}\}\gamma(s)|\eta,\xi,\epsilon;\ s+1,t+1,m+1\rangle$$

(equation continued)

$$+\{(s+t+1)(s-t+1)\}^{\frac{1}{2}}\ 2\frac{\epsilon}{\xi\eta^2}(\xi^2+\eta^2)^{\frac{1}{2}}\ \gamma(s)|\eta,\xi,\epsilon;\ s+1,t,m+1\rangle$$

$$-\{(s-t+1)(s-t+2)\}^{\frac{1}{2}}\ \frac{1}{\xi\eta}\{\xi-(\xi^2-\epsilon^2)^{\frac{1}{2}}\}\ \gamma(s)|\eta,\xi,\epsilon;\ s+1,t-1,m+1\rangle$$

$$+\{(s-t)(s-t-1)\}^{\frac{1}{2}}\ \frac{1}{\xi}\{\xi+(\xi^2-\epsilon^2)^{\frac{1}{2}}\}\ \gamma(s-1)|\eta,\xi,\epsilon;\ s-1,\ t+1,\ m+1\rangle$$

$$+\{(s-t)(s+t)\}^{\frac{1}{2}}\ 2\frac{\epsilon}{\xi\eta^2}(\xi^2+\eta^2)^{\frac{1}{2}}\ \gamma(s-1)|\eta,\xi,\epsilon;\ s-1,t,m+1\rangle$$

$$+\{(s+t)(s+t-1)\}^{\frac{1}{2}}\ \frac{1}{\xi}\{\xi-(\xi^2-\epsilon^2)^{\frac{1}{2}}\}\ \gamma(s-1)|\eta,\xi,\epsilon;\ s-1,\ t-1,\ m+1\rangle \qquad (II.3)$$

$$\mathfrak{R}_-|\eta,\xi,\epsilon;\ s,t,m\rangle$$

$$=\{\epsilon[-\frac{\partial^2}{\partial\eta^2}+\frac{\partial^2}{\partial\xi^2}+\frac{\partial^2}{\partial\epsilon^2}+2\frac{\eta}{\xi}\frac{\partial}{\partial\eta}\frac{\partial}{\partial\xi}+2\frac{\xi}{\epsilon}\frac{\partial}{\partial\epsilon}\frac{\partial}{\partial\xi}+2\frac{\eta}{\epsilon}\frac{\partial}{\partial\epsilon}\frac{\partial}{\partial\eta}]$$

$$+\frac{\eta}{\epsilon}[-2(m-1)-\frac{\epsilon^2}{\eta^2}+2t\frac{1}{\xi}(\xi^2-\epsilon^2)^{\frac{1}{2}}]\frac{\partial}{\partial\eta}$$

$$+\frac{\xi}{\epsilon}[-2(m-1)+3\frac{\epsilon^2}{\xi^2}+2t\frac{1}{\xi}(\xi^2-\epsilon^2)^{\frac{1}{2}}]\frac{\partial}{\partial\xi}-(2m-7)\frac{\partial}{\partial\epsilon}$$

$$+m^2\frac{1}{\epsilon}-6m\frac{1}{\epsilon}+t^2\epsilon[\frac{1}{\eta^2}+2\frac{1}{\xi^2}-\frac{1}{\epsilon^2}]-s(s+1)\frac{\epsilon}{\xi^2}+t\frac{\epsilon}{\xi}(\xi^2-\epsilon^2)^{-\frac{1}{2}}$$

$$+4t\frac{1}{\xi\epsilon}(\xi^2-\epsilon^2)^{\frac{1}{2}}-2t\frac{\epsilon}{\xi\eta^2}(\xi^2+\eta^2)^{\frac{1}{2}}\ \beta(s)+\frac{5}{2}\frac{1}{\epsilon}+\frac{5}{2}\frac{\xi^2}{\epsilon}(\xi^2-\epsilon^2)^{-1}$$

$$-\frac{1}{4}\epsilon(\xi^2+\eta^2)^{-1}-\frac{9}{4}\epsilon(\xi^2-\epsilon^2)^{-1}+\frac{\epsilon}{\eta^2}\alpha(s)\}|\eta,\xi,\epsilon;\ s,t,m-1\rangle$$

$$-\{(s-t)(s+t+1)\}^{\frac{1}{2}}\ \frac{1}{\xi}\{\xi-(\xi^2-\epsilon^2)^{\frac{1}{2}}\}\{\frac{1}{\xi}(\xi^2+\eta^2)^{\frac{1}{2}}[\frac{\partial}{\partial\eta}+\frac{1}{\eta}(t+1)$$

$$+\tfrac{1}{2}\eta(\xi^2+\eta^2)^{-1}]-\frac{1}{\eta}\beta(s)\}\ |\eta,\xi,\epsilon;\ s,t+1,m-1\rangle$$

$$-\{(s+t)(s-t+1)\}^{\frac{1}{2}}\ \frac{1}{\xi}\{\xi+(\xi^2-\epsilon^2)^{\frac{1}{2}}\}\{\frac{1}{\xi}(\xi^2+\eta^2)^{\frac{1}{2}}[\frac{\partial}{\partial\eta}-\frac{1}{\eta}(t-1)$$

$$+\tfrac{1}{2}\eta(\xi^2+\eta^2)^{-1}]+\frac{1}{\eta}\beta(s)\}|\eta,\xi,\epsilon;\ s,t-1,m-1\rangle$$

$$+\{(s+t+1)(s+t+2)\}^{\frac{1}{2}}\ \frac{1}{\xi\eta}\{\xi-(\xi^2-\epsilon^2)^{\frac{1}{2}}\}\ \gamma(s)|\eta,\xi,\epsilon;\ s+1,t+1,m-1\rangle$$

(equation continued)

$$+ \{(s+t+1)(s-t+1)\}^{\frac{1}{2}}\ 2\frac{\epsilon}{\xi\eta^2}(\xi^2+\eta^2)^{\frac{1}{2}}\ \gamma(s)\ |\eta,\xi,\epsilon;\ s+1,t,m-1\rangle$$

$$+ \{(s-t+1)(s-t+2)\}^{\frac{1}{2}}\ \frac{1}{\xi\eta}\{\xi+(\xi^2-\epsilon^2)^{\frac{1}{2}}\}\ \gamma(s)|\eta,\xi,\epsilon;\ s+1,t-1,m-1\rangle$$

$$- \{(s-t)(s-t-1)\}^{\frac{1}{2}}\ \frac{1}{\xi\eta}\{\xi-(\xi^2-\epsilon^2)^{\frac{1}{2}}\}\ \gamma(s-1)|\eta,\xi,\epsilon;\ s-1,t+1,m-1\rangle$$

$$+ \{(s-t)(s+t)\}^{\frac{1}{2}}\ 2\frac{\epsilon}{\xi\eta^2}(\xi^2+\eta^2)^{\frac{1}{2}}\ \gamma(s-1)|\eta,\xi,\epsilon;\ s-1,t,m-1\rangle$$

$$- \{(s+1)(s+t-1)\}^{\frac{1}{2}}\ \frac{1}{\xi\eta}\{\xi+(\xi^2-\epsilon^2)^{\frac{1}{2}}\}\ \gamma(s-1)|\eta,\xi,\epsilon;\ s-1,t-1,m-1\rangle \quad \text{(II.4)}$$

$$\mathfrak{R}_6|\eta,\xi,\epsilon;\ s,t,m\rangle$$

$$= \{-(\xi^2+\eta^2)^{\frac{1}{2}}\left[\frac{\partial^2}{\partial\eta^2}+\frac{\partial^2}{\partial\xi^2}+\frac{\partial^2}{\partial\epsilon^2}+2\frac{\epsilon}{\xi}\frac{\partial}{\partial\xi}\frac{\partial}{\partial\epsilon}\right]$$

$$- [\eta(\xi^2+\eta^2)^{-\frac{1}{2}}+\frac{1}{\eta}(\xi^2+\eta^2)^{\frac{1}{2}}]\frac{\partial}{\partial\eta} - \frac{1}{\xi}[4(\xi^2+\eta^2)^{\frac{1}{2}}-\eta^2(\xi^2+\eta^2)^{-\frac{1}{2}}]\frac{\partial}{\partial\xi}$$

$$- [\epsilon(\xi^2+\eta^2)^{-\frac{1}{2}}+\frac{1}{\epsilon}(\xi^2+\eta^2)^{\frac{1}{2}}]\frac{\partial}{\partial\epsilon}$$

$$+ m^2\frac{1}{\epsilon^2}(\xi^2+\eta^2)^{\frac{1}{2}} - 2mt\frac{1}{\epsilon^2\xi}(\xi^2+\eta^2)^{\frac{1}{2}}(\xi^2-\epsilon^2)^{\frac{1}{2}}$$

$$+ t^2(\xi^2+\eta^2)^{\frac{1}{2}}[\frac{1}{\eta^2}-2\frac{1}{\xi^2}+\frac{1}{\epsilon^2}]$$

$$+ s(s+1)\frac{1}{\xi^2}(\xi^2+\eta^2)^{\frac{1}{2}} - 2t\frac{\xi}{\eta^2}\beta(s) + \frac{1}{4}(\xi^2+\eta^2)^{\frac{1}{2}}(\xi^2-\epsilon^2)^{-1} - \frac{9}{4}(\xi^2+\eta^2)^{-\frac{1}{2}}$$

$$+ \frac{1}{\eta^2}(\xi^2+\eta^2)^{\frac{1}{2}}\alpha(s)\}|\eta,\xi,\epsilon;\ s,t,m\rangle$$

$$+ \{(s-t)(s+t+1)\}^{\frac{1}{2}}\ \frac{\eta}{\xi}\{-\frac{1}{\xi}(\xi^2-\epsilon^2)^{\frac{1}{2}}[\frac{\partial}{\partial\epsilon}+(t+1)\frac{1}{\epsilon}-\tfrac{1}{2}\epsilon(\xi^2-\epsilon^2)^{-1}]$$

$$+ m\frac{1}{\epsilon}\}\ |\eta,\xi,\epsilon;\ s,t+1,m\rangle$$

$$+ \{(s+t)(s-t+1)\}^{\frac{1}{2}}\ \frac{\eta}{\xi}\{\frac{1}{\xi}(\xi^2-\epsilon^2)^{\frac{1}{2}}[\frac{\partial}{\partial\epsilon}-(t-1)\frac{1}{\epsilon}-\tfrac{1}{2}\epsilon(\xi^2-\epsilon^2)^{-1}]$$

$$+ m\frac{1}{\epsilon}\}\ |\eta,\xi,\epsilon;\ s,t-1,m\rangle$$

(equation continued)

$$+ 2 \{(s+t+1)(s-t+1)\}^{\frac{1}{2}} \frac{\xi}{\eta^2} \gamma(s) |\eta, \xi, \epsilon; s+1, t, m\rangle$$

$$+ 2 \{(s-t)(s+t)\}^{\frac{1}{2}} \frac{\xi}{\eta^2} \gamma(s-1) |\eta, \xi, \epsilon; s-1, t, m\rangle \tag{II.5}$$

Equation (II.5) for $\mathfrak{R}_6$ is relatively simpler since it is the sum of five states. Equations (II.2)-(II.4) all contain nine states where $\Delta s = 0, \pm 1, \quad \Delta t = 0, \pm 1$.

III. The Casimir Operators of SU(2,2)

(1) The second rank Casimir operator is

$$C_2 = \tfrac{1}{2} L_{ab} L^{ab} = \tfrac{1}{2} \mathfrak{L}_{\mu\nu} \mathfrak{L}^{\mu\nu} + 4iL_{03} - L_{03}^2 - \mathfrak{R}_{\mu} P^{\mu} \tag{III.1}$$

and if we substitute Eqs. (I.4)-(I.6), (II.1)-(II.5) into

$$C_2 | c_2, c_3, c_4; \eta, \xi, \epsilon; s, t, m\rangle = c_2 | c_2, c_3, c_4; \eta\, \xi\, \epsilon; s, t, m\rangle \tag{III.2}$$

where c_2, c_3, and c_4 are the eigenvalues of the three Casimir operators of SU(2,2), we obtain

$$\alpha(s) = c_2 - 2s(s+1) + 4 \tag{III.3}$$

(2) The third rank Casimir operator is

$$C_3 = -\frac{1}{48} \epsilon^{abcdef} L_{ab} L_{cd} L_{ef} \tag{III.4}$$

and

$$\beta(s) = -\frac{c_3}{s(s+1)} \tag{III.5}$$

(3) The fourth rank Casimir operator is

$$C_4 = \tfrac{1}{4} L_{ab} L^{bc} L_{cd} L^{da} - \tfrac{1}{4} C_2^2 - 2C_2 \tag{III.6}$$

and

$$(2s+1)(2s+3)(s+1)^2 \gamma^2(s) = [(s+1)^2 - A^2][(s+1)^2 - B^2][(s+1)^2 - C^2] \tag{III.7}$$

where

$$c_2 = A^2 + B^2 + C^2 - 5$$

$$c_3 = ABC$$

$$c_4 = \tfrac{1}{4}[A^2 + B^2 + C^2 - 1]^2 - [A^2B^2 + B^2C^2 + C^2A^2] \qquad \text{(III.8)}$$

It has been shown[2] that for degenerate UIR's, $C = B+1$. In general, A is discrete, $0, \frac{1}{2}, 1, 3/2, \ldots$; B and C may be discrete or continuous, or even complex. The determination of A, B, and C has been studied in a separate publication.

IV. "Space-like" and "Light-like" Representations of E(3,1)

(1) "Space-like" Unitary Irreducible Representations of E(3,1)

To change from "time-like" to "space-like" UIR's of E(3,1) we simply replace η by $i\omega$, $s(s+1)$ by $r(r-1)$ in Sections II and III.

$$P_\mu P^\mu |\omega,\xi,\epsilon; r,t,m\rangle = \omega^2 |\omega,\xi,\epsilon; r,t,m\rangle$$

$$W_\mu W^\mu |\omega,\xi,\epsilon; r,t,m\rangle = -\omega^2 r(r-1) |\omega,\xi,\epsilon; r,t,m\rangle \qquad \text{(IV.1)}$$

The little group involved here is $O(2,1) \sim SU(1,1)$, and according to Bargmann[5] has the following four classes of unitary irreducible representations:

a) $-r(r-1) > 0$, $t = \pm$ integer
b) $-r(r-1) > \frac{1}{4}$, $t = \pm$ half-integer
} continuous series

c) $r = 1/2, 1, 3/2 \ldots$, $t = r, r+1, \ldots$
d) $r = 1/2, 1, 3/2 \ldots$, $t = -r, -r-1, \ldots$
} discrete series

(2) "Light-like" Unitary Irreducible Representations of E(3,1)

a) UIR's with discrete "spin"

$$P_\mu P^\mu |\xi,\epsilon; t,m\rangle = 0$$

$$W_\mu W^\mu |\xi,\epsilon; t,m\rangle = 0 \qquad \text{(IV.2)}$$

where the basis vector $|\xi,\epsilon; t,m\rangle$ depends on four parameters only. The representation of the generators is as follows:

$$P_\pm |\xi,\epsilon; t,m\rangle = \epsilon |\xi,\epsilon; t,m\pm 1\rangle$$

$$P_5 |\xi,\epsilon; t,m\rangle = (\xi^2 - \epsilon^2)^{\frac{1}{2}} |\xi,\epsilon; t,m\rangle$$

$$\mathcal{P}_6 |\xi,\epsilon;t,m\rangle = \xi|\xi,\epsilon;t,m\rangle$$

$$\mathcal{L}_5 |\xi,\epsilon;t,m\rangle = m|\xi,\epsilon;t,m\rangle$$

$$\mathcal{L}_{\pm}|\xi,\epsilon;t,m\rangle = \{\mp(\xi^2-\epsilon^2)^{\frac{1}{2}}\frac{\partial}{\partial\epsilon} \pm \tfrac{1}{2}\epsilon(\xi^2-\epsilon^2)^{-\frac{1}{2}}$$

$$+ t\frac{\xi}{\epsilon} - (m\pm1)\frac{1}{\epsilon}(\xi^2-\epsilon^2)^{\frac{1}{2}}\}|\xi,\epsilon;t,m\pm1\rangle$$

$$\mathcal{L}_{56}|\xi,\epsilon;t,m\rangle = -i\{(\xi^2-\epsilon^2)^{\frac{1}{2}}\frac{\partial}{\partial\xi} + \tfrac{1}{2}\frac{1}{\xi}(\xi^2-\epsilon^2)^{\frac{1}{2}}$$

$$+\tfrac{1}{2}\xi(\xi^2-\epsilon^2)^{-\frac{1}{2}}\}|\xi,\epsilon;t,m\rangle$$

$$\mathcal{L}_{\pm6}|\xi,\epsilon;t,m\rangle = -i\{\epsilon\frac{\partial}{\partial\xi} + \xi\frac{\partial}{\partial\epsilon} + \tfrac{1}{2}\frac{\epsilon}{\xi} \pm (m\pm1)\frac{\xi}{\epsilon}$$

$$\mp t\frac{1}{\epsilon}(\xi^2-\epsilon^2)^{\frac{1}{2}}\}|\xi,\epsilon;t,m\pm1\rangle$$

$$L_{03}|\xi,\epsilon;t,m\rangle = i\{\xi\frac{\partial}{\partial\xi} + \epsilon\frac{\partial}{\partial\epsilon} + 2\}|\xi,\epsilon;t,m\rangle$$

$$\mathcal{R}_5|\xi,\epsilon;t,m\rangle = \{(\xi^2-\epsilon^2)^{\frac{1}{2}}\left[\frac{\partial^2}{\partial\xi^2} - \frac{\partial^2}{\partial\epsilon^2}\right] + \left[\frac{1}{\xi}(\xi^2-\epsilon^2)^{\frac{1}{2}} + \xi(\xi^2-\epsilon^2)^{-\frac{1}{2}}\right]\frac{\partial}{\partial\xi}$$

$$+[\epsilon(\xi^2-\epsilon^2)^{-\frac{1}{2}} - \frac{1}{\epsilon}(\xi^2-\epsilon^2)^{\frac{1}{2}}]\frac{\partial}{\partial\epsilon} + (\xi^2-\epsilon^2)^{\frac{1}{2}}[(t^2+m^2)\frac{1}{\epsilon^2} - \tfrac{1}{4}\frac{1}{\xi^2}] - 2mt\frac{\xi}{\epsilon^2}$$

$$+(\xi^2-\epsilon^2)^{-\frac{1}{2}}\delta(t)\}|\xi,\epsilon;t,m\rangle$$

$$\mathcal{R}_{\pm}|\xi,\epsilon;t,m\rangle = \{\epsilon\left[\frac{\partial^2}{\partial\xi^2} + \frac{\partial^2}{\partial\epsilon^2} + 2\frac{\xi}{\epsilon}\frac{\partial}{\partial\epsilon}\frac{\partial}{\partial\xi}\right]$$

$$+\left[\frac{\epsilon}{\xi} \pm 2(m\pm1)\frac{\xi}{\epsilon} \mp 2t\frac{1}{\epsilon}(\xi^2-\epsilon^2)^{\frac{1}{2}}\right]\frac{\partial}{\partial\xi}$$

$$\pm(2m\pm5)\frac{\partial}{\partial\epsilon} - t^2\frac{1}{\epsilon} \mp t\frac{1}{\epsilon}[\xi(\xi^2-\epsilon^2)^{-\frac{1}{2}} + \frac{1}{\xi}(\xi^2-\epsilon^2)^{\frac{1}{2}}] + \frac{1}{\epsilon}(m^2\pm4m+\frac{3}{2})$$

$$+\frac{3}{2}\frac{\xi^2}{\epsilon}(\xi^2-\epsilon^2)^{-1} - \tfrac{1}{4}\frac{\epsilon}{\xi^2} - \epsilon(\xi^2-\epsilon^2)^{-1}\delta(t)\}|\xi,\epsilon;t,m\pm1\rangle$$

$$\mathcal{R}_6|\xi,\epsilon;t,m\rangle = \{-\xi\left[\frac{\partial^2}{\partial\xi^2} + \frac{\partial^2}{\partial\epsilon^2} + 2\frac{\epsilon}{\xi}\frac{\partial}{\partial\xi}\frac{\partial}{\partial\epsilon}\right] - 4\frac{\partial}{\partial\xi} - [\frac{\epsilon}{\xi} + \frac{\xi}{\epsilon}]\frac{\partial}{\partial\epsilon} + t^2\frac{\xi}{\epsilon^2}$$

$$-2mt\frac{1}{\epsilon^2}(\xi^2-\epsilon^2)^{\frac{1}{2}} + m^2\frac{\xi}{\epsilon^2} + \frac{3}{2}\xi(\xi^2-\epsilon^2)^{-1} - \frac{5}{4}\frac{1}{\xi} - \xi(\xi^2-\epsilon^2)^{-1}\delta(t)\}|\xi,\epsilon;t,m\rangle \qquad \text{(IV.3)}$$

where the function $\delta(t)$ is related to the eigenvalues of the Casimir operators.

$$C_2 = 3t^2 - \tfrac{1}{2} - 2\delta(t)$$

$$C_3 = t[\delta(t) - t^2 - \tfrac{1}{4}]$$

$$C_4 = -\tfrac{1}{4}\{3t^4 - t^2[1 + 4\delta(t)] + 12[\delta(t) - 1]\} \qquad \text{(IV.4)}$$

It can be shown that $\delta(t) = \frac{5}{4}$.

b) UIR's with continuous "spin"

It can be shown that in the reduction of SU(2,2) with respect to E(3,1), the continuous "spin" representations of E(3,1) do not contribute.

c) The null representation of E(3,1): $P_\mu = 0$.

Similarly, the null representations of E(3,1) with $P_\mu = 0$ also do not contribute.

V. Brief Discussion of the Maximal Compact Subgroup SU(2) x SU(2) x U(1)

The maximal compact subgroup of SU(2,2) is SU(2) x SU(2) x U(1), which is generated by $\vec{J}$, $\vec{K}$, and R_0.[2] The basis vector is $|j,\mu;k,\mu;\lambda\rangle$, and (for simplicity for the moment we drop the extra label α which removes degeneracy),

$$\begin{pmatrix} \vec{J}^2 \\ J_3 \\ \vec{K}^2 \\ K_3 \\ R_0 \end{pmatrix} |j,\mu;\, k,\nu;\, \lambda\rangle = \begin{pmatrix} j(j+1) \\ \mu \\ k(k+1) \\ \nu \\ \lambda \end{pmatrix} |j,\mu;\, k,\nu;\, \lambda\rangle \qquad \text{(V.1)}$$

The other eight generators of SU(2,2) are $P_\pm$, $Q_\pm$, $S_\pm$, $T_\pm$, with

$$P_+|j,\mu;\, k,\nu;\, \lambda\rangle = [(j+\mu+1)(k-\nu+1)]^{\frac{1}{2}}\, a_1(j,k,\lambda)\,|j+\tfrac{1}{2},\, \mu+\tfrac{1}{2};\; k+\tfrac{1}{2},\, \nu-\tfrac{1}{2};\, \lambda+1\rangle$$

(equation continued)

$$+ [(j+\mu+1)(k+\nu)]^{\frac{1}{2}}\, a_2(j,k,\lambda)\,|j+\tfrac{1}{2},\, \mu+\tfrac{1}{2};\, k-\tfrac{1}{2},\, \nu-\tfrac{1}{2};\, \lambda+1\rangle$$

$$+ [(j-\mu)(k-\nu+1)]^{\frac{1}{2}}\, a_3(j,k,\lambda)\,|j-\tfrac{1}{2},\, \mu+\tfrac{1}{2};\, k+\tfrac{1}{2},\, \nu-\tfrac{1}{2};\, \lambda+1\rangle$$

$$+ [(j-\mu)(k+\nu)]^{\frac{1}{2}}\quad a_4(j,k,\lambda)\,|j-\tfrac{1}{2},\, \mu+\tfrac{1}{2};\, k-\tfrac{1}{2},\, \nu-\tfrac{1}{2};\, \lambda+1\rangle$$

$$P_-|j,\mu;\, k,\nu;\lambda\rangle = -[(j-\mu+1)(k+\nu+1)]^{\frac{1}{2}}\, b_1(j,k,\lambda)\,|j+\tfrac{1}{2},\, \mu-\tfrac{1}{2};\, k+\tfrac{1}{2},\, \nu+\tfrac{1}{2};\lambda-1\rangle$$

$$+ [(j-\mu+1)(k-\nu)]^{\frac{1}{2}}\, b_2(j,k,\lambda)\,|j+\tfrac{1}{2},\, \mu-\tfrac{1}{2};\, k-\tfrac{1}{2},\, \nu+\tfrac{1}{2};\, \lambda-1\rangle$$

$$+ [(j+\mu)(k+\nu+1)]^{\frac{1}{2}}\, b_3(j,k,\lambda)\,|j-\tfrac{1}{2},\, \mu-\tfrac{1}{2};\, k+\tfrac{1}{2},\, \nu+\tfrac{1}{2};\, \lambda-1\rangle$$

$$- [(j+\mu)(k-\nu)]^{\frac{1}{2}}\quad b_4(j,k,\lambda)\,|j-\tfrac{1}{2},\, \mu-\tfrac{1}{2};\, k-\tfrac{1}{2},\, \nu+\tfrac{1}{2};\, \lambda-1\rangle$$

etc. The functions $a_i(j,k,\lambda)$ and $b_i(j,k,\lambda)$ have been given in previous publications.[2)]

Next, we discuss the p-λ diagrams, $p = j+k$, which are convenient to the study of the various types of representations.

(a) Finite-dimensional (nonunitary) representations

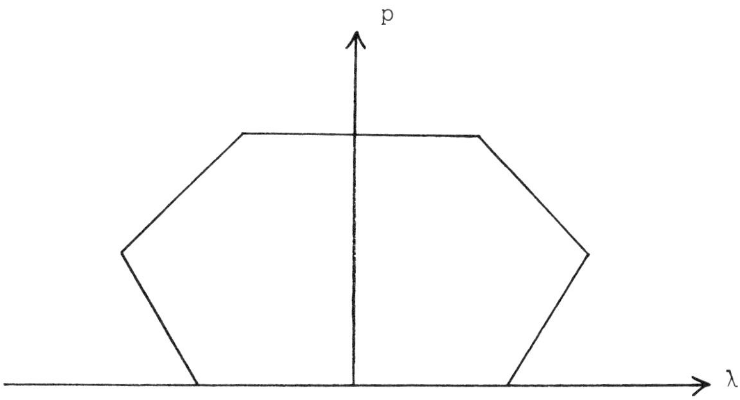

allowed states of SU(2) x SU(2) x U(1) are inside or on the boundary

(b)

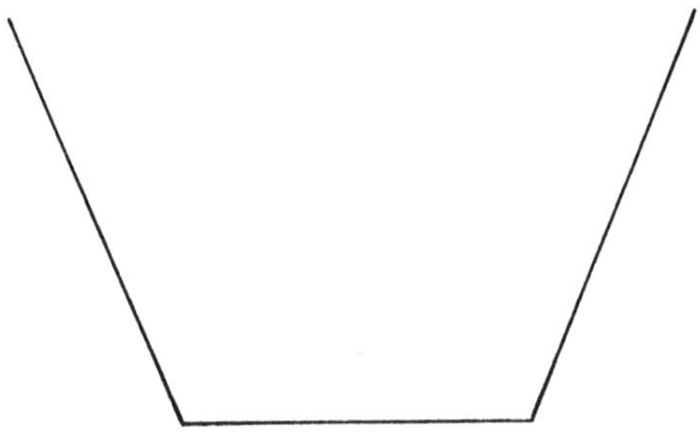

allowed states are inside or on the boundary

(c)

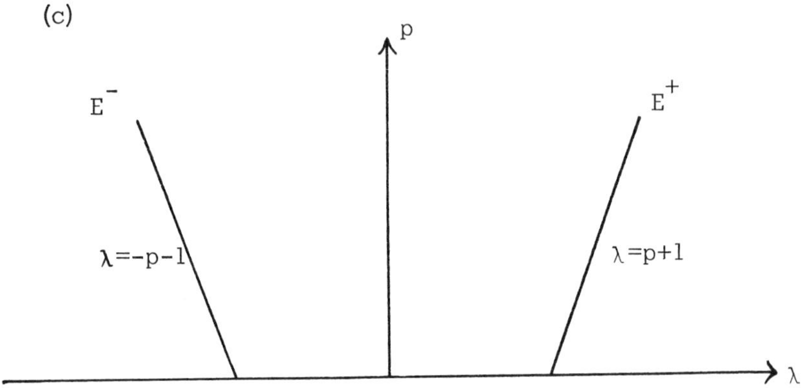

allowed states on the boundary only

(d)

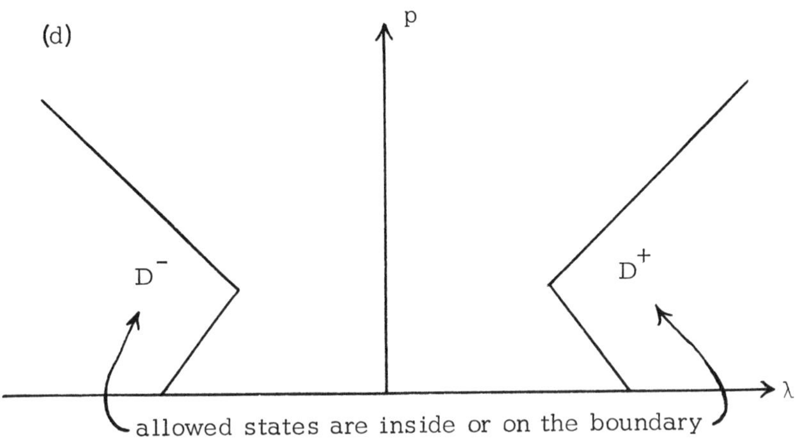

allowed states are inside or on the boundary

(e) λ is unbounded.

VI. Matrix Elements (Overlap Functions) of SU(2,2)

We are interested in the calculation of the matrix elements $\langle j\,\mu;\, k,\nu;\, \lambda|\eta,\xi,\epsilon;\, s,t,m\rangle$ which in general turn out to be confluent hypergeometric functions. Here we shall present only a simple example to illustrate the general technique involved.

We study the "ground state" of the E^{+} series, $|j = p_0, \mu = p_0; k = 0, \nu = 0; \lambda = p_0+1\rangle$ which satisfies the following relation,

$$\mathcal{L}_+|j = p_0, \mu = p_0;\, k = 0,\, \nu = 0;\, \lambda = p_0+1\rangle$$

$$= -i(J_+-K_+)|p_0,\, p_0;\, 0,0;\, p_0+1\rangle = 0 \qquad \text{(VI.1)}$$

Since we are dealing with unitary representations, we have

$$\langle p_0,\, p_0;\, 0,\, 0;\, p_0+1|\mathcal{L}_-|\xi,\epsilon;\, t,m\rangle = 0 \qquad \text{(VI.2)}$$

where "light-like" representations of the iso-Poincaré subgroup have been used. (This choice is dictated by the eigenvalues of the Casimir operators.) If we define

$$f(\xi,\epsilon;\, t) \equiv \langle p_0,\, p_0;\, 0,0;\, p_0+1|\xi,\epsilon;\, t,\, m=p_0\rangle \qquad \text{(VI.3)}$$

eq. (VI.2) with the use of Eq. (IV.1) becomes

$$\left\{ (\xi^2-\epsilon^2)^{\frac{1}{2}} \frac{\partial}{\partial\epsilon} - \tfrac{1}{2}\,\epsilon(\xi^2-\epsilon^2)^{-\frac{1}{2}} + t\,\frac{\xi}{\epsilon} - p_0\,\frac{1}{\epsilon}(\xi^2-\epsilon^2)^{\frac{1}{2}} \right\} f(\xi,\epsilon;\, t) = 0 \qquad \text{(VI.4)}$$

with the solution

$$f(\xi,\epsilon;\, t) = c(\xi,t)\, \epsilon^{p_0-t}\, (\xi^2-\epsilon^2)^{-\frac{1}{4}} \left[\xi + (\xi^2-\epsilon^2)^{\frac{1}{2}}\right]^t \qquad \text{(VI.5)}$$

If, in addition, we also use the expression for $\mathfrak{R}_5$ in Eq. (IV.1) we can determine the function $c(\xi,t)$ in Eq. (VI.5) and obtain

$$f(\xi,\epsilon;\, t=p_0) = c\, e^{-\xi}\, \xi^{-\frac{1}{2}}(\xi^2-\epsilon^2)^{-\frac{1}{4}} \left[\xi + (\xi^2-\epsilon^2)^{\frac{1}{2}}\right]^{p_0} \qquad \text{(VI.6)}$$

Other more general matrix elements can be obtained through the repeated use of Eq. (V.2).

We shall not present all the results we have obtained for the various types of unitary irreducible representations of SU(2,2). The interested readers are referred to a forthcoming article for all the details,[4] which are sufficiently complicated to prevent us from giving any meaningful summary here. They all involve Whittaker functions and their derivatives.

The rest of this talk presents a summary of our results on the reduction of SU(2,2) with respect to the iso-Poincaré subgroup E(3,1), together with the allowed eigenvalues of the three Casimir operators.

Summary of results:

a) UIR's reducible with respect to "light-like" representations of E(3,1)

The $E^{\pm}$ series, $p_0 = 0, \frac{1}{2}, 1, 3/2, \ldots$

$$C_2 = 3(p_0^2 - 1)$$

$$C_3 = \mp p_0(p_0^2 - 1)$$

$$C_4 = -\tfrac{3}{4}(p_0^2 - 1)^2$$

b) UIR's reducible with respect to "time-like" representations of E(3,1)

b1,b2) $D^{\pm}$ series, $j_m, k_m = 0, \frac{1}{2}, 1, 3/2, \ldots$, $\lambda_m = j_m + k_m + s_m$, $s_m = -1, 0, 1, 2, \ldots$ (if either j_m or k_m equals 0, s_m can also be -2)

$$C_2 = 2j_m(j_m+1) + 2k_m(k_m+1) + \lambda_m(\lambda_m+4)$$

$$C_3 = -(\lambda_m+2)(j_m-k_m)(j_m+k_m+1)$$

$$C_4 = \tfrac{1}{4}[(\lambda_m+2)^2 - 4j_m(j_m+1)][(\lambda_m+2)^2 - 4k_m(k_m+1)] - (\lambda_m+4)^2$$

and the $s = 0, \frac{1}{2}, 1, \ldots$ representations of SU(2) are used here.

b3) The most degenerate principal continuous series: $\rho > 0$

$$C_2 = -4 - \rho^2$$

$$C_3 = 0$$

$$C_4 = \tfrac{1}{4}\rho^4 + \rho^2$$

and the $s = 0$ trivial representation of SU(2) is used.

b4) The most degenerate complementary continuous series: $0 \leq \sigma < 1$

$$C_2 = -4 + \sigma^2$$

$$C_3 = 0$$

$$C_4 = \tfrac{1}{4}\sigma^4 - \sigma^2$$

and again the s = 0 representation of SU(2) is used.

c) UIR's reducible with respect to the "space-like" representations of E(3,1)

c1) Principal series: $p_0 = 0, \frac{1}{2}, 1, \ldots$, $\rho > 0$

$$C_2 = p_0^2 - 2\rho^2 - \frac{9}{2}$$

$$C_3 = \pm\, p_0(\rho^2 + \tfrac{1}{4})$$

$$C_4 = \tfrac{1}{4}\, p_0^4 + p_0^2(\rho^2 - \tfrac{3}{4})$$

where the principal series of SU(1,1) is used here. (See diagram b.)

c2) Complementary series: $p_0 = 0, 1, 2, \ldots,$ $-1 < \sigma < 0$

$$C_2 = p_0^2 + 2(\sigma-1)(\sigma+2)$$

$$C_3 = \pm\, p_0\, \sigma(\sigma+1)$$

$$C_4 - \tfrac{1}{4}\, p_0^4 - p_0^2(\sigma^2 + \sigma + 1)$$

where the complementary series of SU(1,1) is used.

c3) For p_0 = half-integer, the $\rho \to 0$ limit of c1) splits up into two inequivalent unitary irreducible representations, with

$$C_2 = p_0^2 - \frac{9}{2}$$

$$C_3 = \pm\, \tfrac{1}{4} p_0$$

$$C_4 = \tfrac{1}{4} p_0^4 - \tfrac{3}{4} p_0^2$$

and $p - \lambda \geq 0$, $p + \lambda \geq p_0 + \frac{1}{2}$. Here, the D^+ series of SU(1,1) is used.

c4) This is the other half of the split representation, with $p + \lambda \geq 0$, $p - \lambda \geq p_0 + \frac{1}{2}$. Here, the D^- series of SU(1,1) is used.

c5) For p_0 = integer = 1,2,3,..., the $\sigma \to 0$ limit of c2) splits up into three inequivalent unitary irreducible representations, with

$$C_2 = p_0^2 - 4$$

$$C_3 = 0$$

$$C_4 = \tfrac{1}{4}p_0^4 - p_0^2$$

and $p - \lambda \geq 0$, $p + \lambda \geq p_0 + 1$. Here, the D^+ series of SU(1,1) is used.

c6) This is the second part of the split representation, with $p + \lambda \geq 0$, $p - \lambda \geq p_0 + 1$. Here, the D^- series of SU(1,1) is used.

c7) This is the third part of the split representation, with $p - \lambda \geq p_0$, $p + \lambda \geq p_0$. Here, the trivial one-dimensional representation of SU(1,1) is used.

c8) For $p_0 = 0$, the $\sigma \to 0$ limit of c2) remains as a single unitary irreducible representation with

$$C_2 = -4$$

$$C_3 = 0$$

$$C_4 = 0$$

and $p - \lambda \geq 0$, $p + \lambda \geq 0$. Again, the trivial one-dimensional representation of SU(1,1) is used.

References

1. See talks given at this Symposium.
2. For other references to the literature, see for instance T. Yao, J. Math. Phys. 8, 1931 (1967); ibid. 9, 1615 (1968).
3. See for instance:
   W. Miller, Jr., Lie Theory and Special Functions (Academic Press, New York, 1968).
   J. D. Talman, Special Functions, A Group Theoretic Approach (Benjamin, New York, 1968).
4. For details see a forthcoming article by T. Yao (to be published).
5. V. Bargmann, Ann. Math. 48, 568 (1947); A. O. Barut and C. Fronsdal, Proc. Roy. Soc. A287, 532 (1965).

Section D: General

COORDINATE TRANSFORMATIONS THAT FORM GROUPS IN THE LARGE†

K. H. Mariwalla‡
Physics Department
Arya-Mehr University of Technology
Tehran, Iran

Abstract

One generally takes geometric symmetry group as motions (metric automorphisms). We show that both in nonrelativistic and relativistic mechanics there exist (well known) examples where the physical symmetry is larger than the group of motions of flat space into itself. Similar situations exist in general relativity. It is pointed out that there exist a wealth of coordinate transformation groups (both linear and nonlinear realizations of groups used in elementary particle theory) in both flat space and curved space-time which arise as automorphisms of some geometric entity such that the distance is left unchanged. Various possible approaches for application to elementary particle theory are suggested.

Since the inception of relativity and quantum hypothesis a great deal of progress has been made in understanding of those phenomena that are related to electromagnetism. The successes of the quantum theory as presently understood have been so far less than spectacular in the field of elementary particles (strong and weak interactions) and almost nil in the field of gravitation. In fact the foundations of the theory of general relativity (G.R.) as a theory of gravitation appear to be quite distinct from those of the quantum theory. In its motivation, elegance and beauty the theory of G.R. is rarely to be surpassed. In spite of what we have said above,

†Presented at the Symposium on De Sitter and Conformal Groups, University of Colorado, Summer 1970.

‡Present address: Institute of Mathematical Sciences, MATSCIENCE, Madras, India.

there have been several efforts to quantize G.R. One of the viewpoints taken is that the present difficulties in the application of quantum theory to elementary particles would somehow disappear if gravitation is also quantized. Among the attempts to quantize G.R. are:

1. B. S. Dewitt[1] uses the concept of infinite dimensional groups (connected with coordinate gauge) to set up a generally covariant field theory in curved space.

2. J. A. Wheeler[2] uses topological considerations to identify the possible elementary structures in space-time which could then be put in some kind of correspondence with elementary particles.

3. P. A. M. Dirac[3] formulates new type of bracket expressions (called Dirac brackets) to introduce Hamiltonian structures in theories where there are gauge groups: e.g. the gauge groups in electrodynamics and coordinate gauge in G.R. On the negative side we mention the work of Salecker and Wigner[4] who point to the great difficulties involved in giving any meaningful concept to quantum mechanical measurement of length and momenta when applied to curved space. The difficulties involved in the structure of quantum field theory even in flat space make it doubtful whether the essential point in the process of quantization is yet understood. It appears that the difficulties originate in the very concept of canonical quantization.

The recent great proliferation of groups in various branches of physics such as classical mechanics, nonrelativistic quantum mechanics and G.R. seems to offer some hope that one can after all put the various offshoots of physics on a unified basis. The hope is founded on the fact that the symmetry groups offer a coordinate free description so crucial to G.R. In the following we attempt to show by simple, well known examples how coordinate transformations that form groups in the large are associated with the symmetry group of the problem, even though these transformation groups are different from what one usually takes as the "geometric" symmetry group[5] (the "motions"). In this way we try to isolate the essential point involved in quantization and suggest a possible method of quantizing G.R. and application to the particle theory.

Nonrelativistic Mechanics

One can describe the trajectory of a particle in three-dimensional space in two different ways.[6] One can conceive of this trajectory as a trajectory in curved space to be described by a geodesic equation; if one further assumes that the space is Riemannian, one can determine the parameters of the curve in terms of the metric tensor components. The maximum number of linearly independent components of the Riemann tensor in this case is just 6. In mechanics, on the other hand, one considers the trajectory to be in absolute (flat) space and in absolute time together with the concept of force. If this force is derivable from a symmetric stress tensor (the symmetry follows from the isotropy of space in the usual way), then the two descriptions are equivalent. For instance, for a spherically symmetric potential, the equivalent curved space metric is

$$ds^2 = f(r)\, dr^2 + r^2 (d\theta^2 + \sin^2\theta\, d\varphi^2) \quad , \tag{1}$$

$$f(r) = -\alpha(k - r^{-2} + 2\Phi \ell^{-2})^{-1} \quad ; \tag{2}$$

where ℓ is related to the angular momentum and α, k are constants to be determined by the boundary conditions. For the Kepler problem

$$\Phi = -MG/r, \quad \alpha > 0, \quad \text{and} \quad k = 2E/m \gtrless 0 \quad ,$$

according as $E \gtrless 0$. It is known[7] that 3-dimensional spherically symmetric metric such as (1) can always be conformally represented in a Euclidean space as follows:

$$ds^2 = H^2(R)\, [\epsilon\, dR^2 + R^2 (d\theta^2 + \sin^2\theta\, d\ ^2)] , \tag{3}$$

where $\epsilon = \pm 1$ denotes the sign of $f(r)$. Since (3) represents a space of constant curvature, its group of motions is a 6-parameter group. In fact it is SO_4 for $E < 0$ and $SO(3,1)$ when $E > 0$. These happen to be the symmetry groups for the Kepler problem in 3-dimensions. However, note that we would obtain the same symmetry for any spherically symmetric potential. Generalizing to n-dimensional space with metric of the type (3), the symmetry will be $SO(n+1)$. People have obtained[8] such results by using the canonical formalism. In this formalism more than one type of symmetry groups arise.

A. <u>The Conjugacy Group</u>

This arises from the very definition of a Poisson bracket (P.B.) which requires the existence of a momentum conjugate to each coordinate

$$\{q_i, p_j\} = \delta_{ij} \quad . \tag{4}$$

For a system with n-degrees of freedom these are n-equations. Each of them corresponds to the 3-dimensional[9] noncompact group d_1^3 of rank 1 with elements q, p, and identity. The conjugacy group is thus a 2n+1 parameter noncompact semisimple group of rank n:

$$d_1^3 \otimes d_1^3 \otimes \ldots \otimes d_1^3 \otimes d_1^3 \quad \text{n-times} . \tag{5}$$

The importance of this group lies in the fact that its elements are the building blocks of all other canonical variables.

B. "The Infinitesimal Canonical Group"

This arises from the symplectic structure of the Poisson bracket. If we put $\eta_1, \eta_2 \dots \eta_n = q_1, q_2 \dots q_n$; $\eta_{n+1} \dots \eta_{2n} = p_1 \dots p_n$, the Hamiltonian equations take the form

$$\frac{d\eta^i}{dt} = \epsilon^{ij} \frac{\partial H}{\partial \eta^j} \quad , \tag{6}$$

and the P.B. is

$$\{f,g\} = \epsilon^{ij} \frac{\partial f}{\partial \eta^i} \frac{\partial g}{\partial \eta^j} \quad . \tag{7}$$

For given η^i, fixed in a small neighborhood, the set of transformations $\underset{\sim}{X}_{ij} = \partial \eta'^i / \partial \eta^j$ leaving the antisymmetric numerical matrix ϵ unchanged constitute the "infinitesimal canonical group";[10] i.e.

$$\underset{\sim}{X} \, \underset{\sim}{\epsilon} \, \underset{\sim}{X}^T = \epsilon \quad . \tag{8}$$

It is clearly isomorphic to the symplectic group in 2n dimensions and on the reals; the rank of this group is n. It has two important subgroups which are also symmetry groups of H (i.e. have vanishing P.B. with H); these are[11] SO(n+1) and SU(n). Since all these groups are local symmetry groups they do not, in general, give the true symmetry group which is necessarily a subgroup of these groups. For a true (physical) symmetry group it is essential that the transformations form a group in the large.[12] For instance in the case of Kepler problem and the harmonic oscillator these transformations arise as nonlinear realizations (in coordinate space) of SO(4) and SU(3) respectively (one can use the generators of SO(4) and SU(3) to explicitly evaluate these transformations). It would therefore be of interest to have a method of obtaining global symmetry group of a problem without having to solve the problem first. However, there does not seem to exist (at least it is unknown to the author) a general method of determining a "global symmetry group."

The system of equations (4) which are the basic building blocks of all canonical transformations have a direct physical significance only if the coordinates q_i are cartesian. In this case the p_i are translation generators in q_i. In transition to quantum mechanics only these are carried over.[13] Even in that case, as was first shown by Wigner,[14] given Heisenberg's equations of motion the canonical commutation relations do not uniquely follow from it.[15] Furthermore, Wigner in a careful analysis of relativistic invariance and quantum phenomena has shown how an unsatisfactory situation results in quantum theory if one insists on treating coordinates as observables. Even in classical relativistic Hamiltonian framework there exist "no interaction theorems"[17] unless one is ready to abandon position as a canonical variable (or the existence of world lines).[18] Thus we conclude that in transition to quantum mechanics what is carried over is not the structure (4) or the symplectic structure of phase space, but only the physical symmetry group, characterized by the coordinate transformations that form groups in the large. This is more in keeping with the spirit of G.R., i.e. a coordinate free description. Before we proceed to consider the relativistic case we review some well known material on continuous transformation groups to show how coordinate transformations that form groups in the large arise naturally in the context of various geometrical objects.[19],[20]

Coordinate Transformation Groups in the Large as Geometric Entities

In the 1870's F. Klein[21] suggested a program of characterizing various geometric entities in terms of the group character of the coordinate transformations as automorphisms of space into itself. But soon mathematicians realized that the group of automorphisms will then be an abstract rather than a transformation group. This was considered as a natural step beyond Klein's own formulation of the program and interest in the studies of transformation groups subsided. But there is an alternative viewpoint. One can consider the coordinate transformation groups as a realization of the abstract group through transformations in the field of coordinates. But since we live in space to which we endow coordinates for convenience, it is essential to know which of our deductions on natural phenomena are dependent on a particular coordinate system employed and which are due to the intrinsic geometrical properties. In this manner the concept of coordinate transformation groups arises naturally in the study of differential geometry and physics, and has recently been used extensively in general relativity. Both in the study of flat and curved space one finds a wide variety of transformation groups which could be of interest in elementary particle theory. With this motivation we summarize in the following some of the well known results on the field of continuous transformation groups and show how coordinate transformations that form groups in the large arise in the context of various geometrical quantities. The principal references used are Eisenhart,[22)-24)] Robertson and Noonan,[24] Yano[26] and a recent work of Katzin et al.[27]

Given an n-dimensional space, if there exists a set of mappings of space into itself such that:

1. The product of two such mappings belongs to the set.
2. The identity map exists.
3. For every map there exists an inverse map, such that these two maps taken in any order give the identity map, then the space is said to admit an automorphism.

The set of all such maps constitute a group. For an r-parameter group one may represent a coordinate transformation as

$$x'^{i} = f^{i}(x^{j}, a_{\alpha}), \quad \alpha = 1,2\ldots r ; \tag{9}$$

and possess "generating vectors"

$$\xi^{i}_{\alpha}(x) = \left.\frac{\partial f^{i}(x^{j}, a_{\beta})}{\partial a_{\alpha}}\right|_{\text{all } a_{\alpha}=0} \tag{10}$$

The necessary and sufficient condition for the existence of these automorphisms is that

$$\frac{\partial x^{i}}{\partial a_{\alpha}} = \xi^{i}_{\beta}(x^{j})\, A^{\beta}_{\alpha}\,(a_{\gamma}) \tag{11}$$

If we introduce the operators

$$\underset{\sim}{X}_{\alpha} = \xi^{i}_{\alpha}\,\frac{\partial}{\partial x^{i}} \quad , \tag{12}$$

then

$$\begin{aligned} x'^{i} &= [\exp(a_{(\alpha)}\, \underset{\sim}{X}_{(\alpha)})]\varkappa^{i} \quad , \\ &\approx x^{i} + a_{(\alpha)}\, \underset{\sim}{X}_{(\alpha)}\, \varkappa^{i} + O(a^{2}) \ . \end{aligned} \tag{13}$$

For this reason $\underset{\sim}{X}_{\alpha}$ are called the infinitesimal generators of coordinate transformations. For the set of r-linearly independent functions ξ^{i}_{α} satisfying (11) one can write

$$\xi^{k}_{\delta}\,\xi^{i}_{\gamma,k} + \xi^{k}_{\gamma}\,\xi^{i}_{\delta,k} = C^{\sigma}_{\delta\gamma}\,\xi^{i}_{\sigma} \tag{14}$$

where comma denotes differentiation. The constants $C^{\sigma}_{\delta\gamma}$ are given in terms of A, by

$$C^{\sigma}_{\gamma\delta} = B^{\alpha}_{\gamma} B^{\beta}_{\delta} (A^{\sigma}_{\alpha,\beta} - A^{\sigma}_{\beta,\alpha}) , \tag{15}$$

where B is a matrix inverse to the matrix A. It is then easy to see that C's satisfy

$$C^{\sigma}_{\alpha\beta} C^{\mu}_{\sigma\gamma} + C^{\sigma}_{\beta\gamma} C^{\mu}_{\sigma\alpha} + C^{\sigma}_{\gamma\alpha} C^{\mu}_{\sigma\beta} = 0 . \tag{16}$$

In terms of the operators $\underset{\sim}{X}_{\alpha}$ one may re-express (14) and (16) as

$$[\underset{\sim}{X}_{\alpha}, \underset{\sim}{X}_{\beta}] = C^{\sigma}_{\alpha\beta} \underset{\sim}{X}_{\sigma} , \tag{17}$$

$$[\underset{\sim}{X}_{\alpha}, [\underset{\sim}{X}_{\beta}, \underset{\sim}{X}_{\gamma}]] + [\underset{\sim}{X}_{\beta} [\underset{\sim}{X}_{\gamma}, \underset{\sim}{X}_{\alpha}]] + [\underset{\sim}{X}_{\gamma}, [\underset{\sim}{X}_{\alpha}, \underset{\sim}{X}_{\beta}]] = 0 \tag{18}$$

Also we find that

$$B^{\nu}_{\lambda,\rho} B^{\rho}_{\mu} - B^{\nu}_{\mu,\rho} B^{\rho}_{\lambda} = C^{\rho}_{\mu\lambda} B^{\nu}_{\rho} . \tag{19}$$

The three fundamental theorems of Lie are

1) The set of transformations $f^i(x;a)$ form a group if they satisfy (11) with det $A(0) \neq 0$ and $f^i(x;0) = x^i$.

2) If we are given a set of r linearly independent functions ξ^i_{α} and the set of constants $C^{\sigma}_{\alpha\beta}$ such that (14) is satisfied, then there exist functions A^{β}_{α} (a) such that (11) is satisfied and yields the solution $f^i(x,a)$ which defines our r-parameter group of automorphisms.

3) If a set of constants $C^{\sigma}_{\alpha\beta}$ satisfy (16), then there exist functions $\xi^i_{\alpha}(x)$ such that (14) is satisfied.

This set of theorems can be applied to a wide variety of systems to determine their global groups of automorphisms. One may therefore apply these to a classical canonical system or to a Riemannian space. In order to apply it it is necessary to decide which geometric object (a set of quantities which transform linearly under coordinate transformations)[22),23)] is under consideration.

Let us apply this to a general Riemannian space. Here the fundamental object is the metric tensor, in terms of which the infinitesimal distance along a curve is given by

$$ds^2 = g_{ij} \, dx^i \, dx^j \tag{20}$$

Let us take g_{ij} as the components of the geometric object whose automorphisms we are interested in finding. The set of transformations $x'^i \to x^i$ such that

$$g'_{ij}(x') = g_{ij}(x') \tag{21}$$

when they exist are called the metric automorphisms or group of motions of space into itself. In terms of the generating function $\xi^i_{(\alpha)}$ the condition for the existence of motions may be expressed by the vanishing of the Lie derivative[26)]

$$\underset{\xi}{\pounds}\, g_{ij} = h_{ij} \equiv \xi_{i;j} + \xi_{j;i} = 0 \tag{22}$$

(killing equations), where the semicolon denotes covariant differentiation. If the equations (22) have r solutions, then the space is said to admit an r-parameter group of motions. The equation (20) is a first integral of the equation of a geodesic

$$\frac{d^2 x^i}{ds^2} + \Gamma^i_{jk} \frac{dx^i}{ds} \frac{dx^j}{ds} = 0 \; ; \tag{23}$$

The Γ^i_{jk} are the christoffel symbols and are called affinities. The automorphisms which leave affinities unchanged in form are called affine-automorphism (or affine-collineations). The condition for their existence is that

$$\pounds\, \Gamma^i_{jk} = 0 \; ; \tag{24}$$

or equivalently

$$h_{ij;k} = 0 \tag{25}$$

has solutions. Similarly one can consider Ricci, Einstein and curvature automorphisms. For curvature automorphisms the necessary condition is the existence of solutions to

$$h_{jm;ih} = h_{jm;hi} \; . \tag{26}$$

One can also consider transformations that are based on a different geometric concept than distance. For instance the mappings which preserve angle (conformal correspondences) and those which map paths (geodesics) into paths (geodesic or projective correspondences).

By projective correspondence we mean those changes in Γ which leave the system of geodesics unchanged; however the metric is changed. One therefore defines a projective affinity (π) and projective curvature tensor (W). The condition for the existence of a projective automorphism then is that the $\pounds\pi = 0$ (which implies that $\pounds W = 0$) or

$$(n+1)\ h_{ij;k} = 2g_{ij}\xi^{m}{}_{;mk} + g_{ik}\xi^{m}{}_{;mj} + g_{jk}\xi^{m}{}_{;mi}\ , \qquad (27)$$

where n is the dimensionality of space.

For conformal correspondences

$$g_{ij} \rightarrow e^{2\sigma} g_{ij} \qquad (28)$$

and one gets in this case a hierarchy of three automorphisms. Let $g = \det(g_{ij})$. Then for the existence of conformal motions the condition is

$$\pounds(g^{-1} g_{ij}) = 0; \quad \text{or}$$

$$h_{ij} = 2\sigma\ g_{ij}, \quad \sigma = \xi^{k}{}_{;k}\ n^{-1} \qquad (29)$$

has solutions. When σ is a constant, these motions are called homothetic. One can similarly define automorphisms arising from the vanishing of the Lie derivative of conformal affinity and the conformal curvature tensor. The various possibilities are summarized by Katzin et al.[27]

Thus we see that there exist a wide variety of coordinate transformations that form groups in the large; these arise as coordinate transformations which leave a given geometric object unchanged in functional form. Since most of the coordinate transformations that form groups in the large (apart from motions and affine automorphisms of flat space) are nonlinear, they are of intrinsic interest as they could possibly be used in a nonlinear theory of elementary particles in which fields act as coordinates thus achieving a true democracy of all the fields. This could be connected with general relativity in two possible ways.

1) Given a cosmology and a geometric object determine its automorphism in the form of nonlinear coordinate transformations. Construct a Lagrangian invariant under these coordinate transformations, where now coordinates are the fields themselves.

2) Given a cosmology and some group of automorphisms associated with it, find the group space[19] of the parameters. The automorphism of the group space of the parameters gives the elementary particle group. The group space is in general Riemannian where the coordinates now are the parameters of the group of automorphisms of the original space. If one identifies the fields and the coordinates such

that the Lagrangian has the symmetry of automorphisms of group space we get another possible theory.

Special Relativity

The relativistic invariance group is the Poincaré group. These transformations leave the expression

$$ds^2 = \eta_{ij} dx^i dx^j = dx_o^2 - dx_1^2 - dx_2^2 - dx_3^2 \tag{30}$$

unchanged in the form. That is, the Poincaré group is a group of metric automorphisms. From a physical and geometrical viewpoint it would appear that the only requirement should be that ds^2 remain unchanged. A possible reason (apart from simplicity) for assigning a preferred role to metric automorphisms is that the Poincaré generators have physical interpretation familiar from the Galilean group. But in the investigations of the past fifty years we have come a long way and have at our disposal more conserved quantities than one can comfortably account for on the old picture. (We remark parenthetically that in general relativity the familiar conservation laws arising from Galilean and Poincaré invariance do not seem to have any relevance, any way.) Some of the possible candidates for larger groups in flat space are:

1. The affine group in flat space. This is the inhomogeneous linear groupL $GL(4,R) \otimes T_4$; Lorentz group and the group $SU(2,1)$ are its subgroups. Its generators are

$$\xi^i = \alpha^i + \beta^i_j x^j . \tag{31}$$

2. The group of projective transformations of flat space. This includes the affine group as a subgroup. It is a 24 parameter group with generators[22)]

$$\xi^i = \alpha^i + \beta^i_j x^j + \gamma_k x^k x^i \tag{32}$$

3. Conformal transformations of flat space. These may be determined as follows. A space of constant curvature can always be brought into the form[23)]

$$ds^2 = \eta_{ij} \Lambda^2 dx^i dx^j , \tag{34}$$

$$\Lambda^{-1} = ax^2 + 2x\cdot b + c \tag{35}$$

$$ac = b^2 + \tfrac{1}{4}K . \tag{36}$$

Since we are interested in flat space, let us put $K = 0$. Now the following cases arise

a. $b^\mu = 0$, $a = 0$, $c = 1$: this gives motions;

b. $b^\mu = 0$, $a = 0$, $c > 1$: this gives dilations

$$x'^i = \alpha x^i . \tag{37}$$

c. $b^\mu = 0$, $a \neq 0$; it follows from (36) that $c = 0$. Hence after a dilation, $\Lambda^{-1} = x^2$. These transformations can be realized as inversions in a unit hypersphere

$$x'^i = (x^2)^{-1} x^i \tag{38}$$

d. $b^\mu = 0$, $a = c = 0$; by (36) b^μ is then a null vector and has therefore only one independent component, which can be made unity by a dilation. If we take $b^\mu = (1; 0,0,1)$, then $\Lambda^{-1} = (x-b) = (x_0 - x_3)$ and for the transformations we get

$$x_0' = \tfrac{1}{2}\Lambda(x^2 - 1) \quad , \quad x_3' = -\tfrac{1}{2}\Lambda(x^2 + 1)$$

$$x_2' = \Lambda y \quad , \quad x_1' = \Lambda x \quad . \tag{39}$$

For $a = 1$, we get (38) combined with translation as

$$x'^i = (\delta^i_j + b^i x_j)(1 + 2b\cdot x + b^2 x^2)^{-1} x^j \tag{40}$$

The set of coordinate transformations consisting of motions, (37) and (40) constitute the so-called "conformal group." Its Lie algebra is isomorphic to the Lie algebra of SO(4,2), or SU(2,2). The Lie algebra of the group of motions of De Sitter space (space of constant curvature $\neq 0$) is isomorphic to the Lie algebra of SO(4,1) or SO(3,2). This has led to the belief that De Sitter group is a subgroup of the "conformal group." However the above analysis makes it clear that the "conformal group" is related to the flat space.

We now turn to an application of the "conformal group" to the problem of irreversible loss of radiation by a charge in acceleration.

The equation of motion of a particle in uniform acceleration is

$$\frac{da^\mu}{d\tau} - a^2 V^\mu = 0 \ , \tag{41}$$

where V^μ and a^μ are velocity and acceleration 4-vectors, $a^2 = a_\mu a^\mu =$ constant. Equation (41) integrates to give

$$x^\mu = x_o^\mu + a^{-2}(a^\mu - a_o^\mu) \ , \tag{42}$$

where x_o^μ and a_o^μ are position and acceleration four-vectors in a given momentary rest frame. If we put

$$y^\mu = x^\mu - x_o^\mu, \quad A^\mu = \tfrac{1}{2}\, a_o^\mu \tag{43}$$

we get from (42): $A_\mu(y^\mu + y^2 A^\mu) = 0$; or the equivalent set

$$y'^\mu = \lambda(A,y)\,(y^\mu + y^2 A^\mu), \qquad A^\mu y'_\mu = 0,$$

which defines a coordinate transformation in y^μ. On squaring we get $y'^2 = y^2\lambda^2(1 + 2A\cdot y + A^2 y^2)$, so that the inverse transformation is

$$y^\mu = \lambda^{-1}\left[y'^\mu - \frac{\lambda^{-1}\, y'^2 A^\mu}{1 + 2A\cdot y + A^2 y^2} \right]$$

In order that these transformations form a group, the direct and inverse transformations must have the same form; this determines λ uniquely as

$$\lambda^{-1}(A,y) = 1 + 2A\cdot y + A^2 y^2$$

Thus transformations from an inertial frame to a uniformly accelerated frame

$$y'^\mu = \frac{y^\mu + y^2 A^\mu}{1 + 2A\cdot y + A^2 y^2} \ , \tag{44}$$

together with (43) and the conditions

$$A^\mu y'_\mu = 0, \qquad A^\mu A_\mu = 0 \ . \tag{45}$$

We see that these transformations are defined for a region around a given (but arbitrary) momentary rest frame. This region is defined by $\lambda^{-1} \geq 0$. As $\lambda^{-1} \to 0$, the velocity of the particle approaches the velocity of light. These transformations when applied to the Coulomb field yield the correct Bondi-Gold fields[28)] for a charge in uniform acceleration. It is known that a point charge moving with uniform velocity or with uniform acceleration does not radiate.[28),29)] In this connection it is amusing to note that only for these two motions (viz. uniform velocity and uniform acceleration) there exist transformations, connecting these frames with inertial frames, that form groups in the large. Consequently the direct and inverse transformations are of the same type and the frames of reference so related are equivalent. On the other hand, in flat space, there exist no transformations connecting an inertial frame to one in nonuniform acceleration such that the direct and inverse transformations are of the same type. Thus a frame of reference in nonuniform acceleration is in no sense equivalent to one in uniform motion (or uniform acceleration) and hence irreversible loss of radiation in this case is to be expected.

In the above we have seen an example of a coordinate transformation group in the large, to wit the "conformal group," which is bigger than the Poincaré group and is a symmetry group in the physical sense as far as classical relativity is concerned. Therefore it must also be the symmetry group in quantum theory.

Discussion

We have discussed the role of coordinate transformations that form groups in the large as physical symmetry groups. We saw that in nonrelativistic mechanics (classical and quantum) and in classical special relativity there exist examples of such coordinate transformation groups which though not motions (of space into itself--metric automorphisms) are symmetry groups of the problem. In general relativity, in the absence of the usual conservation laws of energy and momenta one is led to take seriously the conservation of such geometric entities as curvature. In this case again one has to consider the coordinate transformation groups that are not motions. Furthermore the group of metric automorphisms of a curved space (or the metric automorphisms of the "group space of the metric automorphisms of the given curved space") is in general obtained as a group of nonlinear coordinate transformations unlike the transformations of the usual Poincaré group.

We therefore conjecture that such groups of coordinate transformations are of relevance also to elementary particle theory. More specifically they can arise in elementary particle theories in any one of the following ways and their variations.

1. In a nonlinear theory of elementary particles, an elementary particle group can arise as metric automorphisms of curved space with fields playing the role of coordinates.
2. The elementary particle Lagrangian has the same symmetry (and its realization) as the automorphisms of a geometric entity such as curvature tensor in a given cosmology.
3. The symmetry group of the elementary particle Lagrangian is identical to the metric automorphisms of the group space of the metric automorphisms of a given cosmology.

It would appear that this is possibly the only way of bringing G.R. and short range interactions within the same fold. The point of view adopted in making this assertion is that in transition from the classical to quantum theory what is carried over is the structure of the symmetry group and its relevant realizations rather than the canonical commutation relations.

References

1. B. S. Dewitt, Dynamical Theory of Groups and Fields (Gordon and Breach, New York, 1965).
2. J. A. Wheeler, Geometrodynamics (Academic Press, New York, 1962).
3. P. A. M. Dirac, Can. J. Math. 2, 129 (1950); Lectures on Quantum Mechanics, Belfor Graduate School of Science Monograph Series No. 2 (Yeshiva University, New York, 1964).
4. H. Saleckar and E. P. Wigner, Phys. Rev. 109, 571 (1958).
5. The use of this term is due to Wigner; for discussion and other references see R. M. F. Houtappel et al, Rev. Mod. Phys. 37, 959 (1965).
6. K. H. Mariwalla, to be published in J. Math. Phys. Nov.-Dec. 1970.
7. T. Levi-Civita, The Absolute Differential Calculus (Blackie and Sons, London, 1926), p. 411.
8. D. M. Fradkin, Prog. Theor. Phys. (Kyoto) 37, 798 (1967); H. Bacry et al, Comm. Math. Phys. (Germany) 3, 323 (1966); S. P. Aillileuv, Zh. Eksperim. I Teor. Fiz. 33, 200 (1967) [Engl. transl. Sov. Phys. JETP 6, 156 (1958)].
9. N. Jacobson, Lie Algebras (Interscience Pub., New York, 1962), p. 11-12.
10. C. W. Kilmister, Hamiltonian Dynamics (John Wiley and Sons, Inc., New York, 1964).
11. N. Mukunda, Phys. Rev. 155, 1383 (1967), shows that all classical systems with n degrees of freedom have local invariances of SO_{n+1} and SU_n.
12. For an equivalent viewpoint see P. Stehle and M. Y. Han, Phys. Rev. 159, 1070 (1967). The connection between SU_n and SO_{n+1} in the physical context has been discussed by D. G. Ravenhall et al, Phys. Rev. 164, 1950 (1967) and for SU_2 and SO_3 by A. Cisneros and H. V. McIntosh, J. Math. Phys. 10, 277 (1969).
13. For a careful analysis, see K. H. Mariwalla, "Uncertainty Principle and Canonically Conjugate Variable," to be published.
14. E. P. Wigner, Phys. Rev. 77, 711 (1950).
15. Stated in terms of creation and destruction operators; there exist two classes of generalized commutation relations: the para-Fermi and para-Bose rings. H. S. Green, Phys. Rev. 90, 270 (1953).

The first of these are related to the representations of the orthogonal group--C. Ryan, E. C. G. Sudarshan, Nucl. Phys. 47, 207 (1963)--and the second to those of the symplectic group--S. Kamefuchi and Y. Takahashi, Nucl. Phys. 36, 177 (1962).
16. E. P. Wigner, Rev. Mod. Phys. 29, 255 (1957).
17. D. G. Currie, T. F. Jordan and E. C. G. Sudarshan, Rev. Mod. Phys. 35, 350 (1963).
18. See references in Ref. 5.
19. J. A. Schouten, Ricci-Calculus, 2nd ed. (Springer-Verlag, Berlin, 1954).
20. A. Trautman, in Lectures on General Relativity, Brandeis Summer Institute Lectures, 1964 (Prentice-Hall, Inc., Englewood Cliffs, N. J.,
21. F. Klein, Programm zum Eintritt in die Philosophische Fakultät (Erlangen 1872); reprinted in Math. Ann. 43, 63 (1893).
22. L. P. Eisenhart, Continuous Groups of Transformations (Princeton University Press, Princeton, N.J., 1933).
23. L. P. Eisenhart, Riemannian Geometry (Princeton University, Princeton, N.J., 1926).
24. L. P. Eisenhart, Non-Riemannian Geometry, Amer. Math. Soc. Coll. Publs. N. Y., 1927.
25. H. P. Robertson and T. W. Noonan, Relativity and Cosmology (W. B. Saunders Co., London, 1968).
26. K. Yano, The Theory of Lie Derivatives and Its Applications (North-Holland Publ. Co., Amsterdam, 1957).
27. G. H. Katzin et al, J. Math. Phys. 10, 617 (1969).
28. H. Bondi and T. Gold, Proc. Roy. Soc. A229, 416 (1955); H. Bondi, In Jubillee of Relativity Theory, Helvetica Physica Acta Supp. IV, 1956, p. 98, considers the problem from the point of view of equivalence principle and the true fields.
29. P. A. M. Dirac, Proc. Roy. Soc. A167, 148 (1938), shows radiation reaction to vanish identically in these two cases.

NONCONTINUOUS REPRESENTATIONS OF LIE GROUPS†‡

Bernd Pirrung
Institut für Theoretische Physik
Technische Universität Clausthal
Clausthal-Zellerfeld, Germany

Section I

If we want to describe a physical system by means of group theoretical methods, we do two things: (i) We choose--because of physical or formal reasons--a certain Lie group G, and (ii) we identify the physical system or certain properties of it with a (a) faithful, (b) unitary, (c) irreducible, (d) linear, and (e) continuous representation of G.

The relation between the physical system and the representation is given in the following way: The generators of the representation are identified with observables and a rule is given on how to express all observables by these generators.

This is the way one usually proceeds, and the reasons for the choice of such a representation are the following:

(a) The transformations should be observables, and therefore the representation ought to be faithful, since different observables should correspond to different operators.

(b) If one examines pure symmetry transformations, then the invariance of the theory enforces the (anti-) unitarity by the theorem of Wigner. Nonsymmetry groups contain in general the symmetry group as a subgroup. Thus at least this subgroup has to be represented unitarily, and this unitarity is carried over to the representation of the whole group.

(c) One should suppose that reducible representations contain dynamics which one can only describe in the frame of group theory, if one embeds G into a larger group which is represented irreducibly, then one obtains the physical contents, if one reduces this representation with respect to G.

---

†Presented at the Symposium on De Sitter and Conformal Groups, University of Colorado, Summer 1970.

‡Part of a paper to be published by H. D. Doebner and B. Pirrung.

(d) The linearity of the representation is desirable, as nonlinear representations are more difficult to handle.
(e) Now let us consider for example the 3-dimensional rotation group. An experimentalist can realize every rotation; hence the continuity of the 3-dimensional representation of the rotation group. If one uses noncontinuous representations, one loses the properties which distinguish Lie groups, since there are no generators.

But this is not the whole story!
(a) The faithfulness is violated in very simple and often used examples. Consider the unitary, irreducible representations of the n-dimensional translation group T_n. They are characterized by an n-vector k and have the form: $x \to \exp i(k\cdot x)$. These representations are not faithful because of two reasons: $\exp i(k\cdot x)$ has a period of 2π, and the scalar product $(k\cdot x)$ has the same value for different values of x. Therefore also the often used representations of the Poincaré group in the Hilbert space are not faithful. A faithful representation of T_n--which is of course reducible--is for instance:

$$x \to V_x = \begin{pmatrix} e^{ix_1} & & & & 0 \\ & \ddots & & & \\ & & e^{ix_n} & & \\ & & & e^{i\pi x_1} & \\ & & & & \ddots \\ 0 & & & & e^{i\pi x_n} \end{pmatrix}$$

And obviously this representation is unitary.
(b) The unitarity is--as mentioned above--not always enforced by Wigner's theorem, and the most important argument for the unitarity is the fact that it is very difficult to classify nonunitary representations.
(c) The reasons for the use of irreducible and linear representations are similar, namely of mathematical kind.
(d) In fact one cannot realize transformations which are infinitesimal. Thus the experimentalist cannot verify the existence of generators, and one can prove the continuity only approximately by showing that the mean values, which are measured for a small but not infinitesimal interval, are continuous functions of the group parameters. We shall prove later that these mean values exist for one class of noncontinuous representations. I want to stress that one should avoid noncontinuous representations in any case; there are too many of them. But we shall see later that they appear in some constructions which are reasonable from the physical point of view.

Now we have seen that the choice of faithful, unitary, irreducible, linear and continuous representations is a simplification, which cannot always be justified by physical arguments. We have seen that these properties are sometimes not even compatible, and therefore one should not disregard representations, which do not have these properties.

Section II

Now we shall investigate the continuity of representations. Let us first recall the definition:[8)]

> Consider a Lie group G and a representation $x \to V_x$ of G in a Hilbert space H.
>
> The representation is weakly continuous iff $(V_x f, g)$ is a continuous function of x for all $f, g \in H$.
>
> It is strongly continuous, iff the mapping $x \to V_x$ is continuous for all $f \in H$.
>
> It is weakly λ-measurable, iff $(V_x f, g)$ is a λ-measurable function of x for all $f, g \in H$.

Of course one can construct noncontinuous representations by changing the topology of the group. But we do not want to do that; we always use the normal (Euclidean) topology.

The global properties of noncontinuous representations of locally compact groups are described by the following theorem:[1),6)10)]

> Every weakly continuous, unitary representation of a locally compact group in a Hilbert space is strongly continuous.
>
> Every weakly λ-measurable, unitary representation of a locally compact group in a separable Hilbert space is strongly continuous.

That means: If we want to construct noncontinuous representations in a separable Hilbert space, we obtain only representations which are not weakly λ-measurable. And if we want to obtain measurable, noncontinuous representations, we have to consider non-separable Hilbert spaces, and that is the thing we shall do first. A simple example for such a representation is given in Ref. 4 in connection with a model of a ferromagnet. Another example is given in Ref. 8, §22.20.

Of course there are many nonseparable Hilbert spaces, but we now consider one which is well known and applied in physics, namely the infinite direct product $H_\otimes = \prod_{\alpha=1}^{\infty} \otimes H_\alpha$ of Hilbert spaces H_α, which was defined and investigated by John von Neumann in Ref. 11.†

---

†$H_\otimes$ is the closure of the pre-Hilbert space $H'_\otimes$, where

(footnote continued on bottom of next page)

Now consider a Lie group G with a c, f, u (continuous, faithful, unitary and linear) representation $x \to V_x^\alpha$ in H_α, $\alpha = 1,2,3\ldots$ and define $x \to W_x$ in the following way:

$$W_x \varphi = \sum_{\nu=1}^{p} \prod_{\alpha=1}^{\infty} \otimes V_x^\alpha f_{\alpha,\nu}, \text{ if } \varphi = \sum_{\nu=1}^{p} \prod_{\alpha=1}^{\infty} \otimes f_{\alpha,\nu},$$

$$\text{and } W_x \varphi = \lim_{\varkappa \to \infty} W_x \varphi_\varkappa, \text{ if } \varphi = \lim_{\varkappa \to \infty} \varphi_\varkappa .$$

We can then prove the following statements:

a) $x \to W_x$ is a linear representation of G in $H_\otimes$ (this representation is called $\otimes$-representation)

b) The $\otimes$-representation is faithful and unitary.

c) The $\otimes$-representation is weakly λ-measurable.

<u>Proof:</u>

Obviously W_x is a linear operator in $H_\otimes$. If x and y are two elements of G, then

$$W_{xy}\left(\sum_{\nu=1}^{p} \prod_{\alpha=1}^{\infty} \otimes f_{\alpha,\nu}\right) = \sum_{\nu=1}^{p} \prod_{\alpha=1}^{\infty} \otimes V_{xy}^\alpha f_{\alpha,\nu} = \sum_{\nu=1}^{p} \prod_{\alpha=1}^{\infty} \otimes V_x^\alpha V_y^\alpha f_{\alpha,\nu} =$$

$$W_x\left(\sum_{\nu=1}^{p} \prod_{\alpha=1}^{\infty} \otimes V_y^\alpha f_{\alpha,\nu}\right) = W_x W_y \left(\sum_{\nu=1}^{p} \prod_{\alpha=1}^{\infty} \otimes f_{\alpha,\nu}\right)$$

$$H'_\otimes = \left\{\varphi \,\middle|\, \varphi = \sum_{\nu=1}^{p} \prod_{\alpha=1}^{\infty} \otimes f_{\alpha,\nu};\ p < \infty;\ \prod_{\alpha=1}^{\infty} \|f_{\alpha,\nu}\| < \infty \text{ for } \nu = 1,\ldots,p;\right.$$

$$\left.(\varphi,\Psi) = \sum_{\nu=1}^{p} \sum_{\mu=1}^{q} \prod_{\alpha=1}^{\infty} (f_{\alpha,\nu}, g_{\alpha,\mu})\right\}$$

It is necessary for this construction that one defines $\prod_{\alpha=1}^{\infty} z_\alpha = 0$, if $\prod_{\alpha=1}^{\infty} |z_\alpha|$ converges and $\prod_{\alpha=1}^{\infty} z_\alpha$ does not converge. Here z_α, $\alpha=1,2..$ are complex numbers. That means, for example, $\prod_{\alpha=1}^{\infty} (-1)^\alpha = 0$.

i.e. $W_{xy} = W_x W_y$ on $H'_\otimes$. As a similar calculation can be performed, if φ is a limit element of $H'_\otimes$, we obtain a). b) is obvious. As $x \to V^\alpha_x$ is continuous, it is weakly λ-measurable, and therefore by theorem (11.18) in Ref. 9 $(W_x\varphi, \Psi)$ is measurable for all $\varphi, \Psi \in H_\otimes$, i.e. the $\otimes$-representation is weakly λ-measurable. Obviously $x \to W_x$ is faithful, if $x \to V^\alpha_x$ is faithful for one α. Therefore we define:

> $x \to W_x$ is completely faithful (cf) iff $x \to V^\alpha_x$ is faithful for $\alpha = 1, 2, 3, \ldots$.

We have seen that the $\otimes$-representation is faithful and unitary, if $x \to V^\alpha_x$ is faithful and unitary for all α. But a similar theorem cannot be proved concerning continuity. Let us consider a simple case:

> Let G be a nondiscrete topological group and $x \to V^\alpha_x$ a c,f,u representation of G in H_α, $\alpha = 1, 2, \ldots$ Suppose that $H_\alpha = H_\beta$ and $V^\alpha_x = V^\beta_x$ for all α and β. Then the $\otimes$-representation $x \to W_x$ is not continuous.

<u>Proof:</u>

We show that $(W_x\varphi, \Psi)$ is a noncontinuous function of x, if

$$\varphi = \prod_{\alpha=1}^{\infty} \otimes f_\alpha, \quad \Psi = \prod_{\alpha=1}^{\infty} \otimes V^\alpha_x f_\alpha, \quad f_\alpha = f_\beta, \text{ for all } \alpha, \beta, \text{ and } \|f_\alpha\| = 1.$$

In an arbitrary neighbourhood U_x of x we can always find an element y such that $(V^\alpha_y f_\alpha, V^\alpha_x f_\alpha) := a \neq 1$ for all α. $|a| \leq 1$, because $x \to V_x$ is c, f, u.

Thus:

$$|(W_x\varphi, \Psi) - (W_y\varphi, \Psi)| = \left| \prod_{\alpha=1}^{\infty} (V^\alpha_x f_\alpha, V^\alpha_x f_\alpha) - \prod_{\alpha=1}^{\infty} (V^\alpha_y f_\alpha, V^\alpha_x f_\alpha) \right| =$$

$$= \left| \prod_{\alpha=1}^{\infty} 1 - \prod_{\alpha=1}^{\infty} a \right| = |1 - 0| = 1,$$

as either the absolute value of a is less than 1 or $a = \exp(i\varphi)$ and in both cases the infinite product $\prod_{\alpha=1}^{\infty} a$ is zero. Therefore in every neighbourhood of x there is an element y such that

$$|(W_x\varphi, \Psi) - (W_y\varphi, \Psi)| = 1,$$

which completes the proof.

Thus we have found a new method to construct noncontinuous representations of every nondiscrete topological group.†

Section III

Now we shall prove the following general theorem for non-solvable Lie groups:

Let G be a non-solvable Lie group with a c, f, u representation $x \to V_x^\alpha$ in H_α, $\alpha = 1,2,3\ldots$
Then the $\otimes$-representation $x \to W_x$ of G is not continuous.

Proof:

It is well known that every semisimple and therefore also every non-solvable Lie group contains SO(3) or(and) SO(2,1) as a subgroup.[3] Therefore we have only to prove that the $\otimes$-representation of SO(3), resp. SO(2,1), which is given by the representation of G, is not continuous.

Let us consider first SO(3): According to Ref. 5 we can choose $f_\alpha \in H_\alpha$, with $||f_\alpha|| = 1$, such that

$$(V_x^\alpha f_\alpha, f_\alpha) = V^\alpha_{\ell_\alpha \ell_\alpha}(\varphi_1, \theta, \varphi_2) =$$

$$= \exp(-i\ell_\alpha \varphi_1)\, 2^{-\ell_\alpha}(1 + \cos\theta)^{\ell_\alpha} \exp(-i\ell_\alpha \varphi_2), \quad x \in SO(3),$$

(ℓ_α is a positive integer or half integer, φ_1, θ, φ_2 are the Euler angles). Now let us take $\Phi = \Psi = \prod_{\alpha=1}^{\infty} \otimes f_\alpha$ and investigate the continuity at x = e (e denotes the identity).

$$|(W_y\Phi, \Psi) - (W_e\Phi, \Psi)| = |\prod_{\alpha=1}^{\infty}(V_y^\alpha f_\alpha, f_\alpha) - \prod_{\alpha=1}^{\infty}(f_\alpha, f_\alpha) =$$

$$= |\prod_{\alpha=1}^{\infty} 2^{-\ell_\alpha}(1 + \cos\theta)^{\ell_\alpha} - 1| = |0 - 1| = 1,$$

because y may be chosen such that $\theta \neq 0$ and $\varphi_1 = -\varphi_2$. Then $a =: 2^{-1}(1 + \cos\theta)$ is less than 1, and therefore $\prod_{\alpha=1}^{\infty} a^{\ell_\alpha} = 0$.

---

†If we do not demand complete faithfulness, then it is of course easy to construct continuous $\otimes$-representations. Take $V_x^1 = V_x$ and $V_x^\alpha = 1$, $\alpha = 2,3,4,\ldots$.

In a similar way one can prove that the ⊗-representation of SO(2,1) is not continuous (by using the matrix elements given in Ref. 2), which completes the proof.

Section IV

This result cannot be generalized, because abelian groups and even the Heisenberg group may have c, cf, u ⊗-representations:† This can be easily shown for the translation group T_n, by taking as representation in H_α

$$x \to V_x^\alpha = \begin{pmatrix} e^{i\frac{x_1}{\alpha^4}} & & & & & 0 \\ & \ddots & & & & \\ & & e^{i\frac{x_n}{\alpha^4}} & & & \\ & & & e^{i\pi\frac{x_1}{\alpha^4}} & & \\ & & & & \ddots & \\ 0 & & & & & e^{i\pi\frac{x_n}{\alpha^4}} \end{pmatrix}$$

The proof for the Heisenberg group affords a lengthy calculation. That is all that we want to say about noncontinuous representations in the infinite direct product of Hilbert spaces.

†It is very probable that a Lie group G has a c, cf, u ⊗-representation, if G has the following property:

For every $\delta > 0$ there exist $\epsilon_1, \ldots \epsilon_n > 0$, $\epsilon_i \le \delta$, such that

$$g(\alpha_1', \ldots \alpha_n') \cdot g(\alpha_1'', \ldots \alpha_n'') = g(\alpha_1, \ldots \alpha_n)$$

implies

$$g(\epsilon_1\alpha_1', \ldots \epsilon_n\alpha_n') \cdot g(\epsilon_1\alpha_1'', \ldots \epsilon_n\alpha_n'') = g(\epsilon_1\alpha_1, \ldots \epsilon_n\alpha_n)$$

for all $g \in G$ ($\alpha_1, \ldots \alpha_n$ are the group parameters).

It is obvious that groups with this property--we call them contractible --may be treated in a similar way as the translation group T_n. It can be proved that if a group is contractible, it must be nilpotent (and therefore solvable).

Section V

Now we shall discuss noncontinuous representations in separable spaces.

The 3-dimensional representation of the 3-dimensional rotation group may be written in the following form:

$$g(\varphi_1, \theta, \varphi_2) \to V_g = V_{a_3(\varphi_1)} V_{a_1(\theta)} V_{a_3(\varphi_2)},$$

where, for example, $V_{a_3(\varphi)}$ has the form

$$V_{a_3(\varphi)} = \begin{pmatrix} C(\varphi) & -S(\varphi) & 0 \\ S(\varphi) & C(\varphi) & 0 \\ 0 & 0 & 1 \end{pmatrix}.$$

Here the functions C and S have to fulfill the functional equations:

$$C(\varphi_1 + \varphi_2) = C(\varphi_1)\, C(\varphi_2) - S(\varphi_1)\, S(\varphi_2)$$

$$S(\varphi_1 + \varphi_2) = S(\varphi_1)\, C(\varphi_2) + S(\varphi_2)\, C(\varphi_1)$$

$$C^2(\varphi) + S^2(\varphi) = 1.$$

These equations have of course the solutions $\cos\varphi$ and $\sin\varphi$ for $C(\varphi)$ and $S(\varphi)$, but also $\cos f(\varphi)$ and $\sin f(\varphi)$ are solutions if $f(\varphi_1 + \varphi_2) = f(\varphi_1) + f(\varphi_2)$.[12] Hamel has shown in Ref. 7 that there exist noncontinuous functions f which fulfill these equations. And if we use these functions and replace $\cos\varphi$ by $\cos f(\varphi)$ and $\sin\varphi$ by $\sin f(\varphi)$, we obtain a noncontinuous finite dimensional representation of the rotation group.

In the same way one can construct arbitrary dimensional representations of SO(3), which are not continuous, and thus one can even obtain noncontinuous representations of each group, which contains SO(3) as a subgroup.†

References

1. D. Auslander and C. Moore, Mem. Am. Math. Soc. 62 (1966).
2. A. O. Barut and C. Fronsdal, Proc. Roy. Soc. A287, 532 (1965).
3. E. B. Dynkin, Trans. Am. Math. Soc. 6, 175 (1957).
4. A. Katz and Y. Frishman, Suppl. Nuovo Cimento 5, 749 (1967).

---

† It is even possible to construct noncontinuous representations of the Poincaré group such that the restriction to SO(3) and to the translation group is continuous.

5. I. M. Gel'fand, R. A. Minlos, and Z. Ya. Shapiro, Representations of the Rotation and Lorentz Groups and Their Applications (Pergamon Press, Oxford, 1963).
6. R. Godement, Trans. Am. Math. Soc. 63, 1 (1948).
7. G. Hamel, Math. Ann. 60, 459 (1905).
8. E. Hewitt and K. A. Ross, Abstract Harmonic Analysis I (Springer-Verlag, Berlin, Göttingen, Heidelberg, 1963).
9. E. Hewitt and K. Stromberg, Real and Abstract Analysis (Springer-Verlag, Berlin, Heidelberg, New York, 1965).
10. R. T. Moore, Mem. Am. Math. Soc. 78 (1968).
11. J. von Neumann, Collected Works III, ed. by A. H. Taub (Pergamon Press, New York, 1961), p. 322.
12. L. Vietoris, J. reine angew. Math. 186, 1 (1949).

Section E: Expansions

GENERALIZED O(2,1) EXPANSIONS†

C. Cronström ‡
Institute for Advanced Study
Princeton, New Jersey 08540

Abstract

We give a brief review of the O(2,1) expansion for a square-integrable scattering amplitude and its relation to the Sommerfeld-Watson transform. The conventional O(2,1) expansion is then generalized to cover the case of asymptotically growing but power bounded functions. Certain ambiguities inherent in the generalized O(2,1) expansion are discussed in detail.

For simplicity we discuss only the case of two-body amplitudes with equal mass kinematics and no spin.

I. Introduction

The material presented in this talk is to a large extent based on unpublished work by W. H. Klink and myself, which is still in progress.[1] One of the objectives of this work is to generalize the O(2,1) expansion for square integrable scattering amplitudes, which was first discussed by J. F. Boyce,[2] to asymptotically growing scattering and production amplitudes. Here I will only discuss the simplest possible situation; namely the expansion of a two-body scattering amplitude with equal mass kinematics and without spin. The expansion formula we arrive at has also recently been derived by H. D. I. Abarbanel and L. M. Saunders,[3] and by C. E. Jones et al.[4] The derivation of the generalized O(2,1) expansion formula given here is slightly more complicated than the derivations given in Refs. 3 and 4, but exhibits clearly an ambiguity inherent in the generalization of the standard O(2,1) expansion to non-square integrable functions.

†Presented at the Symposium on De Sitter and Conformal Groups, University of Colorado, Summer 1970.

‡On leave from the Department of Theoretical Physics, University of Helsinki, Finland.

This ambiguity is quite essential in establishing the connection between the generalized O(2,1) expansion and the Sommerfeld-Watson representation[5] of the amplitude.

The group theory underlying the standard O(2,1) expansion has been given in Ref. 2, and in the review articles by P. Winternitz,[6),7] and also in the talk by P. Winternitz at this conference.[8] Let me therefore only briefly mention the relevant arguments in arriving at an O(2,1) expansion of a square-integrable amplitude.

We consider a scattering amplitude F(s,t) (s and t are the usual energy and momentum transfer variables, respectively) for a fixed value of t in the physical region of the s-channel. In the general mass case the variable t can be positive for a range of values of s in the s-channel physical region, provided the masses satisfy a certain inequality. We shall, however, only consider negative fixed values of t. The momentum transfer vector is then space-like, and the little (or stability) group corresponding to a fixed momentum transfer is given by O(2,1). This can most easily be seen by considering the amplitude in the brick-wall system (in which the momentum transfer vector has a component along the third space axis only). In this system the amplitude is parametrized (for fixed t) in terms of a hyperbolic angle β, which is related to s and t as follows (for equal masses m),

$$x \equiv \cosh \beta = \frac{2s}{4m^2 - t} - 1 \qquad (I.1)$$

$$F(s,t) \equiv f(t,x) \qquad (I.2)$$

Any amplitude which is square-integrable (in x) can be expanded in terms of the unitary irreducible representations of O(2,1).

The expansion formula, which is known in classical analysis as the Mehler-Fock representation,[9),10] takes the following form

$$f(x) = \int_0^\infty dq\, a(q)\, P_{-\frac{1}{2}+iq}(x) \qquad (I.3)$$

$$a(q) = q \tanh(q\pi) \int_1^\infty dx\, f(x)\, P_{-\frac{1}{2}+iq}(x) \qquad (I.4)$$

Here we have suppressed the dependence of the functions involved on t (this will also be done in the sequel, whenever expedient). The function $P_\ell(x)$ in Eqs. (I.3) and (I.4) is the Legendre function of the first kind. An account of this and related functions can be found e.g.

in the first volume of the Bateman Manuscript Project[11] (references to this work in what follows will be given as B followed by the appropriate page number).

The representation (I.3) is very similar to a Sommerfeld-Watson background integral (taken along Re $\ell = -\frac{1}{2}$). There are two important differences, however.

In the first place the range of integration in Eq. (I.3) is $(0,\infty)$ and not $(-\infty,\infty)$ as it would be in the case of a SW background integral. However, since Eq. (I.4) defines the expansion coefficient a(q) as an even function in q (because of the symmetry of the function $P_\ell(x)$ under the substitution $\ell \to -\ell-1$) we can simply extend the range of integration in Eq. (I.3) to $(-\infty,\infty)$. In so doing, however, we create a possibility for introducing ambiguities in the integral representation, since one may now add any "reasonable" odd function of q to the expansion coefficient a(q) without affecting the integral representation. This fact will be very important in the sequel.

Secondly, in applying the expansion formulae (I.3) and (I.4) to an amplitude f(x) we are not a priori forced to split up the amplitude into two parts, corresponding e.g. to even and odd signature. However, the splitting of an amplitude into an even and odd signature part in conventional Regge theory is not really a logical necessity but rather a matter of choice.[12] A detailed account of Regge theory without signature has also been given in a recent paper by T. K. Gaisser and C. E. Jones.[13] We shall therefore proceed without introducing signature although one could do so albeit in an ad hoc manner.

The rest of this paper is organized as follows. In Sec. II we show in detail how the expansion formulae (I.3) and (I.4) can be brought into a form in which they coincide with the Sommerfeld-Watson transform of an amplitude which is assumed to satisfy a fixed t dispersion relation. Sec. III is devoted to the generalization of Eqs. (I.3) and (I.4) to non-square integrable functions. It is shown that the generalized O(2,1) expansion can still be made to coincide with the SW transform, provided one makes effective use of the ambiguity mentioned above.

The final Sec. IV gives a summary of the results obtained.

II. Comparison with the Sommerfeld-Watson Transform

By introducing a complex variable

$$\ell = -\tfrac{1}{2} + iq \qquad \text{(II.1)}$$

by dividing Eq. (I.4) by q tanh $(q\pi)$, and by extending the range of integration in Eq. (I.3), we can rewrite the Eqs. (I.3) and (I.4) as follows,

$$f(x) = \frac{1}{2i} \int_{-\frac{1}{2}-i\infty}^{-\frac{1}{2}+i\infty} \frac{d\ell\,(2\ell+1)\,\tilde{a}(\ell)}{\tan \pi\ell} P_\ell(x) \tag{II.2}$$

$$\tilde{a}(\ell) = \tfrac{1}{2} \int_1^\infty dx\, f(x)\, P_\ell(x) \tag{II.3}$$

The Eq. (II.2) can be recognized as a Sommerfeld-Watson background integral, provided the expansion coefficient $\tilde{a}(\ell)$ can be related to the partial wave amplitude $a_\ell(t)$ as follows,

$$a_\ell(t) = -\tilde{a}(\ell) \cos \pi\ell \quad . \tag{II.4}$$

We shall assume that the (square-integrable) amplitude $f(x)$ satisfies a fixed t dispersion relation,

$$f(x) \equiv F(s,t) = \frac{1}{\pi} \int_{4m^2}^\infty \frac{ds'\, A_s(s',t)}{s'-s-i\epsilon} \tag{II.5}$$

Since the presence of a u-channel contribution to $F(s,t)$ plays no decisive role in what follows, we simply neglect it for ease of writing. Inserting Eq. (II.5) in Eq. (II.3) we obtain, using the formula B.140,

$$\tilde{a}(\ell) = -\frac{(4m^2-t)^{-1}}{\cos \pi\ell} \frac{1}{\pi} \int_{4m^2}^\infty ds'\, A_s(s',t)$$

$$\left\{ e^{-i\pi\ell} Q_\ell\left(\frac{2s'}{4m^2-t} -1 -i\epsilon\right) + e^{i\pi\ell} Q_{-\ell-1}\left(\frac{2s'}{4m^2-t} -1-i\epsilon\right)\right\} \tag{II.6}$$

where $Q_\ell(x)$ is the Legendre function of the second kind. Eq. (II.6) defines $\tilde{a}(\ell)$ as manifestly even in $q(\ell = -\frac{1}{2}+iq)$. However, the evenness of $P_\ell(x)$ in Eq. (II.2) guarantees that only the even part of the integrand in Eq. (II.2) contributes to the integral. We can thus add any reasonable odd function of q to $\tilde{a}(\ell)$ defined by Eq. (II.6). In particular, we may replace the expression (II.6) by the function

$$\tilde{a}'(\ell) = \frac{2}{\pi} \frac{(t-4m^2)^{-1}}{\cos \pi\ell} \int_{4m^2}^{\infty} ds' \, A_s(s',t) \, e^{-i\pi\ell} Q_\ell\left(\frac{2s'}{4m^2-t} - 1 - i\varepsilon\right) \tag{II.7}$$

We shall now turn to the Sommerfeld-Watson transform of the amplitude defined by Eq. (II.5). We shall then temporarily consider the variable t fixed in the interval

$$0 < t < 4m^2 \tag{II.8}$$

and

$$0 \leq s < 4m^2 \tag{II.9}$$

We can now define a variable z_t,

$$z_t = \frac{2s}{t-4m^2} + 1 \tag{II.10}$$

with values in $[-1,1]$. The amplitude $F(s,t)$ can then be expanded as follows

$$F(s,t) = \sum_\ell (2\ell+1) \, \hat{a}_\ell(t) \, P_\ell(z_t) \tag{II.11}$$

with

$$\hat{a}_\ell(t) = \tfrac{1}{2} \int_{-1}^{+1} dz_t \, P_\ell(z_t) \, F(s,t) \tag{II.12}$$

Inserting Eq. (II.15) in Eq. (II.12) we obtain

$$\hat{a}_\ell(t) = -\frac{2}{\pi} (t-4m^2)^{-1} \int_{4m^2}^{\infty} ds' \, A_s(s', t) \, e^{-i\pi\ell} \, Q_\ell\left(\frac{2s'}{4m^2-t} - 1\right) \tag{II.13}$$

For t fixed in the interval (II.8) the series (II.11) converges in a Lehmann ellipse with semi-major axis $(4m^2 + t)/(4m^2 - t)$. We can then perform a slightly unorthodox Sommerfeld-Watson transform of the series (II.11), retaining the factor $e^{-i\pi\ell}$ in Eq. (II.13) even for complex values of ℓ. The result is

$$F(s,t) = \frac{i}{2} \int_{-\frac{1}{2}-i\infty}^{-\frac{1}{2}+i\infty} \frac{d\ell (2\ell+1)\, \hat{a}_\ell(t)\, P_\ell(-z_t)}{\sin \pi\ell} \tag{II.14}$$

We note that the argument $-z_t$ of the Legendre function in Eq. (II.14) is defined in terms of s and t by the same expression (I.1) that defines $x = \cosh \beta$ in Eq. (II.2).

Let us also remark that the unorthodox step from Eq. (II.11) to Eq. (II.14) could have been avoided by using $\xi \equiv -z_t$ as an expansion variable instead of z_t. However, the final result, Eq. (II.14), with $\hat{a}_\ell(t)$ defined by Eq. (II.13) would still have been obtained, the unconventional factor $e^{-i\pi\ell}$ now being a result of the analytic continuation of $P_\ell(-\xi)$ to $P_\ell(\xi) = P_\ell(-z_t)$.

We can now (trivially) continue the representation (II.2) in s and t to the ranges of values given by Eqs. (II.8) and (II.9), with the symmetric expansion coefficient $\tilde{a}(\ell)$ replaced by the equivalent expansion coefficient $\tilde{a}'(\ell)$ given by Eq. (II.7). Comparing the expressions (II.7) and (II.13) we see that the functions $\hat{a}_\ell(t)$ and $\tilde{a}'(\ell)$ indeed satisfy the relation (II.4), i.e.

$$\hat{a}_\ell(t) = -\,\tilde{a}'(\ell) \cos \pi\ell \tag{II.15}$$

There is thus a perfect agreement between the O(2,1) result, Eq. (II.2), and the Sommerfeld-Watson representation, Eq. (II.14).

III. The Generalized O(2,1) Expansion

As a natural generalization of the Mehler-Fock representation (I.3) we consider the following ansatz,

$$f(x) = \int_{-\infty}^{\infty} dq\, A(p,q)\, P_{p+iq}(x) \tag{III.1}$$

with

$$-\tfrac{1}{2} \le p \le p_0 < 0 \tag{III.2}$$

For $p = -\frac{1}{2}$ Eq. (III.1) reduces to Eq. (I.3), with $A(-\frac{1}{2},q)$ given by Eq. (I.4). Let us now invert Eq. (III.1) for p fixed in the interval (III.2). We use the integral representation B.156,

$$P_{p+iq}(\cosh \beta) = \frac{\sqrt{2}}{\pi} \int_0^\beta \frac{d\varphi \cosh[(\frac{1}{2} + p + iq)\varphi]}{\sqrt{\cosh \beta - \cosh \varphi}} \tag{III.3}$$

Inserting Eq. (III.3) in Eq. (III.1) and inverting the order of integration we obtain

$$f(\cosh \beta) = \frac{\sqrt{2}}{\pi} \int_0^\beta \frac{d\varphi\, \tilde{A}(p,\varphi)}{\sqrt{\cosh \beta - \cosh \varphi}} \qquad \text{(III.4)}$$

with

$$\tilde{A}(p,\varphi) = \int_{-\infty}^{\infty} dq\, A(p,q) \cosh\left[(\tfrac{1}{2} + p + iq)\varphi \right] \qquad \text{(III.5)}$$

Eq. (III.4) is essentially Abels equation,[14] which can be inverted, with the result,

$$\tilde{A}(p,\varphi) = \frac{d}{d\varphi} \int_0^\varphi \frac{d\beta \sinh \beta\, f(\cosh \beta)}{\sqrt{2(\cosh \varphi - \cosh \beta)}} \qquad \text{(III.6)}$$

Let us now split up Eq. (III.5) into its real and imaginary part, respectively,

$$\tilde{A}_R(p,\varphi) = 2 \cosh r\varphi \int_0^\infty dq\, A_R^+(p,q) \cos q\varphi$$

$$- 2 \sinh r\varphi \int_0^\infty dq\, A_I^-(p,q) \sin q\varphi \qquad \text{(III.7)}$$

$$\tilde{A}_I(p,\varphi) = 2 \sinh r\varphi \int_0^\infty dq\, A_R^-(p,q) \sin q\varphi$$

$$+ 2 \cosh r\varphi \int_0^\infty dq\, A_I^+(p,q) \cos q\varphi \qquad \text{(III.8)}$$

Here the subscripts R and I denote the real and imaginary part of the function, respectively, the superscripts + and - denote the even and odd part (in q), respectively, and we have introduced the abbreviation

$$r = \tfrac{1}{2} + p \quad . \qquad \text{(III.9)}$$

The problem that remains is now to invert the Eqs. (III.7) and (III.8). It is not difficult to verify that the following expressions satisfy these equations,

$$A_R^+(p,q) = \frac{1}{\pi} \int_0^\infty d\varphi\, e^{-r\varphi}\, \tilde{A}_R(p,\varphi) \cos q\varphi$$

$$A_I^-(p,q) = -\frac{1}{\pi} \int_0^\infty d\varphi\, e^{-r\varphi}\, \tilde{A}_R(p,\varphi) \sin q\varphi \qquad \text{(III.10)}$$

and

$$A_R^-(p,q) = \frac{1}{\pi} \int_0^\infty d\varphi\, e^{-r\varphi}\, \tilde{A}_I(p,\varphi) \sin q\varphi$$

$$A_I^+(p,q) = \frac{1}{\pi} \int_0^\infty d\varphi\, e^{-r\varphi}\, \tilde{A}_I(p,\varphi) \cos q\varphi \qquad \text{(III.11)}$$

We then recover the function A(p,q) as

$$A(p,q) = A_R^+(p,q) + A_R^-(p,q) + i\,A_I^+(p,q) + i\,A_I^-(p,q) \qquad \text{(III.12)}$$

i.e.

$$A(p,q) = \frac{1}{\pi} \int_0^\infty d\varphi\, e^{-(p+\frac{1}{2}+iq)\varphi}\, \tilde{A}(p,\varphi) \qquad \text{(III.13)}$$

Before expressing A(p,q) in terms of the function f(x) by using Eq. (III.6) let us consider the question of uniqueness of the solution (III.13). We then have to specify in more detail what we require of a solution A(p,q). It is reasonable to require that A(p,q) should in fact be a function of the complex variable $\ell = p + iq$,

$$A(p,q) = F(p + iq) \qquad \text{(III.14)}$$

where

(i) $F(\ell)$ is analytic and regular in a strip contained in $-\frac{1}{2} < p < 0$
(ii) $F(\ell)$ vanishes to the appropriate order as $|\mathrm{Im}\,\ell| \to \infty$
(iii) $F(\ell)$ reduces to an expression equivalent to Eq. (I.4) in the limit $\mathrm{Re}\,\ell \to -\frac{1}{2}$.

Let us now assume that there are two functions A(p,q) satisfying Eq. (III.5) (or Eqs. (III.7) and (III.8)) and the conditions above. Then their difference $\Delta(p,q)$ has to satisfy the equations,

$$\cosh r\varphi \int_0^\infty dq\, \Delta_R^+(p,q) \cos q\varphi$$

$$- \sinh r\varphi \int_0^\infty dq\, \Delta_I^-(p,q) \sin q\varphi = 0 \qquad \text{(III.15)}$$

and

$$\sinh r\varphi \int_0^\infty dq\, \Delta_R^-(p,q) \sin q\varphi$$

$$+ \cosh r\varphi \int_0^\infty dq\, \Delta_I^+(i,q) \cos q\varphi = 0 \qquad \text{(III.16)}$$

Eqs. (III.15) and (III.16) are satisfied e.g. by the choice

$$\int_0^\infty dq\, \Delta_R^+(p,q) \cos q\varphi = c^-(\varphi) \sinh r\varphi$$

$$\int_0^\infty dq\, \Delta_I^-(p,q) \sin q\varphi = c^-(\varphi) \cosh r\varphi \qquad \text{(III.17)}$$

and

$$\int_0^\infty dq\, \Delta_R^-(p,q) \sin q\varphi = d^-(\varphi) \cosh r\varphi$$

$$\int_0^\infty dq\, \Delta_I^+(p,q) \cos q\varphi = - d^-(\varphi) \sinh r\varphi \qquad \text{(III.18)}$$

The functions $c^-(\varphi)$ and $d^-(\varphi)$ in Eqs. (III.17) and (III.18) should be real, odd, independent of p, and such that $c^-(\varphi) \exp r\varphi$ and $d^-(\varphi) \exp r\varphi$ have Fourier transforms. The requirement that $c^-(\varphi)$ and $d^-(\varphi)$ be independent of p follows from the analyticity condition via the Cauchy-Riemann equations.

Let us consider the example

$$c^-(\varphi) = \lambda_1\, \varphi\, e^{-\frac{1}{2}|\varphi|}$$

$$d^-(\varphi) = - \lambda_2\, \varphi\, e^{-\frac{1}{2}|\varphi|} \qquad \text{(III.19)}$$

where λ_1 and λ_2 are real constants. From Eqs. (III.17) and (III.18) we then obtain

$$\Delta(p,q) = \frac{\lambda}{\pi} \frac{2\ell+1}{[\ell(\ell+1)]^2} \tag{III.20}$$

where $\lambda = \lambda_1 + i\lambda_2$ and $\ell = p+iq$.

We have thus demonstrated by means of an example that the solution to the Eqs. (III.7) and (III.8) is not unique.

A little consideration and abstraction from the example (III.20) allows us to infer that we may in fact add any function $\Delta(\ell)$, which is analytic in the strip $-\frac{1}{2} < \mathrm{Re}\,\ell < p_0$, whose boundary value on $\mathrm{Re}\,\ell = -\frac{1}{2}$ is antisymmetric in q, and which is such that $\Delta(\ell)\,P_\ell(x)$ is integrable on $\mathrm{Re}\,\ell = -\frac{1}{2}$ and $\mathrm{Re}\,\ell = p$ to the expansion coefficient $A(p,q)$ in the integral (III.1) without affecting the value of this integral.

Having established the fundamental nonuniqueness of the expansion coefficient $A(p,q)$ in the ansatz (III.1), let us return to the expression (III.13) which gives one solution for $A(p,q)$. Inserting Eq. (III.6) in Eq. (III.13), integrating by parts and changing the order of integration we find

$$A(p,q) = \frac{(\ell+\frac{1}{2})}{\pi} \int_0^\infty d\beta \sinh\beta \left[\int_\beta^\infty \frac{d\varphi\, e^{-(\ell+\frac{1}{2})\varphi}}{\sqrt{2(\cosh\varphi - \cosh\beta)}}\right] f(\cosh\beta) \tag{III.21}$$

The term within brackets in Eq. (III.21) can be recognized as an integral representation (B.155) for the Q_ℓ-function. The final result is thus

$$A(p,q) = \frac{(\ell+\frac{1}{2})}{\pi} \int_0^\infty d\beta \sinh\beta\, Q_\ell(\cosh\beta)\, f(\cosh\beta) \tag{III.22}$$

Let us introduce the notation

$$b(\ell) = \frac{2\pi}{2\ell+1} A(p,q) \tag{III.23}$$

The pair of formulae, Eq. (III.1) and Eq. (III.22) then take the form

$$f(x) = \frac{1}{2\pi i} \int_{p-i\infty}^{p+i\infty} d\ell\,(2\ell+1)\, b(\ell)\, P_\ell(x) \tag{III.24}$$

$$b(\ell) = \int_1^\infty dx\, Q_\ell(x)\, f(x) \tag{III.25}$$

It should be observed that all the steps in the derivation of the inversion formula (III.25) are valid for functions which (in addition to satisfying certain smoothness conditions) behave asymptotically as

$$f(x) = O(x^{p-\epsilon}) \tag{III.26}$$

where p is fixed in the interval (III.2) and ϵ is an arbitrarily small positive number. It should be emphasized that the result given in Eqs. (III.24) and (III.25) does not mean that we have shown that functions behaving asymptotically as given by Eq. (III.26) can be expanded as in Eq. (III.24) with the expansion coefficient $b(\ell)$ given by Eq. (III.25). What we have done so far is to assume that the functions f(x) we consider can be represented by the expression (III.24), and then shown that a suitable candidate for the expansion coefficient $b(\ell)$ is given by Eq. (III.25). We have also shown that the expansion coefficient $b(\ell)$ in the formula (III.24) is not uniquely determined by the function f(x) so that the particular choice given by Eq. (III.25) is but one member of an equivalence class of coefficients $b(\ell)$ (the members of this equivalence class differing by functions $\Delta(\ell)$ of the type discussed previously).

On this level of sophistication it is not difficult to justify the validity of the Eqs. (III.24) and (III.25) even for asymptotically growing functions. A "rigorous" argument would be somewhat lengthy, so we shall merely give a plausibility argument. The expansion coefficient $b(\ell)$ defined by Eq. (III.25) is analytic in ℓ in the whole right-hand ℓ-plane. We may therefore shift the contour of integration in Eq. (III.24) to the right as far as we please. The formulae (III.24) and (III.25) then make sense for functions behaving asymptotically as given by Eq. (III.26), where p now is any finite positive number.

Let us stress again that the derivation of the Eqs. (III.24) and (III.25) does not constitute a complete proof of the generalized O(2,1) expansion theorem. What remains to be done is to formulate conditions on the class of functions that are to be represented by Eq. (III.24), which would ensure the convergence of the representation. The most general conditions of that kind are global conditions on the functions to be expanded, which would ensure the convergence of the representation (III.24) in the mean (in the sense of a suitably defined norm). For physical applications it is, however, of greater interest to find global and local conditions on the functions to be expanded, which ensure point-wise convergence of the representation (III.24). Considerations of this kind will be left for future communications.[1)]

Let us finally examine the connection between the formulae (III.24) and (III.25), and the Sommerfeld-Watson representation of the amplitude considered in Sec. II. At first sight there appears to be a contradiction between Eq. (III.24) and the Sommerfeld-Watson representation. We can for instance move the integration contour to the right at will in Eq. (III.24) without changing the form of the representation, whereas in the SW-representation we pick up discrete terms (due to the term cosec $\pi\ell$ in the integrand) when we move the integration contour to the right. This contradiction is only apparent, as can be seen by using the freedom of adding non-contributing functions $\Delta(\ell)$ to the integrand in Eq. (III.24) (or to the integrand in the SW-representation). Let us for simplicity demonstrate this fact only for the case when p satisfies the condition (III.2).

We shall then again assume that the amplitude f(x) is given by the fixed t dispersion relation (II.5). Inserting Eq. (II.5) in Eq. (III.25) we obtain

$$b(\ell) = \frac{2}{t - 4m^2} \int_{4m^2}^{\infty} ds' \, A_s(s', t) \frac{1}{\pi} \int_1^{\infty} \frac{dx \, Q_\ell(x)}{x + \xi}$$

$$\xi \equiv 1 - \frac{2s'}{4m^2 - t} + i\epsilon \qquad \text{(III.27)}$$

From the analytic properties of the Q_ℓ-function follows the dispersion relation

$$\frac{1}{\pi} \int_1^{\infty} \frac{dx \, Q_\ell(x)}{x + \xi} = \frac{1}{\sin \pi\ell} \left\{ \frac{1}{2} \int_{-1}^{+1} \frac{dx \, P_\ell(x)}{\xi - x} - Q_\ell(\xi) \right\} \qquad \text{(III.28)}$$

However, the first term on the right-hand side of Eq. (III.28) does not contribute to the integral (III.24) since this term (together with the factor $2\ell+1$) is of the type $\Delta(\ell)$ discussed previously. We may thus replace the expression (III.27) by the equivalent expression,

$$b'(\ell) = \frac{2(4m^2 - t)^{-1}}{\sin \pi\ell} \int_{4m^2}^{\infty} ds' \, A_s(s', t) \, Q_\ell(\xi) \qquad \text{(III.29)}$$

Using the relations (B.140) we get,

$$b'(\ell) = \frac{2(t - 4m^2)^{-1}}{\sin \pi\ell} e^{-i\pi\ell} \int_{4m^2}^{\infty} ds' \, A_s(s', t) \, Q_\ell\left(\frac{2s'}{4m^2 - t} - 1 - i\epsilon\right) \qquad \text{(III.30)}$$

Comparing Eq. (III.30) with (II.13) we see that there is perfect agreement between the generalized O(2,1) expansion and the SW-representation also for Re $\ell \neq -\frac{1}{2}$. The argument outlined above can be carried through for a general positive value of p as well, with unimportant changes in detail.

IV. Summary

The main result of this paper is the derivation of a generalized O(2,1) expansion formula for asymptotically growing amplitudes. It was shown that the generalized O(2,1) expansion coefficient is not unique, but that to each function to be expanded corresponds a whole equivalence class of expansion coefficients. Conditions which would ensure convergence (pointwise or in the mean) of the generalized O(2,1) representation are not given, but will be elucidated in forthcoming communications.[1)]

We have further shown that the generalized O(2,1) expansion agrees with the Sommerfeld-Watson representation of the amplitude, defined without signature, in the sense that the expansion coefficient in the SW-representation (which is essentially the partial wave amplitude) belongs to the equivalence class of O(2,1) expansion coefficients.

Acknowledgments

The author wishes to express his appreciation to Dr. Carl Kaysen for the hospitality of the Institute for Advanced Study.

It is also a pleasure to thank Professor M. L. Goldberger and Professor P. Winternitz for valuable discussions, and Professor S. Adler for his reading of the manuscript.

References

1. C. Cronström and W. H. Klink (in preparation).
2. J. F. Boyce, J. Math. Phys. 8, 675 (1967).
3. H. D. I. Abarbanel and L. M. Saunders, Princeton preprint (March 1970).
4. C. E. Jones, F. E. Low and J. E. Young, MIT preprint (April 1970).
5. V. de Alfaro and T. Regge, Potential Scattering (North-Holland Publishing Co., Amsterdam, 1965), Chap. 9.
6. P. Winternitz, Lectures in the Fourth Finnish Summer School in Theoretical Physics 1968 (Gordon and Breach, New York, 1969).
7. P. Winternitz, Lectures at the Summer School on Group Representations and Quantum Theory, Dublin, Ireland, 1969.
8. P. Winternitz, talk given at this Conference.
9. F. G. Mehler, Math. Ann. 18, 161 (1881).

10. V. A. Fock, CR (Doklady) Acad. Sci. U.R.S.S. (N.S.) 39, 253 (1943).
11. Bateman Manuscript Project, Vol. 1, Higher Transcendental Functions, edited by A. Erdelyi et al (McGraw-Hill, New York, 1953).
12. A. O. Barut, The Theory of the Scattering Matrix (MacMillan, New York, 1967), Section 9.1.
13. T. K. Gaisser and C. E. Jones, Phys. Rev. 184, 1602 (1969).
14. E. T. Whittaker and G. N. Watson, A Course of Modern Analysis (Cambridge, 1927), Chap. 11.

EXPANSION THEORY AND THE LORENTZ GROUPS IN NON-CANONICAL BASES†

N. W. Macfadyen‡
Physics Department
Carnegie-Mellon University
Pittsburgh, Pennsylvania

I. Introduction

There are several reasons for studying the reduction O(3,1) ⊃ O(2,1): the intrinsic interest of the topic; the elucidation of the role of "second-kind" functions;[1] the possible application to Regge theory; and, most pertinent to this symposium, the study of how things can go wrong if we choose an unusual basis. Since there are so many reductions of SO(4,2), it may be of interest to examine the unexpected behavior of such a well known group as SL(2,C) in an SU(1,1) basis.

The results I shall present are by no means all new: the reductions SL(2,C)⊃ SU(1,1)[2] and SL(2,R) ⊃O(1,1)[3] have been treated by many people, and I cannot mention them all here. As far as I am aware, the actual use of the matrix elements of finite transformations in this basis to expand functions defined over the group <u>is</u> new, as is also the brief summary I shall give of the chain SL(2,C) ⊃ SU(1,1) ⊃ O(1,1).[4] This is not the place to give explicit proofs or detailed arguments, and so these will be almost completely absent. They can be found in the references.

II. The Reduction SL(2,R) ⊃ O(1,1)

Because this problem displays so many features of the higher dimensionality without also having its complexities, I shall go into most detail here and simply give an umbrella assurance that proofs for that more interesting case follow the same lines.[1] I apologize to all those to whom this is quite familiar.

Consider then a representation j of SL(2,R). Instead of letting the operators act on special functions defined over a homogeneous space of the group--i.e.,

---

†Presented at the Symposium on De Sitter and Conformal Groups, University of Colorado, Summer 1970.
‡Present address: DAMTP, University of Cambridge, England.

$$T_r:\ Y_j^{\,m}(\Omega) = Y_j^{\,m}(\Omega r)\ , \tag{1}$$

it is much more convenient to carry the j-dependence in the action of the operators T_r, and let the basis functions be simple. Following the Russian school, we set up our representation on a space of functions defined over a hyperbola we shall parametrize by β_τ: the τ-label specifies the sheet, and $\beta \in (-\infty, \infty)$ the position thereon, and the operators $T_r^{\,j}$ are specified by

$$[T_r^{\,j}: f]^\tau (\beta) = |\lambda|^{2j} (\text{sign}\,\lambda)^\nu f^{\tau'} (\beta') \tag{2}$$

where the parameters are defined uniquely by

$$\begin{pmatrix} & 1 \\ -1 & \end{pmatrix}^{\rho} \begin{pmatrix} \text{ch}\,\beta/2 & \text{sh}\,\beta/2 \\ \text{sh}\,\beta/2 & \text{ch}\,\beta/2 \end{pmatrix} \begin{pmatrix} r_{11} & r_{12} \\ r_{21} & r_{22} \end{pmatrix} =$$

$$\begin{pmatrix} \lambda^{-1} & \mu \\ & \lambda \end{pmatrix} \begin{pmatrix} & 1 \\ -1 & \end{pmatrix}^{\rho'} \begin{pmatrix} \text{ch}\,\beta'/2 & \text{sh}\,\beta'/2 \\ \text{sh}\,\beta'/2 & \text{ch}\,\beta'/2 \end{pmatrix} \tag{3}$$

or, in matrix notation, $\epsilon^\rho \beta r = k\epsilon^{\rho'} \beta'$. The parameters τ and ρ are discrete; $\tau = \pm 1$ only and $\rho = \frac{1}{2}(1 - \tau)$. The label ν is either 0 or 1, depending on whether the eigenvalues of J_3 in the given representation are integer or half-integer.

It is easy to check that this is indeed a representation of SL(2,R), and clearly we have diagonalized the generator K_1, conjugate to the boost β: that is,

$$[T^{\,j}_{\beta_1} : f]^\tau(\beta) = f^\tau (\beta + \beta') \quad . \tag{4}$$

The new feature is the appearance of the label τ. We can look at this in two ways, of which the simpler is algebraic: the operators $\Delta = J_3^{\,2} - K_1^{\,2} - K_2^{\,2}$ and K_1 do not form a maximal Abelian set but can be augmented by a reflection R which in the standard homomorphism of SL(2,R) and O(2,1) has the significance of a change of sign of the 2-axis:

$$R: (x_0, x_1, x_2) = (x_0, x_1, -x_2). \tag{5}$$

The other way of regarding this is as a corollary of Gel'fand's horospheric method--it is exactly analogous to the two terms required[4),5)] to expand a function defined over a hyperboloid with a hyperbolic coordinate system. In either case, (2) tells us that provided the representation j is irreducible, each representation μ of SO(1,1) is doubly degenerate therein. We shall treat principal series representations: if Re $j = -\frac{1}{2}$, we can introduce an inner product

$$(f,g) = \frac{1}{2\pi} \sum_{\tau} \int_{-\infty}^{\infty} \overline{f^{\tau}(\beta)}\ g^{\tau}(\beta)\ d\beta \tag{6}$$

and then provided $f(\beta)$ belongs to a certain space of functions $\mathcal{D}_j$, the representation is both unitary and irreducible.

Having thus set up a representation, we must choose a set of basis functions. Since the τ are discrete labels, these are obviously also 2-vectors, and a convenient choice is just

$$\varphi^{+}_{\mu}(\beta) = \begin{pmatrix} e^{i\mu\beta} \\ 0 \end{pmatrix} \qquad \varphi^{-}_{\mu}(\beta) = \begin{pmatrix} 0 \\ e^{i\mu\beta} \end{pmatrix} . \tag{7}$$

Obviously many other choices are possible: a different one with some advantages (particularly for the supplementary series of representations) is to take the sum and difference of these, corresponding to eigenvalues of the reflection R. Clearly, $(\varphi^{\tau}_{\mu}, \varphi^{\tau'}_{\mu'}) = \delta_{\tau\tau'}\delta(\mu-\mu')$.

The vectors actually do not belong to $\mathcal{D}_j$ because they are not square-integrable, but we can regard them as members of the dual space in the usual way.

III. Matrix Elements

Let us first consider what we have to calculate. It is found that the parametrization $r = \beta b \beta'$ covers SL(2,R) only with three choices of b: convenient ones are:

$$b_1 \equiv \xi = \begin{pmatrix} e^{\xi/2} & \\ & e^{-\xi/2} \end{pmatrix} \qquad b_2 \equiv \xi\epsilon = \begin{pmatrix} & e^{\xi/2} \\ -e^{-\xi/2} & \end{pmatrix}$$

$$b_3 \equiv \theta = \begin{pmatrix} \cos\theta/2 & \sin\theta/2 \\ -\sin\theta/2 & \cos\theta/2 \end{pmatrix} \tag{8}$$

with $-\infty < \xi < \infty$, $-2\pi \leq \theta < 2\pi$. The first double coset class is exactly the analogue of the single one in the canonical parametrization $r = \theta\,\xi\,\theta'$, and we calculate its matrix elements in a straightforward manner. Clearly these will have a pair of τ-labels attached; indeed, by (2) we find

$$[T^j_\xi : \varphi^+]^+ (\beta) = (e^{-\xi}\, ch^2\, \beta/2 - e^{\xi}\, sh^2\, \beta/2)^j\, \varphi^+(\beta')$$

$$[T^j_\xi : \varphi^+]^- (\beta) = 0$$

$$[T^j_\xi : \varphi^-]^- (\beta) = (e^{\xi}\, ch^2\, \beta/2 - e^{-\xi}\, sh^2\, \beta/2)^j\, \varphi^-(\beta'')$$

$$[T^j_\xi : \varphi^-]^+ (\beta) = (-\text{sign}\, \beta)^\nu\, (e^{\xi}\, sh^2\, \beta/2 - e^{-\xi}\, ch^2\, \beta/2)^j\, \varphi^-(\beta''') \qquad (9)$$

where

$$th\, \beta'/2 = e^{\xi}\, th\, \beta/2$$

$$th\, \beta''/2 = e^{-\xi}\, th\, \beta/2$$

$$th\, \beta'''/2 = e^{-\xi}\, coth\, \beta/2 \quad . \qquad (10)$$

Notice that although the subspace $\tau = +$ is invariant, its complement $\tau = -$ is not so; for $\xi < 0$ the situation is reversed. Inserting the basis functions we find

$$d^{j--}_{\mu'\mu}(\xi) = \frac{\Gamma(-j-i\mu')\,\Gamma(-j+i\mu')}{\Gamma(-2j)} \cdot \frac{1}{2\pi}\, (sh\, \xi/2)^{2j}\, (th\, \xi/2)^{-i(\mu+\mu')} \times$$

$$\times F(-j+i\mu, -j+i\mu';\; -2j;\; -1/sh^2\, \xi/2) \qquad (11)$$

$$d^{j++}_{\mu'\mu}(\xi) = \frac{\Gamma(j+i\mu+1)\,\Gamma(j-i\mu+1)}{\Gamma(j+i\mu'+1)\,\Gamma(j-i\mu'+1)}\, d^{-j-1--}_{\mu'\mu}(\xi) \qquad (12)$$

$$d^{j-+}_{\mu'\mu}(\xi) = 0$$

$$\cos\pi(j+\nu/2)\, d^{j+-}_{\mu'\mu}(\xi) = \cos\pi(i\mu'-\nu/2)\, d^{j--}_{\mu'\mu}(\xi) - \cos\pi(i\mu-\nu/2)\, d^{j++}_{\mu'\mu}(\xi) \; . \qquad (13)$$

All this is valid for $\xi > 0$; if $\xi < 0$ we find

$$d^{j\tau\tau'}_{\mu\mu'}(-|\xi|) = d^{j-\tau'-\tau}_{\mu\mu'}(|\xi|) = d^{-j-1\ \tau'\tau}_{\mu'\mu}(|\xi|) \quad . \tag{14}$$

Now notice what effect this decomposition has had upon the matrix elements. Consider in particular the integral defining (11):

$$d^{j--}_{\mu'\mu}(\xi) = \frac{1}{2\pi}\int_{-\infty}^{\infty} e^{-i\mu'\beta}\,(e^{\xi}\,\mathrm{ch}^2\,\beta/2 - e^{-\xi}\,\mathrm{sh}^2\,\beta/2)^j\, e^{i\mu\beta''}\, d\beta\ . \tag{15}$$

The term in parentheses is always positive; and since it has a minimum value of unity, the asymptotic behavior in the complex j-plane (or the ξ-plane) comes entirely from the ends of the integration, at both of which it has behavior $e^{j\xi}$. This agrees with (11). Similarly, (12) has specific and simple behavior of the opposite kind: only (13) is of mixed type. Closer examination shows that (12) and (11) are exactly the classical "second kind" functions for the group SL(2,R), analytically continued to imaginary values of $m = i\mu$ (that is, modulo irrelevant phase factors). The significance of (13) we shall see shortly. Notice that for $\xi = 0$ the first two matrix elements degenerate into $\delta(\mu-\mu')$.

We just mention that the matrix elements of the other double-coset classes can be calculated similarly. We find

$$\mathfrak{D}^{j\tau\tau'}_{\mu\mu'}(\xi\epsilon) = (-1)^{\nu(\rho'+1)}\, d^{j\tau-\tau'}_{\mu-\mu'}(\xi)\ , \tag{16}$$

and $\mathfrak{D}^{j++}_{\mu'\mu}(\theta)$ turns out[1] to be just a representation function of SU(2) after analytic continuation in j, μ, μ'. The discrete series k^t, $t = \pm$, of representations of SL(2,R) behave similarly; but O(1,1) is not degenerate in any of these representations and consequently all the matrix elements vanish if sign $(\tau\mu) \neq t$. Those remaining are exactly the continuations of the principal continuous series matrix elements.

Now let us return to (12) and (13). Since these relate expressions involving j and -j-1, which label equivalent representations, we expect that a study of the intertwining operator will be useful. Recall that this is an isometric operator A: $\mathfrak{D}_j \to \mathfrak{D}_{-j-1}$, which satisfies

$$A\, T^j_r = T^{-j-1}_r\, A \qquad \forall\ r \in SL(2,R)\ . \tag{17}$$

We ask how A transforms our pseudo-basis states, and find that the only possibility is

$$A: \varphi_\mu^{j\tau} = \alpha_\mu^\tau \varphi_\mu^{-j-1\ \tau} \quad \oplus \quad \alpha_\mu'^\tau \varphi_\mu^{-j-1-\tau} \tag{18}$$

with the relations

$$\alpha^+_\mu = \alpha^-_{-\mu} \equiv \alpha_\mu$$

$$\alpha'^+_\mu = (-1)^\nu \alpha'^-_{-\mu} \equiv \alpha'_\mu \ . \tag{19}$$

Further, we know an integral representation of A in the reduction $SL(2,R) \supset E_2$:

$$[A:f](x) \propto \int_{-\infty}^{\infty} |x-x'|^{-2j-2} \operatorname{sign}^\nu(x-x')\, f(x')\, dx' \ .$$

Together with the requirement of isometry, this specifies the intertwining coefficients α up to a pure phase; carrying out the integrations we find[1)]

$$\alpha_\mu = \pi^{-1}\,\Gamma(j+i\mu+1)\,\Gamma(j-i\mu+1)\ \cos\pi(i\mu - \nu/2)$$

$$\alpha'_\mu = -\pi^{-1}\,\Gamma(j+i\mu+1)\,\Gamma(j-i\mu+1)\ \cos\pi(j-\nu/2)\ . \tag{20}$$

In the canonical basis $SL(2,R) \supset O(2)$ we should have found $A:\psi^j_m = a_m \psi_m^{-j-1}$, where with this normalization[4)]

$$a_m = \pi^{-1}\,\Gamma(j+m+1)\,\Gamma(j-m+1)\ \sin\pi(m-j)\ e^{-i\pi\ \nu/2} \tag{21}$$

Now return to (17) and calculate the matrix elements when r belongs to the first coset class; we obtain two equations

$$\alpha'_{\mu'}\, d^{j++}_{\mu'\mu}(\zeta) = \alpha'_\mu\, d^{-j-1--}_{\mu'\mu}(\zeta) \tag{22}$$

$$\alpha'_\mu\, d^{-j-1+-}_{\mu'\mu}(\zeta) = \alpha_{\mu'}\, d^{j++}_{\mu'\mu}(\zeta) - \alpha_\mu\, d^{-j-1++}_{\mu'\mu}(\zeta) \tag{23}$$

which upon inserting the values of α, α' are just (12) and (13). Clearly an analogous result holds for the second coset class--we just use (16); but it fails for the third, which we called θ, since then none of the function is of simple behavior in the j-plane and none vanishes--and it was precisely the vanishing of d^{j-+} ($\xi > 0$) that gave the usefulness of (22). Therefore this class, which has no analogue in the decomposition $SL(2,R) \supset O(2)$, has no meaningful second-kind decomposition at all. This is of interest.

IV. The Reduction $SL(2,C) \supset SU(1,1) \supset O(2)$

Everything here follows exactly as in the last section; we define a representation by

$$[T_a^x : \varphi]^\tau (v) = \lambda^{-j_o-\sigma-1} \bar{\lambda}^{j_o-\sigma-1} \varphi^{\tau'}(v') \tag{24}$$

where

$$\epsilon^\rho va = k\epsilon^{\rho'} v' \qquad x \equiv \{j_o, \sigma\} ,$$

$v \in SU(1,1)$, $a \in SL(2,C)$, k is the complexification of the k of the last section, $2j_o$ is integral and σ imaginary. The complicating feature is that $\varphi^\tau(v)$ must now be expanded in terms of the representation functions $\mathcal{D}^j_{\tau j_o, m}(v)$ of SU(1,1)--i.e., in shorthand notation

$$f^\tau(v) = \int dM(j) \sum_m \mathcal{D}^j_{\tau j_o, m}(v)\, \tilde{f}^{j\tau}_m \tag{25}$$

where $\int dM(j)$ stands for summing over the discrete and integrating over the continuous principal series. There are exactly the same double coset classes as before, except for an important caveat we shall mention shortly, and we can calculate the matrix elements exactly as before. Thus, for $\xi > 0$ we obtain

$$\begin{aligned} d^{x--}_{\ell m j}(\xi) = {} & (-1)^{m+j_o} \left\{ \frac{\Gamma(j_o-\ell)\Gamma(m-\ell)\Gamma(j+j_o+1)\Gamma(j+m+1)}{\Gamma(-j_o-\ell)\Gamma(-m-\ell)\Gamma(j-j_o+1)\Gamma(j-m+1)} \right\}^{\frac{1}{2}} \\ & \times \frac{\Gamma(\ell-\sigma+1)}{(m+j_o)!} (2\,\mathrm{sh}\,\xi)^{\ell-j} e^{-\xi(m+j_o-\sigma+\ell-j+1)} \\ & \times \sum \frac{(m+\ell+1)_t\,(j_o+\ell+1)_t}{\Gamma(m+j_o-\sigma+\ell+2+t)t!} \\ & \times F(m+j_o+\ell-j+t+1,\ \ell-\sigma+1;\ m+j_o+\ell-\sigma+2+t;\ e^{-2\xi}) \\ & \times {}_4F_3(m-j,\ j_o-j,\ -m-j_o-t,\ -t;\ -\ell-m-t,\ -\ell-j_o-t,\ m+j_o+1;\ e^{-2\xi}) \end{aligned} \tag{26}$$

and

$$d^{x++}_{\ell mj}(\xi) = d^{-x--}_{jm\ell}(\xi) \quad . \tag{27}$$

Many more relations and identities can be found in Ref. 1; they all bear a strong similarity to those of the last section, although like (27) some are simpler because of the normalization of the basis functions. We need not give them here. Let us proceed at once to examine the asymptotic behavior of (26) as $\xi \to \infty$; from the defining integral we see that this is just $e^{\xi\sigma}$, with no term in $e^{-\xi\sigma}$, and so we can just sum the leading terms of the series to obtain

$$d^{x--}_{\ell mj}(\xi \to \infty) \sim \{\ \}^{\frac{1}{2}} \frac{\Gamma(\ell-\sigma+1)\Gamma(-\ell-\sigma)}{\Gamma(j_o-\sigma+1)\Gamma(m-\sigma+1)(m+j_o)!} e^{-\xi(m+j_o-\sigma+1)} \tag{28}$$

As we expected, this is exactly the form of a classical second-kind function, and once again closer inspection shows that is just the analytic continuation (in ℓ and j) of that function. Therefore (26) and (27) play exactly the role for SL(2,C) that their sisters (11) and (12) played for the real group. We can of course find the function $d^{x+-}(\xi)$ directly by integration, but the result is very complex and it is better to use equivalence. Now the representations $x = \{j_o,\sigma\}$ and $-x = \{-j_o,-\sigma\}$ are equivalent, and we can define intertwining coefficients

$$\alpha^{+}_{j} = \alpha^{-}_{j} \equiv \alpha_j$$

$$\alpha'^{+}_{j} = (-1)^{\nu} \alpha'^{-}_{j} \equiv \alpha'_j$$

and deduce

$$\alpha'_\ell \, d^{x++}_{\ell mj}(\xi) = \alpha'_j \, d^{-x--}_{\ell mj}(\xi) \tag{29}$$

$$\alpha^{x+-}_{\ell mj}(\xi) = \beta_\ell(-x)\, d^{x--}_{\ell mj}(\xi) - \beta_j(-x)\, d^{x++}_{\ell mj}(\xi) \tag{30}$$

where $\beta_j(x) = \alpha_j(x)/\alpha'_j(x)$. The coefficients are very difficult to calculate directly but can be found by examining the asymptotics of the d-functions:

$$\alpha_j = \pi^{-1}\,\Gamma(j+\sigma+1)\,\Gamma(\sigma-j)\,\sin\pi(j-j_o)$$

$$\alpha'_j = \pi^{-1}\,\Gamma(j+\sigma+1)\,\Gamma(\sigma-j)\,\sin\pi(\sigma-j_o) \quad . \tag{31}$$

Knowing these, we can express all of the matrix elements of the first two classes by means of (26). Only the third double coset class θ remains, and it seems impossible to calculate its matrix elements in this basis.

V. $SL(2,C) \supset SU(1,1) \supset O(1,1)$

But in this reduction[4] only the third coset class is amenable to calculation. The problem reduces to the previous reduction with different basis vectors, which are specified by the covariance condition arising from the ambiguity in the phase of λ in (24) together with the requirement that K_1 be diagonalized: this implies that the basis vectors $\varphi^\tau(v)$ satisfy

$$\varphi^\tau(e^{i\alpha J_3}\, v\, e^{i\beta K_1}) = e^{ij_o\tau\alpha}\, \varphi^\tau(v)\, e^{i\mu\beta} \,, \tag{32}$$

which means that φ^τ is a "cross-basis" matrix element of $SU(1,1)$ which we can label schematically as $\langle O(2)|\, e^{i\xi K_2}\, |O(1,1)\rangle$. Because $O(1,1)$ is degenerate in a representation of $SU(1,1)$, we actually need another label $t = \pm 1$ to specify which particular subgroup we mean, as in Section II, so that our pseudo-basis elements can be labeled $\mathfrak{D}^{jt}_{\tau j_o, \mu}(v)$. These functions are properly defined and discussed elsewhere[4]--here we only note that they are complete and orthogonal in all labels. The representation functions of $SL(2,C)$ therefore have two pairs of discrete labels on them, corresponding to the pair of discrete operators

$$\begin{aligned} T &: (x_o, x) = (x_o, x_1, x_2, -x_3) \\ R &: (x_o, \underset{\sim}{x}) = (x_o, x_1, -x_2, -x_3) \,, \end{aligned} \tag{33}$$

and can conveniently be written $\mathfrak{D}^{\chi\tau\tau'}_{\ell\mu t,\, j\mu' t'}(a)$. These can be calculated explicitly for a belonging to the third coset θ, and bear distinct resemblances to (26), but of course are not of "second-kind" behavior at all. The reduction has the interesting feature, however, that apart from the matrix elements of the third coset classes in both $SL(2,C)$ and $SU(1,1)$, all the other d-functions are of simple behavior in the group elements. This would seem to be the ultimate that we can hope for in the way of second-kind decompositions.

VI. Expansion Theorems

At last we mention applications--which means expansion theorems over the group. For $SU(1,1) \supset O(1,1)$ everything is well behaved, although we do of course obtain two terms, exactly as we do in expansions over a hyperboloid. For $SL(2,C) \supset SU(1,1) \supset O(2)$, however, unexpected complications occur: specifically, we find that we must introduce new functions, that are not representation functions, for the coset class θ if we are to obtain orthogonality and completeness relations. The new function occurs only in the generalized partial wave projection formula, not in the inversion; for we still find

$$f(a) = \sum_{\tau\tau'} \int dM(\ell) dM(j) \sum_{mm'} \int dx \, \mathfrak{D}^{x\tau\tau'}_{\ell mjm'}(a) \, \tilde{f} \ , \tag{34}$$

but now

$$\tilde{f} = \int d\mu(a) \, \overline{\tilde{\mathfrak{D}}^{x\tau\tau'}_{\ell mjm'}(a)} \, f(a) \tag{35}$$

where $\tilde{\mathfrak{D}}(a) = \mathfrak{D}(a)$ if a is in the first two cosets, but is our new function if it is in the third. The cause of this remarkable phenomenon is easy to find: it lies in the parametrization of the group. We set $a = v\, b\, v'$, and this is in general a seven parameter set, but reduces to six if b is ξ or $\xi\epsilon$. Therefore a straightforward integral over v, v′ and θ will cover some cosets more than once, and so in deriving (35) we find at one point that a measure enters that is not the invariant measure over SU(1,1)--and that therefore the projection function is not a representation function. Clearly the difficulty does not occur for $SL(2,R) \supset O(1,1)$; but it shows that the orthogonality of representation functions is not something to be taken for granted, and such problems may be expected to occur for SU(2,2) in at least some reductions.

Finally, let us note the suggestive similarity this has to the result[6] that a second-kind transform formula exists only for SL(2,R), whereas the inversion holds for both this and SL(2,C).

References

1. N. W. Macfadyen, "The Lorentz Groups and Second-Kind Functions," Carnegie-Mellon Univ. preprint, to be published.
2. N. Mukunda, J. Math. Phys. 9, 50 (1968); S. Ström, Ark. Fys. 34, 215 (1967); A. Sciarrino and M. Toller, J. Math. Phys. 8, 1252 (1967).
3. N. Mukunda, J. Math. Phys. 8, 2210 (1967); ibid. 10, 2092 (1969). N. Ya. Vilenkin, Group Representations and Special Function Theory (American Math. Soc. Monographs, Providence, R. I., 1969). A. O. Barut and E. C. Phillips, Comm. Math. Phys. 8, 52 (1968).

4. N. W. Macfadyen, "The Reduction $O(3,1) \supset O(2,1) \supset O(1,1)$," Carnegie-Mellon Univ. preprint, to be published.
5. N. Ya. Vilenkin and Y. A. Smorodinskii, Zh. E.T.P. 46, 1793 (1964) [Soviet Physics JETP 19, 1209 (1964)].
6. W. Rühl, The Lorentz Groups and Harmonic Analysis (Benjamin, New York, 1969).

TWO-VARIABLE EXPANSIONS
BASED ON THE LORENTZ AND CONFORMAL GROUPS†

P. Winternitz‡
Department of Physics
Carnegie-Mellon University
Pittsburgh, Pennsylvania 15213

Abstract

A review is given of a scattering theory based on two-variable expansions furnished by the Lorentz group. Arguments are given for using the conformal group in such a program, in order to make as much use as possible of relativistic invariance and crossing symmetry simultaneously.

Introduction

The aim of this paper is to give a short review of the present status of a scattering theory, based on two-variable expansions of scattering amplitudes, furnished by an application of the representation theory of the homogeneous Lorentz group O(3,1) and to indicate how further developments of this approach to particle scattering lead us to a consideration of the conformal group of space-time, or rather the group O(4,2).

Much of what is contained in this report was published in a series of original papers[1)-17)] but was reviewed as a whole only in unpublished lectures,[18)] since which there have been further developments.

The reason for writing two-variable expansions has been discussed in the above publications. Let us just repeat the main arguments which are essentially the same as for performing any type of

†Presented at the Symposium on De Sitter and Conformal Groups, University of Colorado, Summer 1970.

‡On leave from Nuclear Research Institute, Czechoslovak Academy of Sciences, Řež, Czechoslovakia. Present address: Department of Physics, University of Pittsburgh, Pittsburgh, Pennsylvania.

direct channel or crossed channel partial wave analysis. Namely, by using group theory to capitalize on the symmetries of the space-time and momentum framework, in which the reactions occur, we wish to treat as much as possible of the kinematics in a general fashion, to transfer the dynamics to the expansion coefficients, which then serve as a tool for making physical assumptions (similar to the reggeized partial wave amplitudes in complex angular momentum theory). The fact that all the dependence on the kinematic parameters (e.g. the Mandelstam variables s, t and u) is contained in known function should make it possible to describe a large amount of data in terms of a few parameters. Further, such expansions can serve as a tool for imposing general principles, like Lorentz invariance, crossing symmetry, unitarity, etc.

We call the coefficients in the O(3,1) two-variable expansions "Lorentz Amplitudes" and the hope underlying all attempts to develop such a theory is that these Lorentz amplitudes are an adequate tool, having reasonable, and in some sense simple, properties.

In Section I we give a brief exposition of the O(3,1) expansions which have so far been considered and mention some of their implications. Section II is devoted to a comparison between our expansions and other two-variable expansions, existing in the literature,[19)-21)] which are based on an SU(3) group. In Section III we show how the group O(4,2) appears in the context of two-variable, or more generally, multi-variable expansions and give some preliminary discussion of O(4,2) expansions.

I. Lorentz Group Expansions

A. Subgroup and non-subgroup type expansions

The general way in which we obtain expansions of scattering amplitudes consists of three steps:

1. We construct a mapping of the physical region of the Mandelstam plane (for two-body reactions) or its generalization (for many-body reactions) onto a homogeneous space of a certain group, in such a manner as to be able to consider the scattering amplitude to be a function of a single point in this space.

2. We choose convenient coordinates on the homogeneous space and find a complete set of generalized harmonic functions of the group.

3. Making use of completeness and orthogonality relations (or of a generalized Plancherel formula), we expand the scattering amplitude in terms of these harmonic functions.

The first step depends on the frame of reference in which we consider the scattering and of course on the group and homogeneous space which we choose. The second step--choice of harmonic functions--is also in general far from unique. Indeed, given a group there are many different bases in which we can consider the group representations. We obtain basis functions by considering complete sets of commuting operators, containing the invariant operators (Casimir operators) and other operators lying in the enveloping algebra of the group algebra and possibly also some further ones. The basis functions are then obtained as the common eigenfunctions of such a complete set of operators. If we limit ourselves to second order operators in the enveloping algebra then we find[6)] that for any given Lie group only a finite number of such sets of commuting operators exists and that there is a one-to-one correspondence between these sets and coordinate systems, allowing the separation of variables in the Laplace operator on the corresponding homogeneous space. Further, it was shown that the simplest types of separable coordinates each correspond to a set of operators consisting of the Casimir operators of all subgroups figuring in a certain reduction of the considered group to its subgroups, whereas more complicated coordinate systems of the elliptic type correspond to other second order operators.

In the following paragraphs we shall consider the O(3,1) expansions, obtained by considering individual chains of subgroups and also a "non-subgroup" type of O(2,1) expansion. We shall limit ourselves to the two-body scattering of spinless particles, so that we have only one scattering amplitude, depending on four momenta p_1, ... p_4, satisfying

$$p_1 + p_2 = p_3 + p_4 \qquad p_i^2 = m_i^2 \qquad m_i > 0 \qquad i - 1,\dots,4 \tag{1}$$

and the dependence on the momenta is restricted by the requirement of Lorentz invariance. Instead of momenta p_i it is convenient to consider the relativistic velocities $v_i = p_i/m_i$, which for arbitrary masses satisfy

$$v^2 = v_0{}^2 - v_1{}^2 - v_2{}^2 - v_3{}^2 = 1 \qquad v_0 \geq 1 \tag{2}$$

and this upper sheet of a two-sheeted hyperboloid is the homogeneous space under O(3,1) which we shall be using.

B. <u>The O(3,1) ⊃ O(3) ⊃ O(2) Reduction</u>

This chain of group reduction corresponds to an introduction of spherical coordinates on the hyperboloid (2), so that the momenta can be written as

$$p_i = m_i(\cosh a_i,\ \sinh a_i \sin\theta_i \cos\varphi_i,\ \sinh a_i \sin\theta_i \sin\varphi_i,$$
$$\sinh a_i \cos\theta_i)$$

$$0 \le a_i < \infty, \quad 0 \le \theta_i \le \pi, \quad 0 \le \varphi_i < 2\pi \tag{3}$$

In order to construct a mapping of the Mandelstam variables s, t and u onto the hyperboloid, it is convenient to make use of the center-of-mass frame. Thus, we fix a time-like vector, the total energy momentum, to be $p_1 + p_2 = (\sqrt{s}, 0, 0, 0)$. Further we choose the coordinate axes so that $\vec{p}_1$ and $\vec{p}_2$ are parallel to the third axis and the first and third axes lie in the scattering plane.

Imposing these conditions, together with the conservation laws, we find that the components of p_1, p_2, and p_4 in (3) can be expressed in terms of a_3 and θ_3 ($\varphi_3 = 0$) so that the scattering amplitude can be written as

$$f(s,t) \equiv f(a_3, \theta_3) \equiv f(v) \tag{4}$$

i.e. a function of a point on the hyperboloid $v^2 = 1$ (or a function of one of the four momenta). It is easy to see that given the above choice of a frame of reference the above parameter $\theta = \theta_3$ is simply the c.m.s. scattering angle and $a = a_3$ is related to the total energy. The relation of a and θ to s and t for arbitrary masses was given previously.[8),18)] For equal masses $m_1 = \ldots = m_4 = \frac{1}{2}$ the formulae simplify to

$$\cosh a = \sqrt{s}, \qquad \cos\theta = 1 + \frac{2t}{s-1} \quad . \tag{5}$$

The basis functions can now be obtained as the common set of eigenfunctions of the Casimir operators of the group O(3,1) and the subgroups in the considered reduction, i.e.

$$\Delta_L \phi_{\sigma\ell m}(a,\theta,\varphi) = \sigma(\sigma+2)\, \phi_{\sigma\ell m}(a,\theta,\varphi)$$

$$L^2 \phi_{\sigma\ell m}(a,\theta,\varphi) = \ell(\ell+1)\, \phi_{\sigma\ell m}(a,\theta,\varphi)$$

$$L_3 \phi_{\sigma\ell m}(a,\theta,\varphi) = m\, \phi_{\sigma\ell m}(a,\theta,\varphi) \tag{6}$$

Here Δ_L, L^2, L_3 can simply be realized as the Laplace operators on the hyperboloid $v^2 = 1$, the sphere $v_0 = \text{const}$ and circle $v_0 = \text{const}$,

v_3 = const. Using spherical coordinates we can separate the variables in (6) and explicitly find the eigenfunctions. To normalize them is a somewhat more difficult task, performed[1] by using the methods of integral geometry. Finally, for scattering amplitudes, depending on a and θ only, we obtain the expansion:

$$f(a,\theta) = \sum_{\ell=0}^{\infty} (2\ell+1) \int_{\delta-i\infty}^{\delta+i\infty} (\sigma+1)^2 \, d\sigma \frac{\Gamma(\sigma+1)}{\Gamma(\sigma+1-\ell)} A_\ell(\sigma) \frac{1}{\sqrt{\sinh a}} P^{-\ell-\frac{1}{2}}_{\frac{1}{2}+\sigma}(\cosh a) \times$$
$$\times P_\ell(\cos\theta) \tag{7}$$

$$A_\ell(\sigma) = \frac{i}{4} \frac{\Gamma(-\sigma-1)}{\Gamma(-\sigma-1-\ell)} \int_0^\infty \sinh^2 a\,da \int_0^\pi \sin\theta\, d\theta\, f(a,\theta) \times$$
$$\times \frac{1}{\sqrt{\sinh a}} P^{-\ell-\frac{1}{2}}_{\frac{1}{2}+\sigma}(\cosh a)\, P_\ell(\cos\theta)\,. \tag{8}$$

Strictly speaking, these formulae are only valid for functions satisfying

$$\int_0^\infty \sinh^2 a\,da \int_0^\pi \sin\theta\, d\theta\, |f(a,\theta)|^2 < \infty \tag{9}$$

i.e. for amplitudes corresponding to total cross-sections limiting to zero as $s \to \infty$. Such functions can be expanded in terms of the basis functions of the irreducible unitary representations of the principal series,[22] corresponding to $\sigma = -1+ip$ in (7) and (8). In order to incorporate more general amplitudes, the expansions must be generalized to non-unitary representations, e.g. by considering more general integration paths[9] in formula (7). Let us note that the reason why only one of the two invariant operators of O(3,1) figures in the set (6) is that the other one is identically equal to zero for particles with spin zero.

C. <u>The $O(3,1) \supset O(2,1) \supset O(2)$ Reduction</u>

This group reduction corresponds to hyperbolic coordinates on the hyperboloid, in which the momenta are

$$p_i = m_i(\cosh\alpha_i \cosh\beta_i,\ \cosh\alpha_i \sinh\beta_i \cos\varphi_i,$$
$$\cosh\alpha_i \sinh\beta_i \sin\varphi_i,\ \sinh\alpha_i) \tag{10}$$

$$-\infty < \alpha_i < \infty \qquad 0 \le \beta_i < \infty \qquad 0 \le \varphi_i < 2\pi$$

Let us now consider the scattering in a brick wall (or Breit) frame, obtained by aligning a space-like vector--the momentum transfer $p_1 - p_3$ with the third axis $p_1 - p_3 = (0,0,0,\sqrt{-t})$ (we are considering $t < 0$ only). Further, let us choose the space axes such that $\vec{p}_1$ and $\vec{p}_3$ are parallel to the third axis and the first and third axes lie in the scattering plane. Such a choice of the reference frame is again sufficient, together with the conservation laws, to enable us to express all the momenta (10) in terms of one of them and thus to obtain the scattering amplitude as a function of a single point, i.e. $f(s,t) = f(\alpha,\beta) = f(v)$. This time the parameter α will be related to the momentum transfer and β will be related to the c.m.s. scattering angle in the crossed channel (in our case in the t-channel). The general relation between α,β and s,t was given for arbitrary masses elsewhere,[8),18)] for $m_1 = \ldots = m_4 = \frac{1}{2}$ we have

$$\sinh\alpha = \sqrt{-t} \qquad \cosh\beta = -1 + \frac{2s}{1-t} \tag{11}$$

The complete set of commuting operators, determining the basis functions, again consists of the Casimir operators of all groups in the reduction, supplemented this time by a discrete operator, corresponding to a reflection of the third axis. Diagonalizing these four operators as in (6), separating variables in hyperbolic coordinates, solving the equations and normalizing the obtained functions, we arrive at the expansion:

$$f(\alpha,\beta) = \frac{1}{16\sqrt{2}\,\pi} \int_{\gamma-i\infty}^{\gamma+i\infty} (2\ell+1)\cot\pi\ell\, d\ell \int_{\delta-i\infty}^{\delta+i\infty} (\sigma+1)^2\, d\sigma\, \frac{\Gamma(\sigma-\ell+1)\,\Gamma(\sigma+\ell+2)}{\Gamma(\sigma+2)} \times$$

$$\times \frac{1}{\cosh\alpha} \Big\{ A^+(\sigma,\ell) \Big[\mathbb{P}_\ell^{-\sigma-1}(-\tanh\alpha) + \mathbb{P}_\ell^{-\sigma-1}(\tanh\alpha) \Big] +$$

$$+ A^-(\sigma,\ell) \Big[\mathbb{P}_\ell^{-\sigma-1}(-\tanh\alpha) - \mathbb{P}_\ell^{-\sigma-1}(\tanh\alpha) \Big] \Big\}\, P_\ell(\cosh\beta) \tag{12}$$

$$A^{\pm}(\sigma,\ell) = \frac{\Gamma(\sigma-\ell-1)\,\Gamma(-\sigma+\ell)}{\sqrt{2}\,\Gamma(-\sigma)} \int_{-\infty}^{\infty} \cosh^2\alpha\, d\alpha \int_0^{\infty} \sinh\beta\, d\beta\, f(\alpha,\beta)$$

$$\cdot \frac{1}{\cosh\alpha} \Big\{ \mathbb{P}_{-\ell-1}^{\sigma+1}(-\tanh\alpha) \pm \mathbb{P}_{-\ell-1}^{\sigma+1}(\tanh\alpha) \Big\}\, P_\ell(\cosh\beta)\,. \tag{13}$$

Again, if the amplitude $f(\alpha,\beta)$ is square integrable with respect to the invariant measure over the hyperboloid, then only the unitary representations of the principal series will figure, i.e. $\ell = -\frac{1}{2} + iq$, $\sigma = -1+ip$, p,q are real. The functions $\mathbf{P}^{\mu}_{\nu}(x)$ are Legendre functions on the $(-1, +1)$ cut.

D. The $O(3,1) \supset E_2 \supset O(2)$ Reduction

Let us now introduce horospheric coordinates on the hyperboloid, putting

$$p_i = m_i(\cosh\gamma_i + \tfrac{1}{2}r_i^2 e^{-\gamma_i},\ r_i e^{-\gamma_i}\cos\varphi_i,\ r_i e^{-\gamma_i}\sin\varphi_i;$$

$$\sinh\gamma_i + \tfrac{1}{2}r_i^2 e^{-\gamma_i}) \tag{14}$$

$$-\infty < \gamma_i < \infty \qquad 0 \le r_i < \infty \qquad 0 \le \varphi_i < 2\pi \quad .$$

We shall consider the scattering in a frame of reference which we call the "light-velocity system." Since it should be adequate for the reduction into E_2, the group of motions of a Euclidean plane, which is the little group of a light-like vector, we wish to obtain a scattering frame by standardizing a light-like vector. A convenient choice, satisfying $K^2(s,t) = 0$, $K(s,t=0) = p_4 - p_2$ is:

$$K(s,t) = p_4 \frac{m_2}{m_4} e^{-A} - p_2 \qquad \cosh A = \frac{m_2^2 + m_4^2 - t}{2m_2 m_4} \tag{15}$$

Putting $K(s,t) = (\omega,0,0,\omega)$, where ω is an arbitrary scaling constant, choosing O13 as the scattering plane and putting the third axis parallel to $\vec{p}_2$ and $\vec{p}_4$, we find that we can again express all γ_i and r_i in (14) in terms of say γ_1 and r_1, thus obtaining $f(s,t) = f(\gamma,r) = f(v)$. This frame of reference was so constructed as to be meaningful particularly for nonequal mass scattering, when $m_1 \neq m_3$ and/or $m_2 \neq m_4$. The relation between s,t and γ,r is quite complicated and is given in previous references.[8),18)] The complete set of commuting operators consists of

$$\Delta_L,\ \theta = (K_1 + L_2)^2 + (K_2 - L_1)^2,\ L_3,$$

where K_i are boost generators and L_i rotation ones. Writing eigenfunction equations similar to (6), separating variables, normalizing, etc., we finally obtain the expansion formulae

$$f(\gamma,r) = \frac{1}{2\pi}\int_0^\infty kdk \int_{\delta-i\infty}^{\delta+i\infty} (\sigma+1)^2\, d\sigma \frac{1}{\Gamma(\sigma+2)}\; A(\sigma,k)\, e^{\gamma} K_{\sigma+1}(ke^{\gamma})\, J_o(kr) \qquad (16)$$

$$A(\sigma,k) = \frac{i}{\Gamma(-\sigma)}\int_{-\infty}^{\infty} e^{-2\gamma}\, d\gamma \int_0^\infty rdr\, f(\gamma,r)\, e^{\gamma} K_{-\sigma-1}(ke^{\gamma})\, J_o(kr) \qquad (17)$$

where again $\sigma = -1+ip$, p real for square-integrable amplitudes. Here $J_o(x)$ are Bessel functions, $K_\nu(z)$ Macdonald cylindrical functions.

E. Discussion of the Subgroup Type Expansions

The maximal subgroup in all three above expansions plays an important role. Beyond simply being a subgroup of the O(3,1) group, which acts as the group of motions of the space of independent kinematic parameters and thus generates the expansions, the subgroups also appear as little groups of the Poincaré group, leaving a certain timelike, spacelike or lightlike vector invariant--namely that vector, the standardization of which determined the frame of reference. Due to this dual role of the subgroups our two-variable expansions incorporate the O(3), O(2,1) and E_2 little group expansions.[23)-25)] Indeed, formula (7) can be interpreted as the standard direct channel partial wave expansion, supplemented by an integral expansion for the partial wave amplitude $a_\ell(s)$. Formula (12) similarly represents an O(2,1) expansion, i.e. the integral of Regge pole theory, together with an integral expansion for the reggeized partial wave amplitude $a(\ell,t)$ (for $t < 0$). Finally the expansion (16) can be viewed as the E_2 little group expansion for $t = 0$, $(m_1 \neq m_3,\ m_2 \neq m_4)$ and as its generalization for $t \neq 0$, again supplemented by a further expansion of the corresponding partial wave amplitude. The O(3,1) little group expansion of Toller[23)] for $t=0$, $m_1=m_3$, $m_2 = m_4$ (elastic forward scattering) is also contained in the two-variable approach, namely as a special limiting case of expansion (12). The relation between these two different O(3,1) expansions is considered in Refs. 9 and 18.

Thus, the O(3,1) two-variable expansions incorporate the Poincaré little group expansions completely, so that they do make full use of relativistic invariance and in particular should be useful for solving problems connected with various types of kinematical constraints upon amplitudes. For particles of spin zero the only problems of this sort are connected with nonequal mass scattering at $t=0$. For a discussion of these problems we refer to previous publications.[9),18)] In particular, a consideration of nonunitary representations makes it possible to incorporate Regge poles, branch points and cuts, etc. in the expansion (12). The expansion for the reggeized partial wave amplitude

$$a(\ell,t) = -\frac{i\cos\pi\ell}{8\sqrt{2}\,\pi\cosh\alpha}\int_{\delta-i\infty}^{\delta+i\infty}(\sigma+1)^2\,d\sigma\,\frac{\Gamma(\sigma-\ell+1)\,\Gamma(\sigma+\ell+2)}{\Gamma(\sigma+2)}\,\cdot$$

$$\cdot\Big\{A^{+}(\sigma,\ell)\Big[\mathbb{P}_{\ell}^{-\sigma-1}(-\tanh\alpha)+\mathbb{P}_{\ell}^{-\sigma-1}(\tanh\alpha)\Big]+$$

$$+A^{-}(\sigma,\ell)\Big[\mathbb{P}_{\ell}^{-\sigma-1}(-\tanh\alpha)-\mathbb{P}_{\ell}^{-\sigma-1}(\tanh\alpha)\Big]\Big\} \quad (18)$$

following from (12) makes it possible to relate the singularities of $a(\ell,t)$ in the complex ℓ plane, i.e. the divergencies of the σ-integral in (18), to the behavior of the Lorentz amplitudes $A^{\pm}(\sigma,\ell)$. In this formalism Lorentz poles, i.e. singularities of $A^{\pm}(\sigma,\ell)$ at finite values of $|\sigma|$ can only lead to fixed singularities in the ℓ-plane whereas moving singularities, e.g. Regge trajectories depend on the behavior of the Lorentz amplitudes for $\mathrm{Im}\,\sigma \to \pm\infty$, $\mathrm{Re}\,\sigma = \delta = \mathrm{const.}$

Let us just mention that if we add Mandelstam analyticity to the assumptions about $f(s,t)$, then the $O(3,1) \supset O(3)$ expansion in one channel and the $O(3,1) \supset O(2,1)$ expansion in the other can be proved to be analytic continuations of each other and the Lorentz amplitudes in the two channels will have definite analytic properties.[15)]

F. Expansions of the Elliptic Type

All three two-variable expansions considered above were of the subgroup type and the stress was on the incorporation of the little group formalism, i.e. on the utilization of relativistic invariance. In this paragraph we shall mention a different approach in which we make use of group representation theory in an "elliptic" basis, not related to any subgroup.

The motivation for going into such complications is that we wish to write crossing symmetric expansions, i.e. expansions which converge in at least two channels and which have particularly simple properties with respect to crossing symmetry. Thus, let us consider a crossing symmetric reaction--one that coincides in the s and t channels. To ensure properties like

$$f(s,t,u) = \pm f(t,s,u)$$

we shall first construct a specially symmetric mapping of s,t onto some coordinates α, β on a hyperboloid and then expand the amplitude $f(\alpha,\beta)$ into such basis functions of the group representations, that they have the same type of dependence on α and β. For all details we refer to the original paper.[16)]

For mathematical simplicity instead of using an O(3,1) hyperboloid as above, we shall make use of the fact that with our usual choice of the scattering plane, all momenta have zero components in the direction of the second space axis, so that we can consider the velocities as lying on an O(2,1) hyperboloid $v_0^2-v_1^2-v_3^2=1$. Let us parametrize this hyperboloid using elliptic functions[26)]

$$v_0 = -\,\mathrm{cn}(\alpha,k)\,\mathrm{cn}(\beta,k),\quad v_1 = i\,\mathrm{sn}(\alpha,k)\,\mathrm{dn}(\beta,k),\quad v_2 = i\,\mathrm{dn}(\alpha,k)\,\mathrm{sn}(\beta,k) \tag{19}$$

where we put the modulus of the elliptic functions $k = 1/\sqrt{2}$ and take the variables in the region

$$\alpha \in (iK,\ iK+2K) \qquad \beta \in (iK,\ iK+2K) \tag{20}$$

where $K = [\Gamma(\frac{1}{4})]^2/4\sqrt{\pi}$ is the real and the imaginary period of the Jacobi functions.

We shall construct a frame of reference, in which an interchange $\alpha \to \beta$, $\beta \to 2K-\alpha$ corresponds to $p_2 \to -p_3$, $p_3 \to -p_2$, $p_1 \to p_1$, $p_4 \to p_4$. Such a frame turns out to be a specifically oriented brick-wall system, in which we have (for $m_1 = \ldots = m_4 = 1$):

$$p_1 = (-\mathrm{cn}^2\alpha_1,\ i\,\mathrm{sn}\alpha_1\,\mathrm{dn}\alpha_1,\ 0,\ i\,\mathrm{dn}\alpha_1\,\mathrm{dn}\alpha_1)$$

$$p_2 = (-\mathrm{cn}\alpha\,\mathrm{cn}\beta,\ i\,\mathrm{sn}\alpha\,\mathrm{dn}\beta,\ 0,\ i\,\mathrm{dn}\alpha\,\mathrm{sn}\beta)$$

$$p_3 = (-\mathrm{cn}\alpha\,\mathrm{cn}\beta,\ -i\,\mathrm{dn}\alpha\,\mathrm{sn}\beta,\ 0,\ -i\,\mathrm{sn}\alpha\,\mathrm{dn}\beta)$$

$$p_4 = (-\mathrm{cn}^2\alpha_1,\ -i\,\mathrm{sn}\alpha_1\,\mathrm{dn}\alpha_1,\ 0,\ -i\,\mathrm{sn}\alpha_1\,\mathrm{dn}\alpha_1) \tag{21}$$

where

$$\mathrm{dn}\alpha_1 = \frac{i}{\sqrt{2}}\left\{\left[1-\frac{(\mathrm{sn}\alpha\,\mathrm{dn}\beta+\mathrm{dn}\alpha\,\mathrm{sn}\beta)^2}{2}\right]^{\frac{1}{2}}-1\right\}^{\frac{1}{2}} \tag{22}$$

The scattering amplitude is now a function of the variables α and β, i.e. of a point, say p_3 on an O(2,1) hyperboloid. It is a simple matter of algebra to show that

$$\tfrac{1}{2}s = 1 - \frac{x^2}{2} - y\left(1 - \frac{x^2}{2}\right)^{\frac{1}{2}}$$

$$\tfrac{1}{2}t = 1 - \frac{x^2}{2} + y\left(1 - \frac{x^2}{2}\right)^{\frac{1}{2}} \qquad x = \mathrm{sn}\alpha\,\mathrm{dn}\beta + \mathrm{dn}\alpha\,\mathrm{sn}\beta$$

$$y = \mathrm{cn}\alpha\,\mathrm{cn}\beta$$

$$\tfrac{1}{2}u = x^2 \tag{23}$$

and

$$\mathrm{cn}^4\begin{Bmatrix}\alpha\\ \beta\end{Bmatrix} = \frac{(s+t)^2 + 2st(2-s-t)}{4(s+t)} \pm \tfrac{1}{2}\left[\frac{stu(s+t-st)}{s+t}\right]^{\frac{1}{2}} \tag{24}$$

The s-channel physical region now corresponds to $\alpha \in (iK, iK+2K)$, $\beta \in (iK, iK+2K)$, the t-channel to $\alpha \in (iK, iK+2K)$, $\beta \in (-iK, -iK+2K)$ and the u-channel, which does not enter symmetrically, to $\alpha \in (iK, iK+2K)$, $\beta \in (0, 2iK)$.

The Laplace operator for the O(2,1) hyperboloid allows the separation of variables in the coordinates (19) and the separated eigenfunctions will be eigenfunctions of the operators:[6),16)]

$$\Delta\psi = (L_3{}^2 - K_1{}^2 - K_2{}^2)\psi = -\ell(\ell+1)\psi \tag{25}$$

$$L\psi = (K_1{}^2 - \tfrac{1}{2}L_3{}^2)\psi = h\psi$$

$$X\psi = p\psi \qquad \psi \equiv \psi^{pq}_{\ell h}(\alpha,\beta)$$

$$Y\psi = q\psi \tag{26}$$

where $p,q = \pm 1$ and X,Y represent reflections of v_1 and v_2, respectively. Upon separating the variables in (25) and (26) and solving the obtained equations, we find

$$\psi^{pq}_{\ell h}(\alpha,\beta) = \Lambda^{p}_{\ell h}(\alpha)\,\Lambda^{q}_{\ell\tilde{h}}(\beta) \qquad h + \tilde{h} = \ell(\ell+1) \tag{27}$$

where $\Lambda^p_{\ell h}(z)$ are Lamé functions,[27] symmetric or antisymmetric with respect to $z = iK + K$ for $p = +1$ or -1 and standardized as

$$\Lambda^{+}(iK+K) = 1 \qquad \Lambda^{+\prime}(iK+K) = 0$$

$$\Lambda^{-}(iK+K) = 0 \qquad \Lambda^{-\prime}(iK+K) = -1\ . \tag{28}$$

Expansions in terms of these functions can be obtained by making use of the method of horispheres[28] (essentially to obtain normalization and completeness relations). Leaving out all the details,[16] we finally obtain an s-channel expansion (for square integrable amplitudes):

$$f^s(\alpha,\beta) = \frac{1}{8\pi^2 i}\int_{-\frac{1}{2}-i\infty}^{-\frac{1}{2}+i\infty} d\ell\,(2\ell+1)\cot\pi\ell \sum_h \sum_{p,q} N^p_{\ell h}|\lambda_{p,q}(\ell,h)|^2 \times$$

$$\times\left\{A^{pq}(\ell,h)\,\Lambda^p_{\ell h}(\alpha)\,\Lambda^q_{\ell\tilde{h}}(\beta) + A^{qp}(\ell,\tilde{h})\,\Lambda^p_{\ell\tilde{h}}(\alpha)\,\Lambda^q_{\ell h}(\beta)\right\} \qquad (29)$$

$$\alpha \in (iK,\ iK+2K) \qquad \beta \in (iK,\ iK+2K)$$

with

$$A^{pq}(\ell,h) = -\tfrac{1}{2}\iint d\alpha\, d\beta\,(cn^2\alpha + cn^2\beta)\,\Lambda^p_{\ell h}(\alpha)\,\Lambda^q_{\ell\tilde{h}}(\beta)\, f^s(\alpha,\beta)$$

$$\left[N^p_{\ell h}\right]^{-1} = \int_{iK}^{iK+2K} |\Lambda^p_{\ell h}(z)|^2 \frac{dz}{\sqrt{2}}$$

$$\lambda_{++}(\ell,h) = \frac{1}{\sqrt{2}}\int_{iK}^{iK+2K} \Lambda^+_{\ell h}(z)\,(icnz)^{-\ell-1}\,dz$$

$$\lambda_{--}(\ell,h) = \frac{i(\ell+1)(\ell+2)}{2}\int_{iK}^{iK+2K} \Lambda^-_{\ell h}(z)\,(icnz)^{-\ell-3}\,snz\,dnz\,dz$$

$$\lambda_{+-} = \frac{(\ell+1)}{2}\int_{iK}^{iK+2K} \Lambda^+_{\ell h}(z)(icnz)^{-\ell-2}\,snzdz$$

$$\lambda_{-+} = \frac{(\ell+1)}{\sqrt{2}}\int_{iK}^{iK+2K} \Lambda^-_{\ell h}(z)(icnz)^{-\ell-2}(idnz)dz \qquad (30)$$

Again, performing a lot of algebra, we can show that the corresponding t-channel expansion is

$$f^t(\alpha,\tilde{\beta}) = \frac{1}{8i\pi^2}\int_{-\frac{1}{2}-i\infty}^{-\frac{1}{2}+i\infty} d\ell\,(2\ell+1)\cot\pi\ell \sum_h \sum_{p,q} q\, N^p_{\ell h}\,|\lambda_{pq}(\ell,h)|^2$$

$$\left\{B^{pq}(\ell,h)\ \Lambda^p_{\ell h}(\alpha)\,\Lambda^q_{\ell\tilde{h}}(2K-\tilde{\beta}) + B^{qp}_{(\ell,\tilde{h})}\Lambda^p_{\ell\tilde{h}}(\alpha)\Lambda^q_{\ell h}(2K-\tilde{\beta})\right\} \qquad (31)$$

$$\alpha \in (iK,\ iK+2K) \qquad \tilde{\beta} \in (-iK,-iK+2K)$$

$$B^{pq}(\ell,h) = -\tfrac{1}{2}q \int_{iK}^{iK+2K} d\alpha \int_{iK}^{iK+2K} d\beta\,(\mathrm{cn}^2\alpha + \mathrm{cn}^2\beta)\, \Lambda^{p}_{\ell h}(\alpha)\, \Lambda^{q}_{\ell \tilde{h}}(\beta)\, f^{t}(\alpha, 2K-\beta) \quad (32)$$

Note that the sums over h in (29) and (31) are over a discrete infinite set of eigenvalues, determined by the orthogonality conditions for $\Lambda^{p}_{\ell h}(z)$. It is crucial that (29) and (31) are expansions over Lamé functions defined over the same intervals, so that they both converge simultaneously.

For a crossing symmetric reaction we must have

$$f^{s}(s,t) = \pm f^{t}(t,s) \quad \text{i.e.} \quad f^{s}(\alpha,\beta) = \pm f^{t}(\alpha, 2K-\beta) .$$

Thus, we shall have "term-by-term" crossing symmetry, if we simply put

$$qB^{pq}(\ell,h) = \pm A^{pq}(\ell,h). \quad (33)$$

Let us stress that so far no use has been made of any analytic continuation in s and t or α and β. This is a problem currently under investigation.

II. Two-Variable Expansions Based on an SU(3) Group

A. The Expansions and Crossing Symmetry

Completely different two-variable expansions of scattering amplitudes have been suggested by Balachandran et al[19),20)] and by Charap and Minton.[21)] Both of these expansions were written for amplitudes in the nonphysical region, namely the inside of the Mandelstam triangle, and originally for reactions involving four spinless particles of equal mass (recently they have undergone much further development--see Ref. 29 and references contained therein).

The main purpose of these expansions (we shall mainly follow Balachandran's approach)[19),20),29)] is to write expansions by means of which crossing symmetry, i.e. symmetries of f(s,t,u) with respect to permutations of s,t and u, can be imposed in a simple and general manner.

Similarly as the O(3,1) expansions, these expansions are generated by a second order differential operator $\mathfrak{G}$ in the variables s, t and u, which is specifically so constructed as to be symmetric under arbitrary interchanges of s,t and u to commute with the angular momentum operator L_s^2 in the s-channel (and hence in all channels). Such an operator is determined uniquely to be

$$\mathfrak{G} = L_s^2 + L_t^2 + L_u^2 \tag{34}$$

It is now a simple matter[19] to find the common set of eigenfunctions of $\mathfrak{G}$ and L_s^2, putting

$$\mathfrak{G}\, S_n^{\ell}(s,t) = (n+\ell)(n+\ell+2)\, S_n^{\ell}(s,t)$$

$$L_s^2\, S_n^{\ell}(s,t) = \ell(\ell+1)\, S_n^{\ell}(s,t) \tag{35}$$

Solving (35) in the nonphysical region $0 \le s \le 1$, $0 \le t \le 1$, $0 \le u \le 1$ ($m_1 = \ldots = m_4 = \frac{1}{2}$) we find that a complete set of solutions can be written as

$$S_n^{\ell}(s,t) = (1-s)^{\ell}\, P_n^{(2\ell+1,0)}(2s-1)\, P_{\ell}(z_s) \qquad z_s = 1 + \frac{2t}{s-1} \tag{36}$$

i.e. a product of Jacobi and Legendre polynomials (n, ℓ = non-negative integers). Any scattering amplitude, square integrable over the Mandelstam triangle with the measure $(1-s)ds\,dz$ can be expanded as

$$f(s,t) = \sum_{n,\ell=0}^{\infty} 2(n+\ell+1)(2\ell+1)\, a_n^{\ell}\, S_n^{\ell}(t) \tag{37}$$

$$a_n^{\ell} = \tfrac{1}{2}\int_0^1 (1-s)ds \int_{-1}^1 dz\, f(s,t)(1-s)^{\ell}\, P_n^{(2\ell+1,0)}(2s-1)\, P_{\ell}(z_s) \tag{38}$$

The important feature of (37) is that the "partial wave crossing matrices" for such an expansion will be block-diagonal. Indeed, if we write a similar t-channel expansion in terms of the functions $T_N^L(s,t)$ obtained by interchanging s and t in (36), we find that any coefficient a_n^{ℓ} in the s-channel can be expressed in terms of a finite number of coefficients b_N^L in the t-channel expansion, because $S_n^{\ell}(s,t)$ and $T_N^L(s,t)$ are both eigenfunctions of the operator $\mathfrak{G}$. It is precisely this block-diagonality of the crossing matrix which makes it possible to impose crossing symmetry simply.

Further, let us note that from the group theoretical point[20] of view the operator $\mathfrak{G}$ can be interpreted as the second order Casimir operator of SU(3) and the functions (36) are basis functions of irreducible representations of SU(3), corresponding to the reduction $SU(3) \supset SU(2) \supset U(1)$.

Generalizations of thse results to physical regions were suggested using the group SU(2,1)[19] or a transformation from sums to integrals in (37) along the lines of the Sommerfield-Watson transformation.[21]

B. Comparison of Lorentz Group and SU(3) Expansions

The simplest way to make a direct comparison between the O(3,1) and SU(3) expansions is to consider both for equal masses and inside the Mandelstam triangle. Indeed, consider the $O(3,1) \supset O(3) \supset O(2)$ expansions for $m_1 = \ldots m_4 = \frac{1}{2}$ and $0 \le s,t,u \le 1$. We can put

$$\cos a = \sqrt{s} \qquad \cos\theta = 1 + \frac{2t}{s-1}$$

and identify a and θ with the spherical coordinates of a point on a sphere, instead of a hyperboloid. We then obtain an $O(4) \supset O(3) \supset O(2)$ expansion

$$f(s,t) = \sum_{n=0}^{\infty} \sum_{\ell=0}^{n} \left\{ \frac{2\ell+1}{4\pi} \frac{\Gamma(N+L+2)}{\Gamma(N-L+1)} (N+1) \right\}^{\frac{1}{2}} A_{n\ell} \cdot$$

$$\cdot \frac{1}{\sqrt{\sin a}} P^{-\ell-\frac{1}{2}}_{\frac{1}{2}+n}(\cos a)\, P_\ell(\cos\theta) \tag{39}$$

$$A_{n\ell} = 2\pi \left\{ \frac{2\ell+1}{4\pi} \frac{\Gamma(N+L+2)}{\Gamma(N-L+1)} (N+1) \right\}^{\frac{1}{2}} \int_0^\pi \sin^2 a\, da \int_0^\pi \sin\theta\, d\theta \cdot$$

$$\cdot f(s,t) \frac{1}{\sqrt{\sin a}} P^{-\ell-\frac{1}{2}}_{\frac{1}{2}+n}(\cos a)\, P_\ell(\cos\theta) \tag{40}$$

The O(4) and SU(3) expansions can now be directly compared and it is possible to expand the two sets of basis functions in terms of each other.[13] The results are quite complicated and we shall not repeat them here. The partial wave crossing matrices can also be calculated once and for all and expressed in terms of generalized hypergeometric functions. However, since the Laplace operator on the O(3,1) hyperboloid or O(4) sphere, generating our expansions, is not symmetric in s,t, and u and is thus a different operator in each channel, the O(4) crossing matrices will not be block diagonal, so that no simple relationship between the Lorentz amplitudes (or O(4) amplitudes) in the two channels is obtained.

To summarize: The O(3,1) expansions make full use of relativistic invariance by completely incorporating the little group formalism, they are written for arbitrary masses and for s,t,u in the physical regions. Crossing and analyticity can be incorporated either by using "elliptic" expansions, or by making use of the connection between the O(3) and O(2,1) reductions.

The SU(3) expansions have particularly simple properties under crossing; however they are in general not related to little group expansions. They have been generalized to arbitrary masses and spins,[29)] however at the price of giving up their group theoretical interpretation. Most of their useful properties were obtained inside the Mandelstam triangle and a continuation into the physical region involves new complications.

It would obviously be of interest to write expansions, incorporating the useful features of both approaches. If these are to be based on group theory, then the corresponding group must contain all the little groups of the Poincaré group as subgroups on one hand and should have a second order Casimir operator, identifiable with the symmetric operator Θ, discussed above, on the other.

Two remarks are in order at this point. First--the group SU(2,1) cannot be used in such a program, since it does not have an E_2 subgroup,[12)] so that it cannot incorporate the complete little group group formalism. Second--a second order operator Θ, commuting with angular momenta in all three channels, or even in two of them, exists if and only if all four masses are equal.

III. Remarks on Possible Conformal Group Expansions

A. General Remarks

In this section we present some arguments indicating that the conformal group or the group O(4,2) can be used to generate expansions of scattering amplitudes and that these are of interest for several reasons:

1. The group O(4,2) is a candidate for the two-variable expansion program discussed above. Indeed, since it contains the whole Poincaré group as a subgroup, it obviously also contains all the little groups, including O(3,1). Further, as we shall show below, the symmetric operator θ can be related to the Laplace-Beltrami operator Δ_c on an O(4,2) hyperboloid.

2. Two-variable expansions should also be developed for reactions involving particles of zero rest mass. Proceeding along lines analogous to Section I we would obtain the scattering amplitude as a function of a point on the cone $v_0{}^2 - v_1{}^2 - v_2{}^2 - v_3{}^2 = 0$. Scale and pure conformal transformations in this space

$$v_\mu{}' = \lambda v_\mu \qquad v_\mu{}' = \frac{v_\mu + \alpha_\mu v_\lambda v^\lambda}{1 + 2\alpha_\nu v^\nu + \alpha_\nu \alpha^\nu v_\lambda v^\lambda} \tag{41}$$

do not leave any hyperboloid invariant; they do, however, transform the cone $v_\mu v^\mu = 0$ into itself.

Thus for zero mass particles and possibly even for massive particles at very high energies it would be of interest to write expansion in terms of the basis functions of the irreducible representations of the conformal group.

3. The third possible application of O(4,2) expansions which we have in mind concerns five point functions, i.e. production amplitudes for reactions of the type

$$1 + 2 \to 3 + 4 + 5 \tag{42}$$

Indeed, if it was possible to map a physical region of a two-body process depending on two variables, onto an O(3,1) or O(2,1) hyperboloid, a similar mapping for reaction (42) depending on five independent parameters, would require at least a five dimensional space. A homogeneous space of the O(4,2) group would obviously serve this purpose and the Laplace-Beltrami operator on such a space would serve to generate five-variable expansions (or expansions of any lower dimension, if desired).

B. <u>The Reduction $O(4,2) \supset O(4) \times O(2)$</u>

Let us introduce spherical coordinates on the hyperboloid

$$v_0{}^2 + v_5{}^2 - v_1{}^2 - v_2{}^2 - v_3{}^2 - v_4{}^2 = 1 \tag{43}$$

putting

$$\begin{aligned} v_0 &= \cosh A \cos \psi & v_1 &= \sinh A \sin a \sin \theta \cos \varphi \\ v_5 &= \cosh A \sin \psi & v_2 &= \sinh A \sin a \sin \theta \sin \varphi \\ & & v_3 &= \sinh A \sin a \cos \theta \\ & & v_4 &= \sinh A \cos a \end{aligned} \tag{44}$$

With

$$0 \le A < \infty, \quad 0 \le a \le \pi, \quad 0 \le \theta \le \pi, \quad 0 \le \psi < 2\pi, \quad 0 \le \varphi < 2\pi \tag{45}$$

these cover the whole hyperboloid.

The Laplace-Beltrami operator can be written as

$$\Delta_c = -\frac{1}{\cosh A \sinh^3 A}\frac{\partial}{\partial A}\cosh A \sinh^3 A\frac{\partial}{\partial A} + \frac{1}{\cosh^2 A}\frac{\partial^2}{\partial \psi^2} -$$

$$-\frac{1}{\sinh^2 A}\left\{\frac{1}{\sin^2 a}\frac{\partial}{\partial a}\sin^2 a\frac{\partial}{\partial a} + \frac{1}{\sin^2 a \sin\theta}\frac{\partial}{\partial\theta}\sin\theta\frac{\partial}{\partial\theta} +\right.$$

$$\left.+\frac{1}{\sin^2 a \sin^2\theta}\frac{\partial^2}{\partial\varphi^2}\right\} \tag{46}$$

The basis functions corresponding to the considered reduction are the eigenfunctions of the set of operators Δ_c, Δ_s (the Laplace operator on the sphere $v_1{}^2 + v_2{}^2 + v_3{}^2 + v_4{}^2 = \sinh^2 A$, L^2 (the angular momentum) and the generators of one parameter rotations L_{12} and L_{05}. They can be written as

$$\varphi_{LN\ell mk}(A,a,\theta,\varphi,\psi) = N_{LN\ell mk}\tanh^N A \cosh^{-L-4} A\,\cdot$$

$$\cdot\, F\{\tfrac{1}{2}(-k+N+L+4),\ \tfrac{1}{2}(k+N+L+4);\ N+2,\ \tanh^2 A\}\,\cdot$$

$$\cdot\,\frac{1}{\sqrt{\sin a}}\ P^{-\ell-\frac{1}{2}}_{\frac{1}{2}+N}(\cos a)\ \ P^m_\ell(\cos\theta)\, e^{im\varphi}\, e^{ik\psi} \tag{47}$$

where $L = -2-i\Lambda$, Λ real corresponds to unitary representations of the principal continuous series and $L=-1,0,1,2,\ldots$ to discrete series. These functions, their normalization, range of parameters, etc. have been discussed by Limic et al.[30)]

Let us now compare the operator Δ_c of (46) with the symmetric operator $\mathfrak{G}$ of (34). Introducing the usual s-channel c.m.s. variables

$$\cos\theta = 1 + \frac{2t}{s-1} \qquad \cosh A = \sqrt{s} \tag{48}$$

it is easy to check that

$$\mathfrak{G} = \frac{1}{4\cosh A \sinh^3 A}\frac{\partial}{\partial A}\cosh A \sinh^3 A\frac{\partial}{\partial A} +$$

$$+\frac{1}{\sinh^2 A}\ \frac{1}{\sin\theta}\frac{\partial}{\partial\theta}\sin\theta\frac{\partial}{\partial\theta}\ . \tag{49}$$

Thus, we have

$$\Delta_C = -4\Theta$$

provided that we keep a, φ and ψ constant in (46) and put $\sin^2 a = \frac{1}{4}$.

Only a certain subset of the basis functions (47) will also be eigenfunctions of the operator Θ, namely these satisfying $N(N+2) = 4\ell(\ell+1)$, i.e.

$$N = 2\ell \qquad \text{or} \qquad N = -2\ell - 2 \quad . \tag{50}$$

These functions will constitute a complete basis for a Hilbert space of square integrable functions, depending on the variables A and θ only.

The O(3) subgroup of O(4) in this reduction chain, producing the angle θ, identified with the c.m.s. scattering angle, will figure as the little group of $p_1 + p_2$ in the corresponding two variable expansion.

C. The Reduction $O(4,2) \supset O(2,2) \times O(2)$

Let us now introduce coordinates on the hyperboloid (43), corresponding to a reduction into $O(2,2) \times O(2)$ and to the further reduction $O(2,2) \supset O(2,1) \supset O(2)$:

$$\begin{aligned} v_0 &= \cosh A \cos a \cosh \beta & v_1 &= \cosh A \cos a \sinh \beta \cos \varphi \\ v_5 &= \cosh A \sin a & v_2 &= \cosh A \cos a \sinh \beta \sin \varphi \\ & & v_3 &= \sinh A \cos \psi \\ & & v_4 &= \sinh A \sin \psi \end{aligned} \tag{51}$$

With

$$-\infty < A < \infty \quad 0 \le a < 2\pi \quad 0 \le \beta < \infty \quad 0 \le \varphi < 2\pi \quad 0 \le \psi \le \pi \tag{52}$$

these coordinates cover the whole hyperboloid. The Laplace-Beltrami operator in these coordinates is

$$\Delta_C = -\frac{1}{\sinh A \cosh^3 A}\frac{\partial}{\partial A}\sinh A \cosh^3 A \frac{\partial}{\partial A} - \frac{1}{\sinh^2 A}\frac{\partial^2}{\partial \psi^2} + $$
$$+ \frac{1}{\cosh^2 A}\left\{\frac{1}{\cos^2 a}\frac{\partial}{\partial a}\cos^2 a \frac{\partial}{\partial a} - \frac{1}{\cos^2 a \sinh \beta}\frac{\partial}{\partial \beta}\sinh \beta \frac{\partial}{\partial \beta}\right.$$
$$\left. - \frac{1}{\cos^2 a \sinh^2 \beta}\frac{\partial^2}{\partial \varphi^2}\right\} \tag{53}$$

The basis functions in this reduction will be the eigenfunctions of Δ_C, Δ_H (Laplace operator on the space $v_0^2 + v_5^2 - v_1^2 - v_2^2 = \cosh^2 A$), H^2 (Laplace operator on the space $v_0^2 - v_1^2 - v_2^2 = \cosh^2 A \cos^2 a$), of the rotation generators L_{12}, L_{34} and of two additional discrete operators R and S, where R corresponds to an inversion of v_3 and v_4 and S to an inversion of v_5.

Let us compare Δ_C of (53) with the symmetric operator $\mathfrak{G}$. This time we use the t-channel brick-wall system variables

$$\sinh A = \sqrt{-s} \qquad \cosh \beta = -1 + \frac{2t}{1-s} \tag{54}$$

and transforming $\mathfrak{G}$ to these variables we get

$$\mathfrak{G} = \frac{1}{4 \sinh A \cosh^3 A} \frac{\partial}{\partial A} \sinh A \cosh^3 A \frac{\partial}{\partial A} + \\ + \frac{1}{\cosh^2 A \sinh \beta} \frac{\partial}{\partial \beta} \sinh \beta \frac{\partial}{\partial \beta} \tag{55}$$

We now see that (53) and (55) satisfy $\Delta_C = -4\mathfrak{G}$, if we keep a, φ and ψ constant in (53) and put $\cos^2 a = \frac{1}{4}$.

Thus, with similar restrictions as in the previous paragraph we can write O(4,2) expansions for functions depending on A and β only, in terms of the eigenfunctions of operator $\mathfrak{G}$. The O(2,1) subgroup in the reduction figures as the little group of the momentum transfer $p_1 - p_3$ (for $t < 0$) and will thus furnish a Regge-type expansion in terms of the Legendre functions $P_\ell(\cosh \beta)$.

IV. Conclusions

We have given a brief exposition of some recent and older work on two-variable expansions of relativistic scattering amplitudes. In particular we have shown how the desire to make maximal use of relativistic invariance and of crossing symmetry leads us to a consideration of the conformal group, which in this approach also comes up in other connections (zero mass particles, production amplitudes). Thus, to the many reasons why the conformal group is of interest, we add a further one--an interest in harmonic analysis on this group. As we have stressed, the specific form which the harmonic analysis takes depends crucially on the parametrization of the homogeneous space under consideration, and is of greatest physical significance. Since each different reduction of a group, say in our case the group SU(2,2) leads to a different parametrization and thus a different expansion, a detailed study of the subgroup structure is necessary. Indeed, a classification of all subgroups of SU(2,2) has been initiated[17] and such a classification of one parameter subgroups of general U(p,q) groups will be presented by J. G. Belinfante at this conference.

References

1. N. Vilenkin, J. Smorodinsky, Zh. Eksperim. i Teor. Fiz. 46, 1793 (1964) [Soviet Phys. JETP 19, 1209 (1964)].
2. P. Winternitz, J. Smorodinsky, and M. Uhlíř, Yadernaya Fiz. 1, 163 (1965) [Sov. J. of Nuc. Phys. 1, 113 (1965)].
3. P. Winternitz and I. Friš, Yadernaya Fiz. 1, 889 (1965) [Sov. J. of Nuc. Phys. 1, 636 (1965)].
4. G. I. Kuznetsov and J. A. Smorodinsky, Yadernaya Fiz. 3, 383 (1966) [Sov. J. of Nuc. Phys. 3, 275 (1966)].
5. G. I. Kuznetsov and J. A. Smorodinsky, Yadernaya Fiz. 6, 1308 (1967) [Sov. J. of Nuc. Phys. 6, 949 (1968)].
6. P. Winternitz, I. Lukač, and J. Smorodinsky, Yadernaya Fiz. 7, 192 (1968) [Sov. J. of Nuc. Phys. 7, 139 (1968)].
7. M. A. Liberman, J. A. Smorodinsky, and M. B. Sheftel, Yadernaya Fiz. 7, 202 (1968) [Sov. J. of Nuc. Phys. 7, 146 (1968)].
8. P. Winternitz, J. Smorodinsky, and M. Sheftel, Yadernaya Fiz. 7, 1325 (1968) [Sov. J. of Nuc. Phys. 7, 785 (1968)]. See also Phys. Letters 26B, 241 (1968).
9. P. Winternitz, J. Smorodinsky, and M. Sheftel, Yadernaya Fiz. 8, 833 (1968) [Sov. J. of Nuc. Phys. 8, 485 (1969)].
10. M. A. Liberman and A. A. Makarov, Yadernaya Fiz. 9, 1314 (1969) [Sov. J. of Nuc. Phys. 9, 766 (1969)].
11. G. I. Kuznetsov, M. A. Liberman, A. A. Makarov, and J. A. Smorodinsky, Yadernaya Fiz. 10, 644 (1969) [Sov. J. of Nuc. Phys. 10, 370 (1970)].
12. W. Montgomery, L. O'Raifeartaigh, and P. Winternitz, Nucl. Phys. B11, 39 (1969).
13. P. Winternitz, Czech. J. Phys. B19, 1589 (1969).
14. P. Pajas and P. Winternitz, J. Math. Phys. 11, 1505 (1970).
15. P. Pajas and P. Winternitz, Phys. Rev. D1, 1105 (1970).
16. N. W. Macfadyen and P. Winternitz, preprint, Carnegie-Mellon U University, Pittsburgh (1970), to be published in J. Math. Phys.
17. J. G. Belinfante and P. Winternitz, preprint, Carnegie-Mellon University, Pittsburgh (1970), to be published.
J. G. Belinfante, these Proceedings.
18. P. Winternitz, preprint RPP/T/3, Rutherford Lab., England (1969).
19. A. P. Balachandran and J. Nuyts, Phys. Rev. 172, 1821 (1968).
20. A. P. Balachandran, W. J. Meggs, J. Nuyts, and P. Ramond, Phys. Rev. 176, 1700 (1968).
21. J. M. Charap and B. M. Minton, J. Math. Phys. 10, 1823 (1969).
22. M. A. Naimark, Linear Representations of the Lorentz Group (Pergamon Press, Oxford, 1964).

23. M. Toller, Nuovo Cimento 53A, 671 (1968); ibid. 54A, 295 (1968 (1968).
24. H. Joos, Lectures in Theoretical Physics VIIA, edited by W. E. Brittin and A. O. Barut (University of Colorado Press, Boulder, 1964).
25. J. F. Boyce, R. Delbourgo, A. Salam, and J. Strathdee, "Partial Wave Analysis I," preprint 1C/67/9, Trieste (1967).
26. A. Erdelyi et al, Higher Transcendental Functions II (McGraw-Hill, Inc., New York, 1953), Chapter 13.
27. A. Erdelyi et al, Higher Transcendental Functions III (McGraw-Hill, Inc., New York, 1955), Chapter 15.
28. N. J. Vilenkin, Special Functions and the Theory of Group Representations (Amer. Math. Soc., Providence, 1968).
29. A. P. Balachandran, W. J. Meggs, P. Ramond, and J. Nuyts, Phys. Rev. 187, 2080 (1969).
30. N. Limic, J. Niederle, and R. Raczka, J. Math. Phys. 7, 1861, 2026 (1966); ibid. 8, 1079 (1967).

PART II: PHYSICAL APPLICATIONS

Section A: Broken Scale Invariance

ON SOME BASIC PROBLEMS CONNECTED WITH POSSIBLE APPLICATIONS OF THE CONFORMAL GROUP IN PARTICLE PHYSICS†

H. A. Kastrup
Sektion Physik
Universität München
München, Germany

(Manuscript not received for publication.)

†Presented at the Symposium on De Sitter and Conformal Groups, University of Colorado, Summer 1970.

SOME REMARKS ON CONFORMAL QUANTUM FIELD THEORY AND THE MASS PROBLEM†

R. L. Ingraham
Research Center
New Mexico State University
Las Cruces, New Mexico 88001

Summary

The conformal invariant quantum field theory of massive particles uses tensor fields defined on the 5-dimensional manifold of "spheres" in space-time, which allows the usual T-exponential definition of the S operator. These irreducible unitary representations have continuous mass spectra, typically $(0, \infty)$. It is suggested how the "mass selection rules" given by the Feynman diagram rules might pick out certain values of the incoming and outgoing masses in a dynamics-dependent way, and thus explain the observed quasi-discrete mass spectrum. Exactly the same mathematical mechanism is responsible for avoiding causality troubles coming from the fact that time-like and space-like intervals can be exchanged by the conformal group.

I. Preliminary

In my opinion, in spite of its many tantalizing promises of physical relevance, the conformal group on space-time has not yet made contact with physics. The central problem is mass. It has been known for a long time that the theory of a massless field admits conformal invariance. But so do massive particles, provided they are grouped into one of the massive conformal IUR's (irreducible unitary representation), each of which has a continuous mass spectrum (either $0 < m < \infty$ or $-\infty < m < 0$ or both) in full accord with O'Raifeartaigh's Theorem. It is just these continuous mass spectra which pose

†Presented at the Symposium on De Sitter and Conformal Groups, University of Colorado, Summer 1970.

the central problem: granted that the conformal group is physical, how is the observed, apparently discrete, mass spectrum to be reconciled to it?

Massive conformal IUR's can be treated in momentum space without introducing new physical concepts. But to treat them in position space, space-time alone is inadequate, one must enlarge it to the 5-dimensional space of all "spheres" in space-time, characterized by their centers x^μ and radii λ. That sphere space is the natural domain of the conformal group is old mathematical news, dating from the classical work of Felix Klein, Sophus Lie, Liouville, and other geometers about a hundred years ago. What is meant is that if one treats the spheres as points X^a $(a = 0,1,\ldots 5)$ in a 5-dimensional projective space, then the conformal group is characterized as that subgroup of the projective group which preserves a certain quadric

$$G_{ab}X^aX^b = 0 \qquad \text{(I.1)}$$

The 4-dimensional locus (I.1) represents the null spheres ($\lambda = 0$), i.e., space-time itself. Non-null spheres are carried into other non-null spheres such that the angle under which they intersect is preserved. The new physical concept so introduced is the fifth coordinate λ.

One can solve for IUR's of $\underset{\sim}{C}$ (the conformal group) as tensor fields $\psi_a\ldots(x,\lambda)$ comprising a set of spin values over this 5-dimensional manifold. The main virtue of these objects is that one has a natural way to build the $\underset{\sim}{C}$-invariant S operator out of them, namely as the T-exponential of an interaction Lagrangian polynomial in these (quantized) fields integrated over the whole 5-dimensional manifold. This permits a study of $\underset{\sim}{C}$-invariant particle theory in perturbation theory via the battery of Feynman diagrams. The problematical nature of λ then gives no trouble since it is integrated out; the external lines of the diagrams bear the familiar five labels p_μ (4-momentum) and m (mass) connected by $p^2 + m^2 = 0$.[1)]

I do think that this 5-dimensional sphere space will eventually find a physical use. The fifth "degree of freedom" is sort of a gauge space, or space of all lengths, and λ should mean something like the ratio of the length unit one has chosen to some standard meter stick chosen once and for all. The dilatation subgroup

$$x^\mu \to \sigma x^\mu, \qquad \lambda \to \sigma\lambda; \qquad \sigma = \text{const.} > 0 \qquad \text{(I.2)}$$

then means a uniform change of unit over all space-time, while the subgroup of special conformal transformations represents a non-uniform, or local, change of unit restricted only by the demand that light propagation in the new, regauged coordinates continues to be governed by $x^2 \equiv \underset{\sim}{x}^2 - t^2 = 0$.

In view of the popularity of studying "scale invariance" at the present moment, it is worthwhile to make a few distinctions which are usually lost in the current brouhaha. First, the group (I.2) is a change of unit. We know this because for example the field equations for the $\psi_a\ldots(x,\lambda)$ have the mass operator $\partial\lambda$ in place of the various fixed masses of ordinary (Poincaré-invariant) field theory. Thus (I.2) induces the change $m \to \sigma^{-1} m$ on all masses (and correspondingly for any dimensional quantity, where we always hold $\hbar = c = 1$). Again, the explicit solutions of these field equations show that p_μ and m are conjugate to x^μ and λ respectively, and one can explicitly verify that (I.2) induces the transformation $p_\mu \to \sigma^{-1} p_\mu$, $m \to \sigma^{-1} m$.

Secondly, change of unit is an exact symmetry of all our physical laws; i.e., every physical field has a definite dimension, and dimensions balance on both sides of an equation.

Thirdly this exact symmetry leads, à la Noether's Theorem, to a conserved current[2] $\mathscr{D}^\alpha(x,\lambda)$ $(\alpha=1,\ldots 5)$

$$\partial_\alpha(\sqrt{g}\ \mathscr{D}^\alpha) = 0 \ , \tag{I.3}$$

which yields a detailed constraint on the behavior in x and λ of massive fields.

Consider now the different group

$$x^\mu \to \sigma x^\mu \ , \qquad m_i \to m_i \quad (i = 1,2,\ldots) \tag{I.4}$$

(m_i any dimensional constants) which underlies the current "scale invariance" investigations. First, it is clear that (I.4) does not mean change of unit. Only in the case that all $m_i = 0$ does it mean that.[3]

Secondly, the group (I.4) is not a symmetry of our physical laws. Whether it is an approximate symmetry (in some sense) and thus can yield physical information (e.g., in high energy scattering) is far from a priori obvious and remains to be demonstrated.

Thirdly, since (I.4) is not a symmetry, Noether's Theorem can only yield the nonconservation law

$$\partial_\mu \mathscr{D}^\mu = \text{terms involving the } m_i \neq 0 \ . \tag{I.5}$$

This is less information than (I.3); for whereas (I.5) gives no way of building a conserved quantity in the case of massive fields, according to (I.3)

$$D \equiv \int_{x^4=t} d^3x d\lambda \lambda^{-5} \mathscr{D}^4(x,\lambda) \qquad (I.6)$$

in time-independent[4] for a system of interacting massive fields.

Therefore in view of these differences in the two types of "scale invariance," we urge that they be clearly distinguished, for clarity of thinking on the subject, and furthermore that the first kind, which is an exact symmetry and is guaranteed to give a conserved quantity in $\underset{\sim}{C}$-invariant QFT, be not so thoroughly ignored.[5]

To return to the subject of this talk after digressing on the physical meaning of λ and the consequent symmetry under changes of unit, the $\underset{\sim}{C}$-invariant QFT of massive particles made possible by the use of sphere space tensor fields contains families of conventional particles with all masses from 0 to ∞. The central problem, unsolved to date, is then by what mechanism the observed quasi-discrete spectrum is picked out. In the following sections I want to remind you of two consequences of this λ-dependence (equivalently: the continuous mass spectra) which look physically promising. One bears on the mass problem noted above, and one, on the causality problem for the conformal group. They both spring from the fact that the elementary wave functions contain the m and λ dependence in Bessel functions $J_n(m\lambda)$ just as the p_μ and x^μ dependence is contained in exponentials $e^{\pm ip\cdot x}$.

II. Conformal Tensors on Sphere Space

If one looks for conformal IUR's as tensor fields over the manifold (x,λ), i.e., if one chooses the generators of the "orbital" plus "spin" form

$$M_{ab} = \overset{o}{M}_{ab} + S_{ab} \qquad (a,b = 0,1,\ldots 5)$$

$$\overset{o}{M}_{ab} \equiv -i(X_a \partial/\partial X^b - X_b \partial/\partial X^a) \qquad (II.1)$$

where the homogeneous coordinates X^a are connected to x^μ and λ via a well known formula and the spin part S_{ab} is some finite-dimensional (matrix) representation of the $\underset{\sim}{C}$-Lie algebra, then the solutions take the form

$$\tilde{\psi}(x,\lambda) = U^{-1}(x)\, \psi(x,\lambda), \quad U(x) \equiv \exp iS_{5\mu} x^\mu \qquad (II.2a)$$

$$\psi(x,\lambda) = e^{-ik\cdot x}\, \varphi(\lambda;k) \qquad (II.2b)$$

$$\varphi(\lambda;k) = \sum_{\delta} \lambda^{\delta+c} Z_{n(\delta)}(m\lambda)\, U_{\delta}(k) \quad \text{(II.2c)}$$

$$\overset{s}{\Delta}\, U_{\delta}(k) \equiv iS_{05} U_{\delta}(k) = \delta U_{\delta}(k) \quad \text{(II.2d)}$$

$\delta = \alpha + m(m=0,1,\ldots m_{max})$;

$$k^2 + m^2 = 0 \quad \text{(II.2e)}$$

$$n(\delta) = n_0, \ \ m \text{ even}; \quad = n_0 - \operatorname{Sgn} n_0, \ \ m \text{ odd}$$

$$n_0 \equiv n(\alpha), \ \ |n_0| \leq 1 \quad \text{(II.2f)}$$

and the spin wave functions $U_\delta(k)$ satisfy a set of Dirac type momentum space wave equations in the space of the matrices S_{ab}. $U_\delta(k)$ is a direct sum of parts each of definite spin. The solutions[7] of these wave equations will fix the values of the various parameters c, α, m_{max}, n_0 (which include the values of the three invariants) for any given IUR. To summarize: the "reduced" wave function $\psi(x,\lambda)$ has the momentum dependence in exponentials $\exp(-ik\cdot x)$ and mass dependence in a linear combination of cylinder functions $Z_n(m\lambda)$ of just two orders multiplied by various powers of λ. (We know m is really the mass by Eq. (II.2e).)

I remark in passing that the x-dependent change of spin frame (II.2a) is a vitally important practical aid in solving the field equations. For it reduces the original translation operator $P_\mu \equiv M_{5\mu}$[8] to the familiar one without a spin part:

$$P'_\mu \equiv U^{-1}(x)(i\partial_\mu + S_{5\mu})U(x) = i\partial_\mu \quad \text{(II.3)}$$

This has the result that the transformed field equations now involve only differential operators independent of x^μ and therefore of a reasonably familiar, manageable type. Proof: the field equations are $Q'_i\,\psi(x,\lambda) = q_i\psi(x,\lambda)$ (q_i = eigenvalues of the invariants Q_i, $i = 2,3,4$) where the transformed invariants Q'_i commute with the $P'_\mu = i\partial_\mu$ and thus are x-independent, Q.E.D. In addition, Q'_2, Q'_4 and Q'_3 are differential operators of orders only 2 and 1, respectively.

Particular examples of these equations have been solved in the literature,[9] for example the scalar ($S_{ab} \equiv 0$), "spinor" ($S_{ab} \equiv (2i)^{-1}\beta_{[a}\beta_{b]}$, where $\beta_a\beta_b + \beta_b\beta_a = 2G_{ab}\underset{\sim}{1}$) and "vector" ($S_{ab} \equiv$ adjoint representation of the $\underset{\sim}{C}$-Lie algebra) IUR's.

III. Causality

Consider the "vector" IUR for definiteness; after quantization the causality of the resulting QFT will be determined by the field commutators in position space. (This is typical of any of the conformal IUR's discussed in Section II.) It comprises (in the inhomogeneous formalism) a Lorentz vector $A_\mu(x,\lambda)$ and Lorentz scalar $A_5(x,\lambda)$ field. One finds

$$[A_5(x_1,\lambda_1), A_5(x_2,\lambda_2)] = i\lambda_1\lambda_2\Delta_0(\lambda_1,\lambda_2,(-x^2)^{\frac{1}{2}}) \qquad \text{(III.1)}$$

where $x \equiv x_1 - x_2$ and

$$\Delta_0 = \int_0^\infty dm m J_0(m\lambda_1) J_0(m\lambda_2)\Delta(x;m) \qquad \text{(III.2)}$$

where $\Delta(x;m)$ is the usual commutator invariant function for mass m:

$$\begin{aligned}\Delta(x;m) &= (4\pi r)^{-1}\partial_r\{\epsilon(x^4)J_0[m(-x^2)^{\frac{1}{2}}]\}, \quad x^2 < 0\\ &= 0 \qquad\qquad\qquad\qquad\qquad\quad, \quad x^2 > 0 \qquad \text{(III.3)}\end{aligned}$$

For $A_\mu(x,\lambda)$, $\Delta_0 \to \Delta_1$ defined with two Bessel functions $J_1(m\lambda)$, while A_μ and A_5 commute.

Thus the commutators are determined by the $g_n(\lambda_1,\lambda_2,(-x^2)^{\frac{1}{2}})$, $n = 0, 1$, where we define the function

$$g_n(z_1,z_2,z_3) \equiv \int_0^\infty dy y J_n(yz_1) J_n(yz_2) J_0(yz_3) \qquad \text{(III.4)}$$

These integrals were evaluated, in the classical age of analysis, by McDonald[10] in 1909. One gets

$$\begin{aligned}g_0(\lambda_1,\lambda_2,(-x^2)^{\frac{1}{2}}) &= \frac{1}{\pi\lambda_1\lambda_2}\csc\theta, \quad |\cos\theta| < 1\\ &= 0 \quad, \quad |\cos\theta| > 1\end{aligned} \qquad \text{(III.5)}$$

in terms of the fundamental finite invariant, the angle θ under which the two spheres intersect:

$$\cos\theta \equiv [(x_1 - x_2)^2 + \lambda_1{}^2 + \lambda_2{}^2]/2\lambda_1\lambda_2 . \qquad \text{(III.6)}$$

g_1 has $\csc\theta \to \cot\theta$ in (III.5). Thus the fields commute at two points if the corresponding spheres do not intersect at all ($|\cos\theta| > 1$), while if they do intersect (θ a real angle), the commutator is a simple trigonometric function of θ. The supports of these commutators for fixed "gauges" λ_1, λ_2 are crescent shaped regions inside the light cone, as shown in the figure.

This shows that for the conformal QFT acausality troubles do not arise, even though conformal transformations can take space-like into time-like intervals and vice versa. The clarification of this puzzle lies in the compensating behavior of λ under $\underset{\sim}{C}$. Even though $(x_1 - x_2)^2 > 0$ may go into $(x_1' - x_2')^2 < 0$, the formulas (III.1) to (III.5) show that the causality depends not on $(x_1 - x_2)^2$ but on the extended quadratic invariant $\cos\theta$, Eq. (III.6), and this indeed is invariant under $\underset{\sim}{C}$. Or, said another way, one has the whole set of Poincaré fields for $0 < m < \infty$, not just those for a few isolated mass values. These arrange to interfere in just such a way that for those time-like intervals which can be transformed to space-like intervals (namely x_1, λ_1 and x_2, λ_2 such that $(x_1 - x_2)^2 < 0$ and $|\cos\theta| > 1$) the commutator is zero anyway. Those time-like intervals for which some signal is possible ($|\cos\theta| < 1$) can never be transformed into space-like intervals (see figure).

IV. Selection Rules on Mass

Consider a typical case of conformal quantum fields in interaction, the Yukawa coupling of the "vector" and "spinor" fields defined in Section II. The general form (II.2) shows that at a vertex there will be the $\delta(k_1 + k_2 - k_3)$ of momentum conservation, while the masses of the three lines will be constrained, not by a δ function, but by the product of three Bessel functions integrated over λ.[11)] But these are just our functions $g_n(m_1, m_2, m_3)$ already defined in Eq. (III.4), which here crop up in a new context.

Rewriting e.g. g_0 in a form more suitable to the present arguments, one gets

$$g_0(m_1, m_2, m_3) = \frac{2}{\pi}\left[\{m_3^2 - (m_1 - m_2)^2\}\{(m_1 + m_2)^2 - m_3^2\}\right]^{-\frac{1}{2}}$$

$$m_1 + m_2 > m_3 > |m_1 - m_2|,$$

$$= 0 \quad \text{otherwise}. \tag{IV.1}$$

These curious "continuous Clebsch-Gordan coefficients" therefore restrict the coupling between Poincaré fields of various masses: if the outgoing mass m_3 is too different from the ingoing masses m_1, m_2 (precisely, if it does not lie between their sum and difference) the coupling is zero. Note that the coupling is maximally strong at the

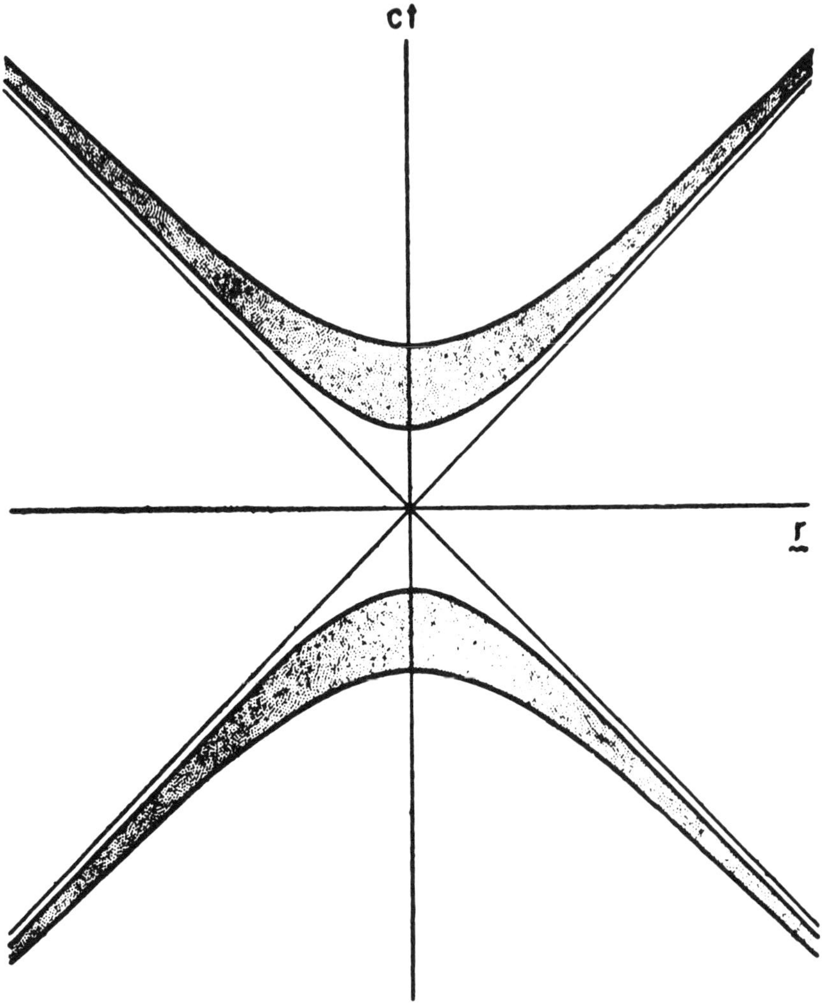

Support of the functions $g_n(\lambda_1, \lambda_2, (-x^2)^{\frac{1}{2}})$ for typical values of $\lambda_1 \neq \lambda_2$.

endpoints of the interval: if $m_3 = |m_1 - m_2| -$ or $(m_1 + m_2)+$, g_0 is actually infinite.

Another interesting feature is that the "coupling constant" for a second order process is not simply the product of the "coupling constants" for the first order processes. For if the two vertices are connected by an internal line, the integration over the internal line mass in Eq. (III.2) produces an effective coupling of the four masses m_1, m_2, m_1', m_2'

$$g_{eff}(m_1, m_2; m_1', m_2') = \int_0^\infty dm\, m g_0(m_1, m_2, m)\, g_0(m_1', m_2', m) \tag{IV.2}$$

where the integration is actually over the finite overlap between $(|m_1 - m_2|, m_1 + m_2)$ and $(|m_1' - m_2'|, m_1' + m_2')$. Thus a theory of "matrix coupling constants" results.

These elementary observations suggest that the conformal QFT at least has the mechanism of emphasizing some configurations[12] of initial and final masses over others. For example, the amplitude may have tremendous resonances, even blow up, at some sets of mass values. And it is clear that this depends on the dynamics, i.e., the topology of the Feynman graphs. One might then guess that these particular values of the masses should be those observed in nature. Of course, there are technical problems to settle first, e.g., how is one to treat a discrete mass value in a formalism with continuous mass spectra.

In any case I feel that the dynamical mechanism afforded by interacting conformal QFT gives a hope of solving the central mass problem alluded to in Section I, and is worth pursuing to a definite conclusion.

References and Footnotes

1. Our metric is diag (+++-)1. x^4 = time, p^4 = energy, etc.
2. $d\theta^2 = g_{\alpha\beta} dx^\alpha dx^\beta = \lambda^{-5}(dx^2 \pm d\lambda^2)$, $x^5 \equiv \lambda$. $d\theta$ ≡ angle between nearby spheres x^α and $x^\alpha + dx^\alpha$. This is a 5-dimensional Riemannian space of constant curvature.
3. In these theories it is assumed that fields transform like $\varphi(\sigma x) = \sigma^d \varphi(x)$ (1) for some d under (I.4). But it is clear that, strictly speaking, this "dimension" d does not exist for any field whose determining equations (field equations, commutation relations, boundary conditions, etc.) contain at least one dimensional constant. For one obviously then has $\varphi(\sigma x; m_i) = \sigma^D \varphi(x; \sigma m_i) \neq \sigma^d \varphi(x; m_i)$ for any d, where D is the ordinary dimension and the m_i are taken typically as masses. In particular (1) does not hold for any interacting field in a meaningful (nondivergent) formalism,

since one needs at least a high-momentum cut-off Λ to make the mathematics (e.g., renormalization) legitimate. Hence it is doubtful that results like "the change of d under an interaction" can be given any consistent sense.

4. Tacit boundary conditions: $\lambda^{-5}\wp^i$ (i=1,2,3) and $\lambda^{-5}\wp^5$ vanish at spatial ∞ and at $\lambda = 0$ and ∞, respectively.
5. The usual reason given is that the 1-dimensional symmetry group (I.2) is "trivial" or "mere dimensional analysis," as if this made it any different in kind from, say, the 3-dimensional rotation group, which is not considered to be "trivial."
6. R. L. Ingraham, in Lectures in Theoretical Physics, Vol. VIIIB (Univ. of Colo. Press, Boulder, 1966), pp. 375-406.
7. These equations have not yet been solved in the general case.
8. We use coordinates such that the fundamental quadric takes the form $G_{ab}X^aX^b = X^2 - 2X^0X^5 = 0$.
9. Y. Murai, Nucl. Phys. 6, 489 (1958).
   R. L. Ingraham, Nuovo Cimento 16, 104 (1960).
   R. L. Ingraham, Phys. Rev. 106, 1324 (1957).
10. H. M. McDonald, Proc. Math. Soc. (London) 7, 142 (1909).
11. The weight λ will always occur in these integrals.
12. Naturally only the ratios of these masses will be determined. But this is exactly what we need, remembering that the dilatation group (I.2) means changes of unit.

CONFORMAL-INVARIANT PHASE SHIFT ANALYSIS†

L. Castell
Max-Planck Institut für Physik und Astrophysik
K. Ringhofer
Universität München
München, Germany

Abstract

We define the conformal-invariant S-matrix for elastic scattering of spin zero particles with mass $m = 0$ and with mass $m > 0$. For the S-matrix a partial wave expansion is given. The comparison with the usual scattering phases $(e^{2i\delta_\ell(s)} - 1)$ allows us to make definite predictions about their s dependence and certain angular momentum correlations.

I. Physical Representations of the Conformal Group for Scalar Particles

In this paragraph we introduce the unitary irreducible representations of the conformal group which describe spinless particles with mass $m > 0$ and mass $m = 0$.

It is well known that the identity component of the conformal group $SO_0(4,2)/C_2$ contains the identity component of the Poincaré group $SO_0(3,1)x)T_4$ as a subgroup. Furthermore it contains the dilatations

$$y^{\mu\prime} = \rho\, y^{\mu}, \quad 0 < \rho < \infty, \tag{1a}$$

and the special conformal transformations

$$y^{\mu\prime} = \frac{y^{\mu} - b^{\mu} y^2}{\sigma(y,b)}\ , \quad \sigma(y,b) = 1 - 2by + b^2 y^2\ , \ \text{metric } {+}{+}{+}{-},$$

$$-\infty < b_{\mu} < +\infty \quad . \tag{1b}$$

---

†Presented by L. Castell at the Symposium on De Sitter and Conformal Groups, University of Colorado, Summer 1970.

The Klein-Gordon equation for massless particles

$$\Box\,\varphi(y) = 0\ , \tag{2}$$

and also the scalar product

$$(\varphi,\psi) = i\int \varphi^*(y)\,\frac{\overleftrightarrow{\partial}}{\partial y^4}\,\psi(y)\,d^3\vec{y}\ , \tag{3}$$

are invariant under the conformal transformations; the wavefunction transforms like

$$\varphi'(y) = \frac{1}{\rho}\,\varphi(y/\rho) \qquad \text{under dilatations} \tag{4a}$$

and

$$\varphi'(y^*) = \frac{1}{\sigma(y,-b)}\,\varphi\left(\frac{y^\mu + b^\mu y^2}{\sigma(y,-b)}\right) \qquad \text{under special conformal transformations.} \tag{4b}$$

From this (and from the Poincaré transformations) we find the infinitesimal operators of the conformal group for zero mass particles:

$$M_{\mu\nu} = \frac{1}{i}\left(y_\mu\frac{\partial}{\partial y^\nu} - y_\nu\frac{\partial}{\partial y^\mu}\right),\quad P_\mu = \frac{1}{i}\,\frac{\partial}{\partial y^\mu}\ ,$$

$$D = \frac{1}{i}\left(y^\mu\frac{\partial}{\partial y^\mu} + 1\right),\quad K_\mu = \frac{1}{i}\left(-y^2\frac{\partial}{\partial y^\mu} + 2y_\mu\left(y^\rho\frac{\partial}{\partial y^\rho} + 1\right)\right). \tag{5}$$

The Fourier transform $\hat{\varphi}(\vec{p})$ in momentum space is defined by

$$\varphi(y) = (2\pi)^{-3/2}\int\hat{\varphi}(p)\,e^{ipy}\,\frac{d^3\vec{p}}{2p^4}\,,\quad p^4 = |\vec{p}|\,, \tag{6}$$

and the corresponding infinitesimal operators in momentum space are given by

$$M_{ij} = \frac{1}{i}(p_i\partial_j - p_j\partial_i),\quad M_{4i} = \frac{1}{i}\,p^4\partial_i,\quad P_\mu = p_\mu\ ,$$

$$D = i(p^j\partial_j + 1),\quad K_i = p_i\partial^j\partial_j - 2(p^j\partial_j + 1)\partial_i,\quad K_4 = p_4\partial_j\partial^j\ ,$$

$$\partial_j = \frac{\partial}{\partial p^j}\ ,\quad i,j = 1,2,3. \tag{7}$$

For the scalar product we obtain in momentum space

$$\langle \varphi, \psi \rangle = \int \hat{\varphi}^*(\vec{p})\, \hat{\psi}(\vec{p})\, \frac{d^3\vec{p}}{2p^4} \quad . \tag{8}$$

We normalize the improper eigenstates $|\vec{p}\rangle$ of the three momentum $\vec{p}$ to

$$\langle \vec{p}|\vec{p}'\rangle = 2|\vec{p}|\delta^3(\vec{p}-\vec{p}') \ . \tag{9}$$

Which representation of the conformal group shall we use to describe scalar particles with mass? Let us make the assumption that a particle with mass transforms in the same way as a certain state of a system of two massless particles. (One could even go so far as to describe a particle with mass as a scattering state of two massless particles, but we do not want to make such a specific interpretation at the moment.)[1] What we have to do is to take the direct product of two representations (7) and to pick out the representation with relative angular momentum zero.

Particles one and two have the four-momenta $p^{\mu}_{(1)}$ and $p^{\mu}_{(2)}$, respectively. Moreover, we have $p^4_{(1)} > 0$, $p^4_{(2)} > 0$ and $p^2_{(1)} = p^2_{(2)} = 0$. Therefore, there are 6 independent variables $p^i_{(1)}$, $p^i_{(2)}$. For the time being, we are only interested in that part of the product representation with the relative angular momentum zero. So we can assume that the functions in the product Hilbert space depend only on the four components of the total momentum $p^{\mu} = p^{\mu}_{(1)} + p^{\mu}_{(2)}$.

$$p^i = p^i_{(1)} + p^i_{(2)} \ , \qquad p^4 = |\vec{p}_{(1)}| + |\vec{p}_{(2)}| \ , \tag{10}$$

$$\frac{\partial}{\partial p^i_{(1)}} = \frac{\partial}{\partial p^i} + \frac{p^i_{(1)}}{|\vec{p}_{(1)}|}\frac{\partial}{\partial p^4} \ , \quad \frac{\partial}{\partial p^i_{(2)}} = \frac{\partial}{\partial p^i} + \frac{p^i_{(2)}}{|\vec{p}_{(2)}|}\frac{\partial}{\partial p^4} \ . \tag{11}$$

The infinitesimal operators of the product representation are then found to be

$$M_{\mu\nu} = M^{(1)}_{\mu\nu} + M^{(2)}_{\mu\nu} = \frac{1}{i}(p_{\mu}\partial_{\nu} - p_{\nu}\partial_{\mu}) \ , \quad P_{\mu} = P^{(1)}_{\mu} + P^{(2)}_{\mu} = p_{\mu} \ ,$$

$$D = D^{(1)} + D^{(2)} = i(p^{\mu}\partial_{\mu} + 2) \ ,$$

$$K_{\mu} = K^{(1)}_{\mu} + K^{(2)}_{\mu} = p_{\mu}\partial^{\rho}\partial_{\rho} - 2(p^{\rho}\partial_{\rho} + 2)\partial_{\mu} \ ,$$

$$\mu,\nu,\rho = 1,2,3,4 \ . \tag{12}$$

From Lorentz invariance it follows that the measure in the scalar product must have the form $g(m^2)\, d^4 p$. Because the representation (12) is by construction self-adjoint, $g(m^2)$ has to be a constant, which we chose to be one. Therefore

$$\langle \varphi | \psi \rangle = \int_{\substack{p^4 > 0 \\ m^2 > 0}} \varphi^*(p)\, \psi(p)\, d^4 p \ , \tag{13}$$

where the integration has to be performed over the forward cone. We will normalize the states $|p\rangle$ with four-momentum p according to

$$\langle p | p' \rangle = \delta^4 (p - p') \ . \tag{14}$$

Actually, the representation (12) of the conformal group, which is supposed to describe particles with mass, is only the first member of a whole series of similar representations[2] of the group $SO_0(4,2)$. Each of these representations is characterized by an integer $\nu \geq 0$, and its infinitesimal operators are

$$M_{\mu\nu} = \frac{1}{i}(p_\mu \partial_\nu - p_\nu \partial_\mu) \ , \quad P_\mu = p_\mu \ ,$$

$$D = i(p^\mu \partial_\mu + 2) \ ,$$

$$K_\mu = p_\mu (\partial_\rho \partial^\rho + \frac{\nu^2}{p^2}) - 2(p^\rho \partial_\rho + 2)\, \partial_\mu \ . \tag{15}$$

The scalar product is again given by (13).

Representations with $\nu = 1, 2, 3 \ldots$ are obtained if one studies the transformation behavior of two-particle states, where each particle is massless but is allowed to carry the same helicity $\nu/2$ or $-\nu/2$. Therefore the quantum number ν contains the information which pair of massless particles has to be used to build up a certain scalar particle with mass. As the π^0 couples electromagnetically to 2γ it has been suggested[3] that the π^0 should be described by the massive representations (15) with the conformal quantum number $\nu = 2$. (A massless particle cannot couple conformal-invariantly to two mass zero particles.) It should be mentioned that a generalized Goldstone argument[3),4)] leads to the same representation for a conformal Goldstone particle. Therefore this massive Goldstone particle can be identified with the π^0. After having assigned the mass zero particles like the neutrino and the γ-quantum, and the massive π^0 to definite irreducible unitary representations of the group $SO_0(4,2)$ or its spin-covering group $SU_0(2,2)$ the question arises, under which conditions could conformal symmetry be physically relevant?[3] A preliminary answer to this problem is that in an extremely relativistic situation, where the energy and the momentum transfer are high compared with the masses of the incoming (and perhaps outgoing) particles $s, -t, -u \gg \Sigma m_i^2$, the scattering matrix may show conformal symmetry. We shall see that the result of the limit $s, -t, -u \to \infty$ will turn out to be different, if we are dealing with a mass zero or a $0 < m^2 < \infty$ representation of the group $SO_0(4,2)$. The additional assumption which some people have proposed that one can neglect all particle masses right from the beginning (in this introduction we have already given some comments which do not justify this assumption) is much more specific and would restrict conformal invariance far too much. Even in the limit of asymptotic high energies phase shifts for massive particle scattering behave quite differently from the phase shifts for mass zero scattering.

II. Clebsch-Gordan Coefficients and S-Matrices

A. Group Theoretical Preliminaries

We start the analysis of the S-matrix elements by the description of the ingoing (or outgoing) two particle states. They will be simply a direct product of two one-particle states with three-momenta $\vec{p}_1$ and $\vec{p}_2$, positive energy $p^4_{(1)} > 0$, $p^4_{(2)} > 0$, with real masses m_1 and m_2, and with conformal quantum numbers ν_1 and ν_2, respectively. (For mass zero particles there is no new conformal quantum number.)[5] Therefore the incoming states are labeled by $|\nu_1, p_1; \nu_2, p_2\rangle$ and the outgoing ones by $\langle\nu_1, p_1'; \nu_2, p_2'|$ for elastic scattering.

The two-particle product representation will always be reducible, its irreducible components will be specified by one or several quantum numbers α. In our examples we shall always find that α is discrete and that one of the α's is given by the maximal spin of a spin multiplet.

Within an irreducible representation α of the direct product the basis states will be labeled by the quantum numbers ℓ, ℓ_3, p^μ, where ℓ is the value of the relative angular momentum or total spin (usually more than one ℓ occurs within one irreducible representation) and ℓ_3 is the value of the third component of the angular momentum. Therefore the basis states can be written as $|\alpha, p, \ell, \ell_3\rangle$ and the normalization can be chosen to be

$$\langle\alpha, p, \ell, \ell_3|\alpha', p', \ell', \ell_3'\rangle = \delta_{\alpha,\alpha'}\, \delta^4(p-p')\, \delta_{\ell,\ell'}\, \delta_{\ell_3,\ell_3'}\,. \tag{16}$$

Using the completeness relation, the matrix element of the operator S-1 is given by

$$\langle \nu_1, p_1'; \nu_2, p_2' | S-1 | \nu_1, p_1; \nu_2, p_2 \rangle = \int d^4p\, d^4p' \sum_{\substack{\alpha,\alpha' \\ \ell,\ell',\ell_3,\ell_3'}}$$

$$\langle \nu_1, p_1'; \nu_2, p_2' | \alpha', p', \ell', \ell_3' \rangle \langle \alpha', p', \ell', \ell_3' | S-1 | \alpha, p, \ell, \ell_3 \rangle$$

$$\langle \alpha, p, \ell, \ell_3 | \nu_1, p_1; \nu_2, p_2 \rangle \quad . \tag{17}$$

From the assumption that the S-matrix is invariant under conformal transformations, and from the fact that there is no degeneracy in our examples, follows that

$$\langle \alpha', p', \ell', \ell_3' | S-1 | \alpha, p, \ell, \ell_3 \rangle = (f(\alpha)-1)\, \delta^4(p-p')\delta_{\ell,\ell'}\, \delta_{\ell_3,\ell_3'}\, \delta_{\alpha,\alpha'} \tag{18}$$

Unitarity means in this generalized framework that the energy-independent reduced S-matrix elements $f(\alpha)$ obey the relation $|f(\alpha)| \leq 1$.

The Clebsch-Gordan coefficient in (17) has the form

$$\langle \alpha, p, \ell, \ell_3 | \nu_1, p_1; \nu_2, p_2 \rangle = g^{\alpha}_{\nu_1,\nu_2}(M^2, \ell, m_1^2, m_2^2) Y_{\ell,\ell_3}(\vec{e})\delta^4(p-p_1-p_2), \tag{19}$$

where $M^2 = -(p_1+p_2)^2$ and the unit vector $\vec{e}$ specifies the direction of $\vec{p}_1$ in the center of mass system. Our task is to determine $g^{\alpha}_{\nu_1,\nu_2}$. The two other factors in (19) follow from Lorentz invariance. Inserting (18) and (19) into (17) we obtain

$$\langle \nu_1, p_1'; \nu_2, p_2' | S-1 | \nu_1, p_1; \nu_2, p_2 \rangle =$$

$$\frac{1}{4\pi} \sum_{\alpha,\ell} (f(\alpha)-1)\, g^{\alpha}_{\nu_1,\nu_2}(s, \ell, m_1^2, m_2^2)\, g^{\alpha *}_{\nu_1,\nu_2}(s, \ell, m_1'^2, m_2'^2)(2\ell+1)$$

$$P_\ell(\cos\theta)\, \delta^4(p_1+p_2-p_1'-p_2') \ , \tag{20}$$

where we have used the relation $\sum_{\ell_3} Y_{\ell,\ell_3}(\vec{e}) Y^*_{\ell,\ell_3}(\vec{e}') = \frac{2\ell+1}{4\pi} P_\ell(\cos\theta)$,

$\cos\theta = \vec{e}\cdot\vec{e}'$. The connection between the $f(\alpha)$ and the usual reduced S-matrix element $e^{2i\delta_\ell(s)}$ will be treated in the examples. Now we shall present the results in detail.

B. Calculations

In this paragraph we determine the Clebsch-Gordan coefficients[3] for two spin zero representations. The product Hilbert space[6] for the two particle states is given by

$$\int \psi^* \psi \frac{d^3\vec{p}_1}{2p_1^{\,4}} \frac{d^3\vec{p}_2}{2p_2^{\,4}} dm_1^2\, dm_2^2 = \int \psi^*\psi\, d^4p^+ \frac{\lambda^{\frac{1}{2}}}{8M^2} d^2\vec{e}\, dm_1^2\, dm_2^2 \quad ,$$

$$p^{\pm} = p_1 \pm p_2\,, \quad M^2 = -(p^+)^2\,, \quad q = \frac{M}{\lambda^{\frac{1}{2}}}\left(p^- - \frac{m_1^2 - m_2^2}{M^2}\, p^+\right) \,,$$

$$\vec{e} = \vec{q} - (q^4/M + p^{4+})\, \vec{p}^{\,+}\,, \qquad \vec{e}^{\,2} = 1\,,$$

$$\lambda = M^4 - 2M^2(m_1^2 + m_2^2) + (m_1^2 - m_2^2)^2 \quad . \tag{21}$$

In case m_1 or m_2 or both are zero, we have to omit the corresponding integration in (21).

In order to calculate the Clebsch-Gordan coefficients we have to diagonalize the Casimir operators of the product representations. In all the cases we shall consider, it is sufficient to deal only with the first operator C_I, because this operator already distinguishes the representations. The operator C_I is given by

$$C_I = \tfrac{1}{2} M_{\mu\nu} M^{\mu\nu} + \tfrac{1}{2}(P_\mu K^\mu + K_\mu P^\mu) - D^2 \ .$$

We shall not give the details of the calculation in the first two cases where the results can be checked directly.

1) First we consider the case where $m_1 = m_2 = 0$. The norm (21) is now simply given by

$$\tfrac{1}{8} \int \psi^* \psi\, d^4 p^+\, d^2\vec{e} \quad . \tag{22}$$

The Casimir operator C_I of the product representation turns out to be

$$C_I = 2\ell(\ell+1) - 4. \tag{23}$$

Each angular momentum state belongs to a different irreducible representation of the conformal group. The Clebsch-Gordan coefficient is given by

$$\langle C(\ell), p, \ell_3 | \vec{p}_1 ; \vec{p}_2 \rangle = 2^{3/2} \, Y_{\ell, \ell_3}(\vec{e}) \, \delta^4(p - p_1 - p_2) \, . \tag{24}$$

It is easy to check that this Clebsch-Gordan coefficient obeys the completeness and orthogonality relations with respect to the norm (22). For the S-matrix element we obtain finally

$$\langle p_1' ; p_2' | S-1 | p_1 ; p_2 \rangle = \frac{2}{\pi} \sum_{\ell} (f_\ell - 1)(2\ell+1) \, P_\ell(\cos\theta) \, \delta^4(p_1 + p_2 - p_1' - p_2') . \tag{25}$$

The reduced S-matrix elements $f_\ell \equiv e^{2i\delta_\ell}$ are constants $|f_\ell| \leq 1$.

2) Next consider the case where $m_1 > 0$ and $m_2 = 0$. If we introduce the variable $x = m_1^2/M^2$, we can express (21) by

$$\frac{1}{8} \int \psi^* \psi \, d^4 p^+ \, M^2 d^2 \vec{e} \, (1-x) dx \, . \tag{26}$$

The Casimir operator C_I is given by

$$C_I = 2(n+\ell)(n+\ell+\nu_1+2) + (\nu_1+1)^2 - 4 , \tag{27}$$

and the Clebsch-Gordan coefficient has the form

$$\langle C(n+\ell), p, \ell, \ell_3 | \nu_1, p_1 ; \vec{p}_2 \rangle = \frac{1}{\sqrt{N}} (1-x)^{\ell} x^{\nu_1/2} P_n^{(1+2\ell, \nu_1)}(2x-1)$$

$$Y_{\ell, \ell_3}(\vec{e}) \frac{1}{M} \delta^4(p - p_1 - p_2) \, .$$

$$n = 0, 1, 2, \ldots \, . \tag{28}$$

The normalization factor is given by

$$N = \frac{2^{-3}(n+2\ell+1)!(n+\nu_1)!}{(2n + 2\ell + \nu_1 + 2)n!(n + 2\ell + \nu_1 + 1)!} \tag{29}$$

The important point is that one irreducible representation is described by the quantum number $n + \ell$, which equals the maximum spin which is contained in one irreducible representation. This has to be proved separately and does not follow from (27).

The S-matrix is given by

$$\frac{2}{\pi}\frac{1}{\sqrt{1-x}}\frac{1}{\sqrt{1-x'}}\sum_{\ell}\Bigg[\sum_{n} f_n^{\ell}\,\frac{1}{8sN}\,P_n^{(1+s\ell,\nu_1)}(2x-1)\,P_n^{(1+2\ell,\nu_1)}(2x'-1)\cdot$$

$$\cdot((1-x)(1-x'))^{\ell+\frac{1}{2}}\,(xx')^{\nu_1/2}-\delta(m_1{}^2-m_1{}'^2)\Bigg]\;(2\ell+1)P_{\ell}(\cos\theta)\cdot$$

$$\cdot\,\delta^4(p_1+p_2-p_1{}'-p_2{}')\,. \qquad (30)$$

In this expression we have replaced M^2 by the total energy s. Correspondingly we have $x = m_1{}^2/s$, $x' = m_1{}'^2/s$. From the irreducibility of the product representation follows $f_n^{\ell} = f_{n-1}^{\ell+1} = f_{n-2}^{\ell+2} = f_{n+1}^{\ell-1} = \ldots$ etc. In order to compare the matrix element (30) with the usual S-matrix elements for elastic scattering where the one-particle states are normalized to $2p^4\delta^3(\vec{p}'-p)$, we have to form a wave packet with respect to the mass. We integrate the incoming and outgoing states over $\int D(m_1{}^2-\bar{m}^2)dm_1{}^2$ and $\int D(m_1{}'^2-\bar{m}^2)dm_1{}'^2$, respectively. The function $D(m^2-\bar{m}^2)$ is different from zero only near $\bar{m}^2$, the actual mass of the particle, and is normalized to one.

$$\int |D(m^2-\bar{m}^2)|^2\;dm^2 = 1. \qquad (31)$$

Instead of

$$\Bigg[\sum_{n} f_n^{\ell}\,\frac{1}{8sN}\,P_n^{(1+2\ell,\nu_1)}(2x-1)\,P_n^{(1+2\ell,\nu_1)}(2x'-1)((1-x)(1-x'))^{\ell+\frac{1}{2}}$$

$$\cdot\,(xx')^{\nu_1/2}-\delta(m_1{}^2-m_1{}'^2)\Bigg] \qquad (32)$$

we obtain

$$\sum_{n}(f_n^{\ell}-1)\;\Big|\int D(m_1{}^2-\bar{m}^2)\frac{1}{\sqrt{8sN}}P_n^{(1+2\ell,\nu_1)}(2x-1)(1-x)^{\ell+\frac{1}{2}}\,x^{\nu_1/2}\,dm^2\Big|^2$$

$$\equiv (e^{2i\delta_{\ell}(s,\bar{m}^2)}-1)\,. \qquad (33)$$

From the generalized unitarity condition $|f_n^{\ell}| \le 1$ we derive that the usual necessary condition for unitarity $|e^{2i\delta\ell}| \le 1$ holds.

This follows from the absolute convergence of the sum

$$\sum_{n=0}^{\infty}\Big|D(m_1{}^2-\bar{m}^2)\,\frac{1}{\sqrt{8sN}}\,P_n^{(1+2\ell,\nu_1)}(2x-1)(1-x)^{\ell+\frac{1}{2}}\,x^{\nu_1/2}\,dm^2\Big|^2 = 1. \qquad (34)$$

If ϵ^2 is the half width of $D(m^2 - \bar{m}^2)$ we can approximate the expression (33) $e^{2i\delta_\ell} - 1$ for the phases δ_ℓ by

$$e^{2i\delta_\ell} - 1 \approx$$

$$\epsilon^2 \sum_{n=0}^{s/\epsilon^2} (f_n^\ell - 1) \frac{1}{8sN} (P_n^{(1+2\ell,\nu_1)} (2\frac{\bar{m}^2}{s} - 1))^2 (1 - \frac{\bar{m}^2}{s})^{2\ell+1} (\frac{\bar{m}^2}{s})^{\nu_1} . \tag{35}$$

The high energy limit of this expression is given by $e^{2i\delta_\ell} - 1 \approx (\frac{\bar{m}^2}{s})^{\nu_1} (\frac{\epsilon^2}{s})$ for $s \to \infty$. Here we have assumed that the sum (35) converges uniformly. Quite unexpected is the fact that in the expression for the phase shift (35) the threshold behavior $(e^{2i\delta_\ell} - 1) \approx (1 - \frac{m^2}{s})^{2\ell+1}$ for $s \to \bar{m}^2$ turns out to be correct. This suggests that one determines for each partial wave the constants f_n^ℓ; $n = 0,1,2...$ from the experimental data. If conformal invariance is satisfied the relations

$$f_n^\ell = f_{n-1}^{\ell+1} = f_{n+1}^{\ell-1} = \ldots \quad \text{etc.}$$

have to hold.

3) Now we consider the case where each of the two incoming particles has mass different from zero. The norm in the product space is given by (21). In the variables $p_\mu{}^+$ and $p_\mu{}^-$ the Casimir operator C_I is given by

$$C_I \psi = \Big[(p^{+2} - p^{-2})(\partial_- \partial_- + \frac{\nu_1^2}{(p^+ + p^-)^2} + \frac{\nu_2^2}{(p^+ - p^-)}) - 2(p^+ \partial_-)^2$$
$$+ 2(p^- \partial_-)^2 + 6(p^- \partial_-) + \nu_1^2 + \nu_2^2 \Big] \psi . \tag{36}$$

In this case we shall not give the exact solution of theproblem, because it is very cumbersome to separate the variables of the first Casimir operator. Instead we shall give only the high energy, and the threshold behavior of the Clebsch-Gordan coefficients. If we introduce the variables $x = m_1^2/M^2$ and $y = m_2^2/M^2$ and solve the equation (36) for $x, y \ll 1$, we see that the solution is independent of C_I. The condition that (36) is a self-adjoint operator selects the square-integrable solutions. We obtain

$$\langle C,p,\ell,\ell_3|\nu_1,p_1;\nu_2,p_2\rangle \approx \frac{1}{M^2}\left(\frac{m_1}{M}\right)^{\nu_1}\left(\frac{m_2}{M}\right)^{\nu_2} Y_{\ell,\ell_3}(\vec{e})\delta^4(p-p_1-p_2). \tag{37}$$

From this we can calculate the high energy behavior of the phase shifts under similar conditions than in the case 2), and obtain

$$(e^{2i\delta_\ell(s,\bar{m}^2)} - 1) \approx \left(\frac{\bar{m}_1^2}{s}\right)^{\nu_1}\left(\frac{\bar{m}_2^2}{s}\right)^{\nu_2}\left(\frac{\epsilon_1^2}{s}\right)\left(\frac{\epsilon_2^2}{s}\right) . \tag{38}$$

To calculate the threshold behavior we transform (36) into the center of mass system, and solve the eigenvalue equation for small values of $|\vec{p}^-|/M$. The first term in the expansion gives us the threshold behavior of the Clebsch-Gordan coefficient. We obtain

$$\langle c,p,\ell,\ell_3|\nu_1,p_1;\nu_2,p_2\rangle \approx \frac{1}{M^2}\left(\frac{\lambda^{\frac{1}{2}}}{M^2}\right)^{\ell} Y_{\ell,\ell_3}(\vec{e})\ \delta^4(p-p_1-p_2),$$

$$\lambda = M^4 - 2M^2(m_1^2 + m_2^2) + (m_1^2 - m_2^2)^2 . \tag{39}$$

From this follows for the phase shifts for elastic scattering

$$(e^{2i\delta_\ell(s,\bar{m}_1^2,\bar{m}_2^2)} - 1) \approx \left(\frac{\epsilon_1^2}{s}\right)\left(\frac{\epsilon_2^2}{s}\right)\left[1 - 2\left(\frac{\bar{m}_1^2}{s}+\frac{\bar{m}_2^2}{s}\right) + \left(\frac{\bar{m}_1^2}{s} - \frac{\bar{m}_2^2}{s}\right)^2\right]^{\ell+\frac{1}{2}} . \tag{40}$$

Equation (40) exhibits the correct threshold behavior $|\vec{p}|^{2\ell+1}$ in a relativistic invariant form. This result gives the hint that a conformal phase shift analysis might also describe some aspects of low energy scattering.

Acknowledgment

One of the authors (L.C.) would like to thank Dr. P. Breitenlohner for interesting discussions.

References

1. Different physical motivations for the representations of the conformal group describing massive particles can be found in: J. A. Schouten and J. Haantjes, Proc. Ned. Acad. Wet. 39, 1059 (1936); Y. Murai, Nucl. Phys. 6, 489 (1958); R. Ingraham, Nuovo Cimento 12, 825 (1954); 16, 104 (1960); L. Castell,

Nuovo Cimento 46, 1 (1966); M. Flato, J. Simon and D. Sternheimer, "Conformal Covariance of Field Equations," to be published; J. Mickelsson and J. Niederle, "On Representations of the Conformal Group Which When Restricted to its Weyl Subgroup Remain Irreducible," to be published.
2. The mathematical properties of these representations have been discussed: R. Raczka, N. Limić and J. Niederle, J. Math. Phys. 7, 1861 (1966); L. Castell, Nucl. Phys. B4, 343 (1967); J. Math. Phys., to be published.
3. L. Castell, Phys. Rev. D2, 1161 (1970).
4. L. Castell, "Spontaneously Broken Symmetry and Conformal Invariance," to be published.
5. G. Mack and I. Todorov, J. Math. Phys. 10, 2078 (1969) L. Castell, Nucl. Phys. B13, 231 (1969).
6. A. J. Macfarlane, J. Math. Phys. 4, 490 (1963); H. Joos, Fortschr. Phys. 10, 65 (1962).

RESTRICTIONS ON INELASTIC CHANNELS FROM CONFORMAL INVARIANCE†

L. Castell
Max-Planck-Institut für Physik und Astrophysik
München, Germany

Abstract

It is shown group-theoretically that conformal invariance restricts all final states of two incoming mass zero particles very strongly. We apply these results to a conformal-invariant scalar field theory and to γ-γ scattering, and conclude that there could be at most elastic scattering. The paper ends with a comment on deep inelastic electron-proton scattering.

I. Mass Zero Representations of $SU_0(2,2)$

Let us consider all irreducible unitary mass $m = 0$, positive energy $p^4 > 0$ representations[1)] of the spin-covering group $SU_0(2,2)$ of the identity component of the conformal group $SO_0(4,2)/C_2$. All these representations are contained in the exceptional degenerate discrete series[2)] E^+. Each irreducible representation in this series is completely characterized by a certain value of the helicity λ, where $\lambda = 0$, $\pm\frac{1}{2}$, ± 1, $\pm 3/2, \ldots$ etc. The states of one irreducible representation can be specified by the quantum numbers of the maximal compact subgroup of $SU_0(2,2)$, namely $SU(2) \times SU(2) \times U(1)$. The quantum numbers are given by $s_{(1)}$, $s_{(1),3}$, $s_{(2),3}$, and n. The most important relation between these quantum numbers are

$$n = s_{(1)} + s_{(2)} + 1, \tag{1}$$

and

$$\lambda = s_{(1)} - s_{(2)} . \tag{2}$$

---

†Presented at the Symposium on De Sitter and Conformal Groups, University of Colorado, Summer 1970.

This leaves three discrete integer or half-integer quantum numbers $s_{(1),3}$, $s_{(2),3}$, and $n \geq 1$ to describe the states of one irreducible representation. They correspond to the three-vector $\vec{p}$ in the momentum basis[3]. We can represent the structure of these irreducible representations in a j - n diagram, $j = s_{(1)} + s_{(2)}$.

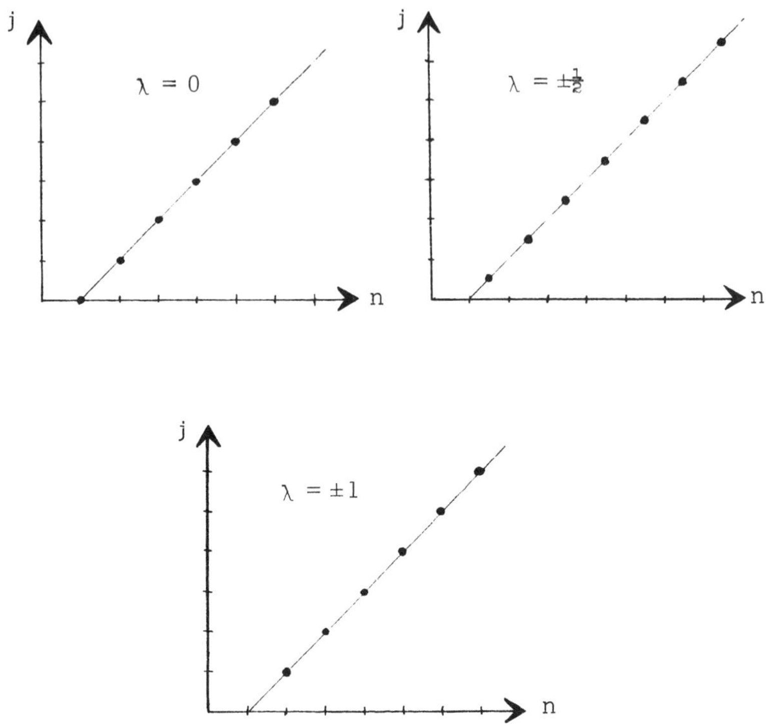

Fig. 1. Mass m = 0, $p^4 > 0$ representations.

II. The General Structure of the Direct Product of Mass m = 0, $p^4 > 0$ Representations

For simplicity we now omit $s_{(i),3}$, $i = 1,2$, and describe the states of the representations by the quantum number $(s_{(1)}, s_{(2)}, n)$, or more conveniently by $[j, \lambda, n]$, where $j = s_{(1)} + s_{(2)}$, $\lambda = s_{(1)} - s_{(2)}$. For mass m = 0, $p^4 > 0$ representations we have $j = n - 1$, and λ is the helicity. The general product state is given by

$$(s'_{(1)}, s'_{(2)}, n') \times (s''_{(1)}, s''_{(2)}, n'') = \sum (s_{(1)}, s_{(2)}, n) \quad ,$$

$$s_{(i)} = s'_{(i)} + s''_{(i)}, \; s'_{(i)} + s''_{(i)} - 1, \ldots \; |s'_{(i)} - s''_{(i)}|, \quad n = n' + n''. \tag{3}$$

Note that for all states in the product of two mass $m = 0$, $p^4 > 0$ representations the maximum j obeys the equation

$$\max j = j' + j'' = n' + n'' - 2 = n - 2. \tag{4}$$

Differently expressed

$$n \geq j + 2 \quad . \tag{5}$$

For the direct product of α mass $m = 0$, $p^4 > 0$ representations the general structure is given by the relation

$$n \geq j + \alpha \quad . \tag{6}$$

This equation is expressed in Fig. 2.

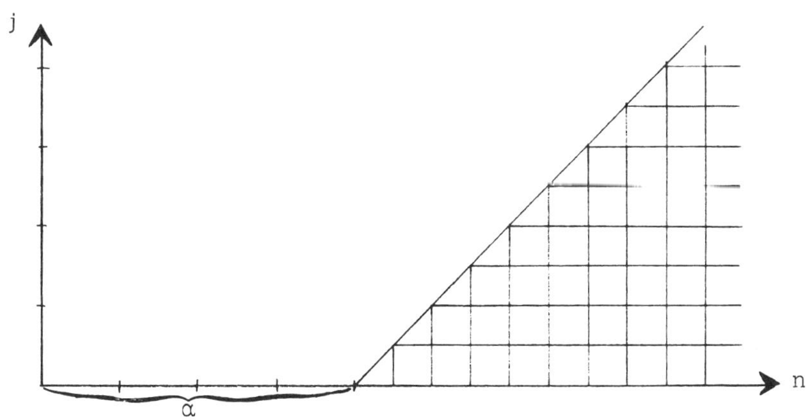

Fig. 2. The direct product of α mass $m = 0$, $p^4 > 0$ representations.

The restriction of the representations of Fig. 2 to the Poincaré group shows that they are characterized by $0 < m^2 < \infty$ and $p^4 > 0$.

III. The Physical Massive Representations of $SU_o(2,2)$

We define the physical representations of $SU_o(2,2)$ by their restriction with respect to the Poincaré group. The physical representations are characterized by the mass spectrum $m = 0$ or $0 < m^2 < \infty$ and by the sign of the energy $p^4 > 0$. In the discrete basis (for more details see Section IV) this corresponds to $n > 0$, where n is the eigenvalue of M_{46}. We have seen that for physical $m = 0$ representation $n = j+1$, and that for the direct product of two $m = 0$ representations $n \geq j + 2$. Yao[2] has shown that for all degenerate representations of $SU_o(2,2)$ for which $n > 0$ we have either the mass $m = 0$ case, or $n \geq j + 2$ for the massive representations. For this we are led to the following conjecture.

All irreducible unitary representations of $SU_o(2,2)$ characterized by a mass spectrum $0 < m^2 < \infty$ and positive energy $p^4 > 0$ are contained in the direct product of a finite number of mass $m = 0$, $p^4 > 0$ representations of $SU_o(2,2)$.

In a scattering process of mass zero particles the investigation of all inelastic massive channels is thus reduced to the investigation of all inelastic mass $m = 0$ channels. And this is the problem we are going to solve.

IV. The Direct Product of Two Mass $m = 0$, $p^4 > 0$, spin 0 Representations

The irreducible representations of the direct product[4] of two mass $m = 0$, $p^4 > 0$, spin $\lambda = 0$ are specified by the mass spectrum $0 < m^2 < \infty$, and by the fixed angular momentum $\ell = 0, 1, 2, 3 \ldots$. This follows from dilatational invariance, and from the value of the first Casimir operator

$$C_I = 2\ell(\ell + 1) - 4 . \tag{7}$$

For the following we shall use the local isomorphism $SU_o(2,2) \approx SO_o(4,2)$, and correspondingly for the maximal compact subgroup $SU(2) \times SU(2) \times U(1) \approx SO(4) \times SO(2)$. If we use the metric +++-+- the first Casimir operator is given by

$$C_I = \tfrac{1}{2} M_{ik}M_{ik} + M_{46}^2 - (M_{i4}M_{i4} + M_{i6}M_{i6}),$$

$$ik = 1, 2, 3, 5. \tag{8}$$

Or if we introduce the rising and lowering operators $M_k^{\pm} = M_{k6} \pm i M_{k4}$ with respect to the eigenvalues n of the operator M_{46}, we obtain

$$C_I = \tfrac{1}{2} M_{ik} M_{ik} + M_{46}^2 - M_i^+ M_i^- - 4M_{46}. \tag{9}$$

The lowest eigenstate with respect to the eigenvalue n is defined by $M_k^-(0, 0, 1) = 0$. For the direct product $M_{\mu\nu}$ is given by

$$M_{\mu\nu} = M'_{\mu\nu} + M''_{\mu\nu}, \quad \mu,\nu = 1,2\ldots 6. \tag{10}$$

The question is now what is the minimum eigenvalue of M_{46} for each irreducible representation of the direct product characterized by the angular momentum ℓ. Consider the product state with maximum j for each fixed n. These states are of the form $[j, 0, j+2]_\rho$, j = 0, 1, 2,... . The index ρ specifies the j + 1 fold degeneracy. If we apply the lowering operator to the j + 1 linear independent combinations of these states, we get in general states of the form $[j - 1, 0, j + 1]$; for <u>one</u> certain linear combinations we get zero, as there are only j linear independent combinations. Now we calculate C_I for this state. From (9) we obtain

$$C_I = j(j+2) + (j+2)^2 - 4(j+2) = 2j(j+1) - 4 . \tag{11}$$

This is the state with the lowest value of n for the irreducible representation characterized by the angular momentum $\ell = j$. We exhibit the reduction into irreducible representations graphically in Fig. 3.

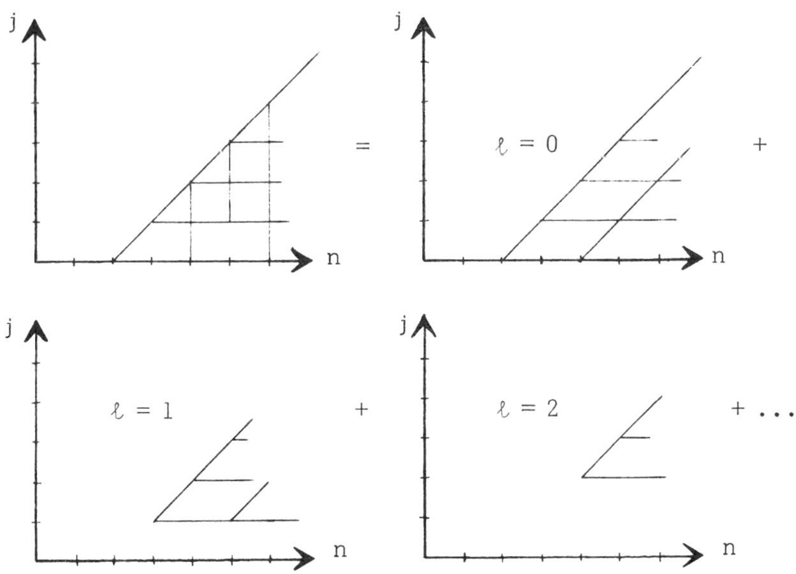

Fig. 3. Reduction of the direct product of two spin 0, mass 0, $p^4 > 0$ representations into irreducible components.

If we compare this reduction with the general structure of the direct product of α mass $m = 0$ representations $\alpha > 2$ we see that the direct product of two mass $m = 0$, $p^4 > 0$, spin 0 representations is not contained in them. Two mass $m = 0$, $p^4 > 0$, spin 0 representations (initial state) can couple conform-invariantly only to two mass 0, $p^4 > 0$ representations. It will turn out in Section V that the "inelasticity" consists in a helicity transfer. The possible final states in each angular momentum channel ℓ are given by two mass $m = 0$, $p^4 > 0$ particles with helicity $\lambda = s$ and $\lambda' = -s$, $s = 0, \frac{1}{2}, 1, 3/2 \ldots \ell/z$

IV.1 Application to a Conformal Invariant Scalar Field Theory

Consider a conformal-invariant quantum field theory which contains for simplicity only a scalar field $\varphi(y)$, e.g. the $\varphi^4(y)$ theory. The one-particle wave function $\langle p|\varphi(y)|0\rangle$ transforms[5] like the direct sum of a mass 0, $p^4 > 0$ representation, and of a physical massive spin 0 representation characterized by the Casimir operator $C_I = \nu^2 - 4 = -3$. This massive state can only be interpreted as the lowest angular momentum $\ell = 0$ component of the direct product of three of the mass 0, $p^4 > 0$, spin 0 states in the sense of Section III. From these restrictions[+] of the basic states and the previous results follows that there can be only elastic mass 0, spin 0 scattering. The reduced S-matrix elements $\exp 2i\delta_\ell$ are constants,[4] for which $|e^{2i\delta_\ell}| = 1$. This is the group theoretical result. If we impose in addition unitarity and crossing, we find that there is no scattering at all.[6]

V. The Direct Product of Two Mass $m = 0$, $p^4 > 0$ Representations of Helicity λ and λ'

The direct product of two mass 0, $p^4 > 0$ representations of arbitrary helicity λ and λ' contains representations of the Poincaré group characterized by $0 < m^2 < \infty$, $p^4 > 0$ and spin

$$\ell = |\lambda - \lambda'| + k, \quad k = 0, 1, 2, \ldots \qquad (12)$$

Each representation occurs once.[7] The first Casimir operator of $SU_o(2,2)$ is given by

$$C_I = (\lambda + \lambda')^2 + 2\ell(\ell + 1) - 4. \qquad (13)$$

Therefore one irreducible representation of $SU_o(2,2)$ is characterized by $0 < m^2 < \infty$, $p^4 > 0$, ℓ and $\lambda + \lambda'$. In Fig. 4 we give the lowest weights of each irreducible representation for the case $\lambda \geq 0$ and $\lambda' \leq 0$.

+It remains to be proved that there are no states with $m = 0$ and $\lambda = \pm 1, \pm 2, \ldots$ in the scalar theory.

In parentheses we denote the quantum numbers (s_1, s_2) for the lowest weight.

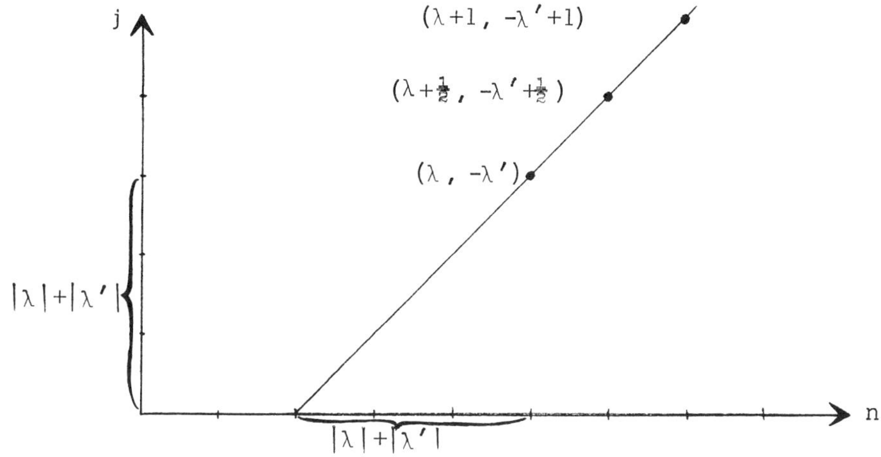

Fig. 4. Lowest weights for the irreducible representations of the direct product, $\lambda \geq 0$, $\lambda' \leq 0$.

In Fig. 5 we treat correspondingly the case where $\lambda \geq 0$, $\lambda' \geq 0$.

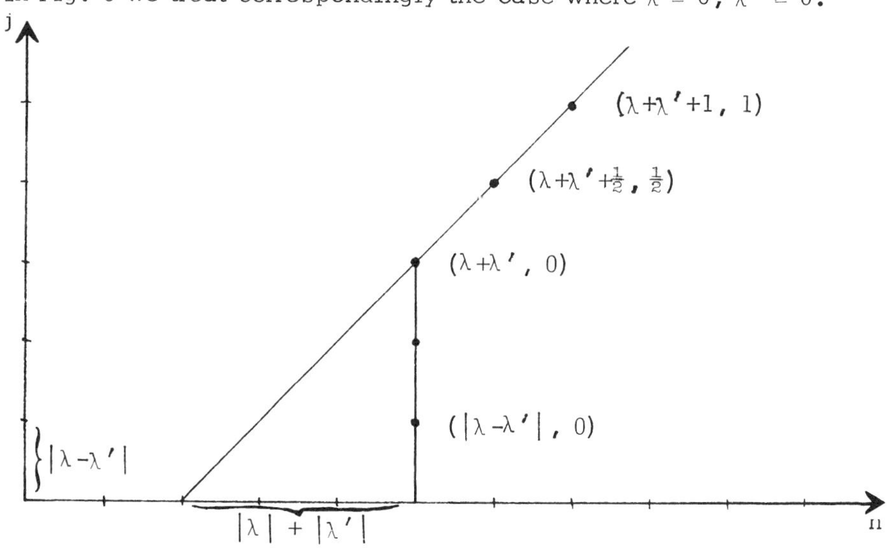

Fig. 5. Lowest weights for the irreducible representations of the direct product $\lambda \geq 0$, $\lambda' \geq 0$.

From the comparison of Fig. 5 and Fig. 4, or more abstractly from the value of C_I follows that for conformal-invariant mass 0 particle-scattering there is no helicity-flip $\lambda \to \lambda$, $\lambda' \to -\lambda'$. The corresponding result has been obtained in field theory[8] for spin 0 scattering off spin $\frac{1}{2}$ and spin 1 massless particles.

V.1. Application to Photon-Photon Scattering

If at high energy a photon-photon scattering becomes conformal-invariant our analysis shows that the inelastic channels are highly restricted. A final state which consists of two mass $m = 0$ particles with helicity $\lambda \neq \pm 1$ is not a realistic possibility, as we are dealing with electromagnetic interactions. Moreover, only if the helicity of the incoming photons is the same (e.g. $\lambda = \lambda' = +1$) we can expect inelastic channels. In the angular momentum $\ell = 0$ channel we can obtain: 1) four mass 0, spin 0 states (which can combine to two massive states), 2) two neutrinos, and one mass 0, spin 0 state, 3) two antineutrinos, and one mass 0, spin 0 state. In the $\ell = 1$ channel the only inelastic final state is given by two antineutrinos, and a mass 0, spin 0 state. From the assignment of massive elementary particles to irreducible representations[4] of the conformal group, we can conclude that none of the inelastic channels corresponds to any state, which contains a massive elementary particles. So there remains only elastic scattering.

V.2. Comment on Deep Inelastic Electron-Proton Scattering

We should like to show that the following approximation (skeleton-theory) which some people propose, in order to explain scaling of deep inelastic electron-proton scattering does not lead to reasonable results. If one replaces the incoming e and p and the outgoing e by neutrinos and studies the conformal-invariant reaction

$$\nu + \nu \to \nu + x,$$

we obtain again only elastic scattering $x = \nu$.

The only inelasticity, which is in the $\ell = 0$ channel (three mass 0, spin 0 states), is excluded as one ν has to be in the final state. This shows that massive elementary particles have to be assigned to massive representations of the conformal group.

Finally it should be mentioned that a corresponding analysis[9] for the two-dimensional Thirring-model shows very elegantly that there is no scattering.

Acknowledgment

The author is indebted to Prof. H. D. Doebner who stressed the importance of the restriction of inelastic channels by group theoretical methods.

References

1. G. Mack and I. Todorov, J. Math. Phys. 10, 2078 (1969);
   L. Castell, Nucl. Phys. B13, 231 (1969);
   J. Michelson and J. Niederle, to be published.
2. T. Yao, J. Math. Phys. 9, 1615 (1968); NYO-3829-52, June 1970.
3. L. Castell, J. Math. Phys., 11 2999(1970).
4. L. Castell, Phys. Rev. D2, 1161 (1970).
5. L. Castell, "Spontaneous Broken Symmetry and Conformal Invariance," M.P.I. preprint, Sept. 1970.
6. In this connection see N. F. Bali, D. D. Coon, and A. Katz, J. Math. Phys. 10, 1939 (1969).
7. M. Schaaf, Lecture Notes in Physics 5 (Springer Verlag, 1970).
8. D. J. Gross and J. Wess, CERN preprint T.H. 1076 (1969).
9. L. Castell, Comm. Math. Phys. 17, 127 (1970).

CANONICAL AND NONCANONICAL
SCALE SYMMETRY BREAKING†

R. Jackiw‡
Massachusetts Institute of Technology
Cambridge, Massachusetts

We have undertaken a study of scale symmetry breaking both on the formal, canonical level, and in detailed perturbative calculations. Our main result is that in perturbation theory mass terms are not the only objects which break scale symmetry, even though this might be the conclusion of a formal canonical investigation. Most probably, this state of affairs persists in the complete theory as well.[1)]

We first summarize here the consequences of the formal theory of scale symmetry breaking. It is shown how one may derive theorems about, e.g. the high energy behavior of Green's functions. These theorems are then demonstrated to be false, and their failure is explained by showing that (1) scale dimensions of fields are affected by interaction and (2) mass terms are not the only objects which break scale symmetry.[2)]

I. Formal Theory

Consider a renormalizable field theory. It is possible to introduce a new, improved energy momentum tensor $\theta^{\mu\nu}$, such that the scale current D^{μ} and the conformal current $K^{\alpha\mu}$ have the form[3)]

$$D^{\mu} = x_{\nu}\, \theta^{\nu\mu} \tag{1}$$

$$K^{\alpha\mu} = 2x^{\alpha} x_{\nu}\, \theta^{\nu\mu} - x^2\, \theta^{\alpha\mu} \tag{2}$$

It is assumed that only masses break dilatation invariance, on the Lagrangian level.

---

†Presented at the Symposium on De Sitter and Conformal Groups, University of Colorado, Summer 1970.

‡Alfred P. Sloan Fellow

$$\partial_\mu D^\mu = \theta^\mu{}_\mu = \text{mass terms} \tag{3}$$

$$\partial_\mu K^{\alpha\mu} = 2x^\alpha \partial_\mu D^\mu \tag{4}$$

To extract the consequences of these hypotheses, we derive Ward identities satisfied by matrix elements of $\theta^{\mu\nu}$. In order to do this, we need to know the commutator of $\theta^{\mu\nu}$ with a renormalized field φ of scale dimensionality d (d = 1 for Bosons, 3/2 for Fermions). Under very general hypotheses, one can show that

$$i[\theta^{oo}(o,\underline{x}), \varphi(0)] = \partial^o \varphi(0)\delta(\underline{x}) + \Sigma^{oi} \varphi(0)\, \partial_i\, \delta(\underline{x}) \tag{5}$$

$$i[\theta^{oi}(o,\underline{x}), \varphi(0)] = \partial^i \varphi(0)\delta(\underline{x}) - \frac{d}{3}\varphi(0)\, \partial^i\, \delta(\underline{x}) + \tfrac{1}{2}\Sigma^{ij} \varphi(0)\, \partial_j\, \delta(\underline{x}) \tag{6}$$

($\Sigma^{\mu\nu}$ is the spin matrix appropriate to the field φ.) These are the formal, canonical commutators. No statement is being made concerning their validity in perturbation theory. Since the commutator of $\theta^{\mu\nu}$ with φ necessarily contains gradient terms, the T product of $\theta^{\mu\nu}$ with φ is not covariant. In order to arrive at the covariant T* product, a covariantizing seagull $S^{\mu\nu}$ must be added. Hence we are led to consider

$$\begin{aligned} F^{\mu\nu}_{ij}(p,q) &= \int d^4x\, d^4y\, e^{iqx}\, e^{ipy} \langle 0|T^*\theta \quad (x)\,\varphi_i(y)\,\varphi_j(0)|0\rangle \\ &= \int d^4x\, d^4y\, e^{iqx}\, e^{ipy} \langle 0|T\,\theta^{\mu\nu}(x)\,\varphi_i(y)\,\varphi_j(0)|0\rangle \\ &\quad + S^{\mu\nu}_{ij}(p,q) \end{aligned} \tag{7}$$

$$F_{ij}(p,q) = \int d^4x\, d^4y\, e^{iqx}\, e^{ipy} \langle 0|T\theta^\mu{}_\mu(x)\,\varphi_i(y)\,\varphi_j(0)|0\rangle \tag{8}$$

In the above i,j label the fields; the labels may be space-time or internal indices. It is assumed that matrix elements of $\theta^\mu{}_\mu$ require no seagull. The covariantizing seagull $S^{\mu\nu}_{ij}$ may be explicitly constructed from the known commutators (5) and (6). Hence a Ward identity may be derived. Its form is

$$q_\mu F^{\mu\nu}_{ij}(p,q) = i\, p^\nu G(p) - i(p+q)^\nu G(p+q)$$
$$+\frac{i}{2} q_\mu \Sigma^{\mu\nu}_{ii'} G_{i'j}(p+q) + \frac{i}{2} q_\mu \Sigma^{\mu\nu}_{jj'} G_{ij'}(p) \qquad (9)$$

Also a trace identity is obtained, from the explicit form of $S^{\mu\nu}_{ij}$

$$g_{\mu\nu} F^{\mu\nu}_{ij}(p,q) = F_{ij}(p,q) - i\, d\, G_{ij}(p+q) - i\, d\, G_{ij}(p) \qquad (10)$$

In Eq. (9) and (10) G_{ij} is the renormalized propagator

$$G_{ij}(p) = \int d^4x\, e^{ipx} \langle 0 | T\, \varphi_i(x)\, \varphi_j(0) | 0 \rangle \qquad (11)$$

The formulae (9) and (10) contain all the restrictions that the various space-time transformations (Lorentz, scale, and conformal) impose on the propagator. (Had we wished to study the n particle Green's function, we would consider the matrix element of $\theta^{\mu\nu}$ with n fields.) Once a model for scale symmetry breaking, e.g. mass terms, is adopted, then one may deduce theorems about G(p). We now show explicitly how these restrictions are contained in Eq. (9) and (10).

1. Lorentz transformations Differentiate (9) with respect to q_α and set q to zero. This gives

$$F^{\alpha\nu}_{ij}(p,0) = -i\, g^{\alpha\nu} G_{ij}(p) - i\, p^\nu \frac{\partial}{\partial p_\alpha} G_{ij}(p)$$
$$+\frac{i}{2} \Sigma^{\alpha\nu}_{ii'} G_{ij}(p) + \frac{i}{2} \Sigma^{\alpha\nu}_{jj'} G_{ij'}(p) \qquad (12)$$

Since $F^{\alpha\nu}_{ij}$ is symmetric in α and ν, we learn from (12) that

$$\left[p^\nu \frac{\partial}{\partial p_\alpha} - p^\alpha \frac{\partial}{\partial p_\nu} \right] G_{ij}(p) = \Sigma^{\alpha\nu}_{ii'} G_{i'j}(p) + \Sigma^{\alpha\nu}_{jj'} G_{ij'}(p) \qquad (13)$$

This is the trivial and well known constraint of Lorentz covariance.

2. Scale transformations Form the trace of (12). We have (suppressing indices)

$$g_{\mu\nu} F^{\mu\nu}(p,0) = -4i\, G(p) - i\, p^\alpha \frac{\partial}{\partial p^\alpha} G(p) \qquad (14)$$

Combining (14) with (10) at q = 0 leaves

$$F(p,0) = i(2d-4)G(p) - i\, p^{\alpha} \frac{\partial}{\partial p^{\alpha}} G(p) \tag{15}$$

This provides a constraint on G(p), once F(p,0) is known, i.e. once we have a model for scale symmetry breaking.

3. Conformal transformations Differentiate (9) by

$$2 \frac{\partial}{\partial q^{\alpha}} \frac{\partial}{\partial q^{\nu}} - g_{\alpha\nu} \frac{\partial}{\partial q^{\beta}} \frac{\partial}{\partial q_{\beta}}$$

and set q = 0. This gives

$$2 \frac{\partial}{\partial q_{\alpha}} g_{\mu\nu} F^{\mu\nu}_{ij}(p,q)\Big|_{q=0} = -8i \frac{\partial}{\partial p_{\alpha}} G_{ij}(p) - 2i\, p_{\beta} \frac{\partial}{\partial p_{\beta}} \frac{\partial}{\partial p_{\alpha}} G_{ij}(p)$$

$$+ i\, p^{\alpha} \frac{\partial}{\partial p^{\beta}} \frac{\partial}{\partial p_{\beta}} G_{ij}(p) + 2i\, \Sigma^{\alpha\beta}_{ii'} \frac{\partial}{\partial p^{\beta}} G_{i'j}(p) \tag{16}$$

The left hand side of (16) may be evaluated from (10). After a rearrangement of terms, we are left with

$$2 \frac{\partial}{\partial q_{\alpha}} F_{ij}(p,q)\Big|_{q=0} = i\,(2d-8) \frac{\partial}{\partial p_{\alpha}} G_{ij}(p) - 2i\, p_{\beta} \frac{\partial}{\partial p_{\beta}} \frac{\partial}{\partial p_{\alpha}} G_{ij}(p)$$

$$+ i\, p^{\alpha} \frac{\partial}{\partial p^{\beta}} \frac{\partial}{\partial p_{\beta}} G_{ij}(p) + 2i\, \Sigma^{\alpha\beta}_{ii'} \frac{\partial}{\partial p^{\beta}} G_{ij}(p) \tag{17}$$

This equation may be simplified by using (13) and (15). First d is eliminated between (17) and (15). This gives

$$2 \frac{\partial}{\partial q_{\alpha}} F_{ij}(p,q)\Big|_{q=0} - \frac{\partial}{\partial p_{\alpha}} F_{ij}(p,0) =$$

$$i \frac{\partial}{\partial p^{\beta}} \Big[p^{\alpha} \frac{\partial}{\partial p_{\beta}} G_{ij}(p) - p^{\beta} \frac{\partial}{\partial p_{\alpha}} G_{ij}(p) + 2\Sigma^{\alpha\beta}_{ii'} G_{ij}(p) \Big] \tag{18}$$

Next we use (13).

$$2\frac{\partial}{\partial q_\alpha} F_{ij}(p,q)\Big|_{q=0} - \frac{\partial}{\partial p_\alpha} F_{ij}(p,0) = i\frac{\partial}{\partial p_\beta}\left[\Sigma^{\alpha\beta}_{ii'} G_{i'j}(p) - \Sigma^{\alpha\beta}_{jj'} G_{ij'}(p)\right] \tag{19}$$

Eq. (19) determines the constraint on G which follows from a model for conformal symmetry breaking.

II. False Theorem

It is now shown that the constraint equations for scale and conformal transformations cannot be interpreted naively. Consider for definiteness the propagator for a theory of spin zero fields with mass μ and a quartic self-interaction. The propagator may be written in the form

$$G(p) = \frac{i}{p^2}\, g(p^2/\mu^2) \tag{20}$$

We find from (15) that g satisfies

$$\frac{p^2}{2}\, F(p,0) = \frac{p^2}{\mu^2}\, g'(p^2/\mu^2) + (1-d)\, g(p^2/\mu^2) \tag{21}$$

Consider the limit as $\mu^2 \to 0$. One might expect that the left hand side vanishes, since F is the matrix element of θ^μ_μ which formally is $\mu^2\varphi^2$. On the right hand side, this limit is equivalent to $p^2 \to \infty$. Hence we find

$$\lim_{p^2\to\infty} g(p^2/\mu^2) \propto (p^2/\mu^2)^{d-1} \tag{22}$$

Since d = 1 for Boson fields, we further conclude that the Boson propagator goes as $1/p^2$ for large p^2.

This result, a weak form of Lehmann's theorem, is manifestly false in perturbation theory where it is known that logarithmic terms are present in the asymptotic domain. Thus we must abandon the steps which lead from the true (by definition) Eq. (21) to the false result. Specifically we cannot conclude that d = 1 and that F vanishes with the mass.

III. True Theorems

Detailed calculation in perturbation theory in lowest non-trivial order of the interaction yields the following conclusions. It remains possible to assert that F vanishes with the mass. However d changes from its canonical value of 1. To exhibit the change in d, we consider the definition of that object

$$i[D(0), \varphi(0)] = i \int d^3x\, x_i\, [\theta^{oi}(0,\underline{x}), \varphi(0)]$$
$$= d\, \varphi(0) \qquad (23)$$

The commutator is evaluated by the Bjorken-Johnson-Low prescription. Specifically an application of this technique to $F^{\mu\nu}(p,q)$, gives by definition

$$\lim_{q_o \to \infty} \frac{\partial}{\partial q^i}\, q_o\, F^{oi}(p,q)\Big|_{\underset{\sim}{q}=0} = i\, d\, G(p) \qquad (24)$$

Hence the true value of d may be computed from the high energy behavior of $F^{\mu\nu}$. Explicit calculation in lowest order gives

$$d = 1 + c\lambda^2 \qquad (25)$$

where c is a well defined numerical constant, and λ is the coupling strength of the quartic self interaction. Substituting this value of d into Eq. (22) (which remains valid to lowest order, since F does vanish with the masses), we find

$$\lim_{p^2 \to \infty} g(p^2/\mu^2) \propto (p^2/\mu^2)^{c\lambda^2} \approx 1 + c\lambda^2 \log p^2/\mu^2 \qquad (26)$$

Explicit calculation of the propagator to the same order verifies (26) with precisely the same coefficient. Perturbative calculations for several models have been performed, and the conclusion is always the same, in lowest order: although the scale breaking term vanishes with the masses, the dimension changes, and the resulting theorem about high energy behavior is verified by comparison with a calculation of the propagator.

Although perturbative calculations beyond lowest order have not been performed, it is possible to obtain answers by another method--that of the renormalization group. The crucial question is whether or not the propagator, in the high energy domain, behaves as a power of (p^2/μ^2). In this case one could say that scale breaking effects disappear with vanishing mass, but d changes from its canonical value. On the other hand if a power behavior for the propagator is not found, then the scale breaking effects do not go away as the mass goes to zero. The renormalization group indicates that the latter behavior is true.

From the renormalization group for the $\lambda\varphi^4$ theory, one can deduce the asymptotic form of the propagator[4)]

$$g(p^2/\mu^2) \xrightarrow[p^2 \to \infty]{} \frac{D[Q(\lambda) + \log p^2/\mu^2]}{D[Q(\lambda)]} \tag{27}$$

D and Q are two undetermined functions. $Q(\lambda)$ is related to the Gell-Mann-Low eigenvalue function $\psi(\lambda)$ by

$$\frac{1}{\psi(\lambda)} = \frac{\partial}{\partial\lambda} Q(\lambda) \tag{28}$$

According to the renormalization group, $\psi(\lambda)$ has a zero at $\lambda = \lambda_0$ if the unrenormalized coupling constant is finite and equal to λ_0. From (27) it is easy to deduce the following equation, which forms the basis for our subsequent discussion.[5)]

$$\frac{p^2}{\mu^2} g'(p^2/\mu^2) \xrightarrow[p^2 \to \infty]{} \beta(\lambda)\, g(p^2/\mu^2) + \psi(\lambda) \frac{\partial}{\partial\lambda} g(p^2/\mu^2) \tag{29}$$

Here $\beta(\lambda)$ is an undetermined function of λ. It is evident from (29) that power behavior for g is in general not obtained due to the presence of the second term on the right hand side. Only if $\psi(\lambda)$ has a zero and λ is chosen to be the zero of this function, does (29) yield a power law for g. Thus we conclude that if the unrenormalized coupling is finite, then scale invariance becomes exact as masses go to zero, but the dimensions change. On the other hand if $\psi(\lambda)$ has no zeros and the unrenormalized coupling constant is infinite (as it is in perturbation theory) scale invariance remains broken when the masses go to zero. Since the renormalization group can be formulated independently of perturbation theory, this analysis applies to the complete theory, as well as to perturbative approximations.

A comparison between (21) and (29) shows that $\beta(\lambda) = d-1$, while $\psi(\lambda)\, \partial/\partial\lambda\, g(p^2/\mu^2)$ is the residual scale breaking term when masses go to zero. We can understand the presence of these "anomalous," noncanonical scale breaking terms in the following way. In calculating matrix elements of $\theta^{\mu\nu}$ it is necessary to insure their conservation. However in specific calculations these matrix elements are not conserved, and conservation is achieved for example by Pauli-Villars regularization. One defines $\theta^{\mu\nu}_{Reg} = \theta^{\mu\nu} - \theta^{\mu\nu}_M$, where $\theta^{\mu\nu}_M$ is formed from regulator fields $\tilde{\varphi}$ carrying mass M. Physical, conserved matrix elements are obtained by letting $M \to \infty$. Consider now the trace of $\theta^{\mu\nu}_{Reg}$, which according to (3) breaks scale invariance. Evidently we have

$$g_{\mu\nu}\,\theta^{\mu\nu}_{\text{Reg}} = \mu^2\varphi^2 - M^2\tilde{\varphi}^2 \tag{30}$$

Thus if matrix elements of $\tilde{\varphi}^2$ behave as M^{-2} for large M, the regulator contribution to (30) survives, even in the physical limit $M \to \infty$. Specific calculation shows that $\tilde{\varphi}^2$ does indeed behave in this fashion. Therefore even when μ^2 is zero, $g_{\mu\nu}\,\theta^{\mu\nu}_{\text{Reg}}$ does not vanish.[6]

References

1. This work was done in collaboration with S. Coleman, see "Canonical and Noncanonical Scale Symmetry Breaking," MIT preprint.
2. Anomalous dimensions and anomalous scale symmetry breaking have also been discussed by K. Wilson, "Broken Scale Invariance and Anomalous Dimensions," SLAC PUB 737 and "Anomalous Dimensions and the Breakdown of Scale Invariance in Perturbation Theory," SLAC PUB 741.
3. The new, improved energy momentum tensor has been discussed in these proceedings; see also C. G. Callan Jr., S. Coleman and R. Jackiw, "A New Improved Energy-Momentum Tensor," Annals of Physics (N.Y.) 59, 42 (1970).
4. The renormalization group argument for the $\lambda\varphi^4$ theory has been detailed in Ref. 1. These results are also easily obtained from the discussion by N. N. Bogoliubov and D. V. Shirkov, Introduction to the Theory of Quantized Fields (Interscience, New York, 1959), pp. 511-552.
5. Results similar to those in Eq. (29) have been derived by K. Symanzik, "Small Distance Behaviour in Field Theory and Power Counting," DESY 70/20 and C. G. Callan, "Broken Scale Invariance in Scalar Field Theory," Cal Tech preprint CALT 68-207 and "Broken Scale Invariance and Scaling Laws in Elementary Particle Physics," Cal Tech preprint CALT 68-259.
6. This anomalous, noncanonical behavior has also been discovered in conventional current algebra; for a summary see R. Jackiw, "Non-Canonical Behaviour in Canonical Theories," CERN preprint, TH. 1065.

A NEW IMPROVED ENERGY MOMENTUM TENSOR†

R. Jackiw‡
Massachusetts Institute of Technology
Cambridge, Massachusetts

In order that the energy momentum tensor be an observable, as surely it is due to its weak coupling to gravity, it is necessary that it possess finite matrix elements. We have shown that for any renormalizable field theory it is possible to find an energy-momentum tensor whose matrix elements are finite.[1] However this object is not always the conventional Belinfante tensor, although the total energy and angular momentum of course remain unchanged.

Before exhibiting our tensor, we turn to the topics of scale and conformal transformations, since our tensor arises very naturally in this context.[2] A Poincaré covariant Lagrangian theory is scale invariant when the following relation is true:

$$4\mathcal{L} = \frac{\delta\mathcal{L}}{\delta\partial^\mu\varphi}(d+1)\,\partial^\mu\varphi + \frac{\delta\mathcal{L}}{\delta\varphi}\,d\varphi\ . \tag{1}$$

Here $\mathcal{L}$ is the Lagrangian function of the theory, assumed to depend on a set of fields φ, and on single derivatives of these fields $\partial^\mu\varphi$. The quantity d is the scale dimension of the field φ, chosen to be 3/2 for Fermions and 1 for Bosons. Moreover a Poincaré covariant Lagrangian describes a conformally invariant theory when two conditions are met:

(1) Scale invariance must hold; i.e. Eq. (1) is true.

(2) The field virial, V^μ, defined by

$$V^\mu \equiv \frac{\delta\mathcal{L}}{\delta\partial^\nu\varphi}(g^{\nu\mu}d - \Sigma^{\nu\mu})\,\varphi\ , \tag{2}$$

must be a total divergence:

†Presented at the Symposium on De Sitter and Conformal Groups, University of Colorado, Summer 1970.
‡Alfred P. Sloan Fellow

$$V^{\nu} = \partial_{\mu} \sigma^{\mu\nu} \tag{3}$$

(In Eq. (2) $\Sigma^{\nu\mu}$ is the spin matrix for the field φ.) It is remarkable that for all renormalizable theories Eq. (3) is true, though of course for none of these theories is scale invariance exact. (Condition (3) is also true for theories involving fields of spin not greater than one, without derivative interactions.)

Once (3) is satisfied, we may introduce our tensor $\theta^{\mu\nu}$ by the following procedure: Form the Belinfante tensor $\theta_B^{\mu\nu}$ in the usual way. Form also the object

$$X^{\lambda\rho\mu\nu} = g^{\lambda\rho}\sigma_+^{\mu\nu} - g^{\lambda\mu}\sigma_+^{\rho\nu} - g^{\lambda\nu}\sigma_+^{\mu\rho} + g^{\mu\nu}\sigma_+^{\lambda\rho} - \tfrac{1}{3}g^{\lambda\rho}g^{\mu\nu}\sigma + \tfrac{1}{3}g^{\lambda\mu}g^{\rho\nu}\sigma \tag{4a}$$

where $\sigma_+^{\mu\nu}$ is the symmetric part, and σ the trace of $\sigma^{\mu\nu}$. The new improved tensor is now given by

$$\theta^{\mu\nu} = \theta_B^{\mu\nu} + \tfrac{1}{2}\partial_\lambda\partial_\rho X^{\lambda\rho\mu\nu} \tag{4b}$$

In terms of $\theta^{\mu\nu}$ the dilatation current D^{μ} and the conformal current $K^{\alpha\mu}$ are given by the simple formulas

$$D^{\mu} = x_{\nu}\theta^{\mu\nu} \tag{5a}$$

$$K^{\alpha\mu} = 2x^{\alpha}x_{\nu}\theta^{\nu\mu} - x^2\theta^{\alpha\mu} \tag{5b}$$

$$\partial_\mu D^{\mu} = \theta^{\mu}_{\ \mu} \tag{6a}$$

$$\partial_\mu K^{\alpha\mu} = 2x^{\alpha}\partial_\mu D^{\mu} \tag{6b}$$

Eq. (6) exhibits the fact that for the theories under consideration, i.e. for theories where (3) is true, conformal invariance is broken by the same mechanism as scale invariance.

Explicit computation shows that $\sigma^{\mu\nu}$ is identically zero except for spin zero fields.

$$\sigma^{\mu\nu} = \frac{g^{\mu\nu}}{2} \sum_{\substack{\text{spin}\\ \text{zero}\\ \text{fields}}} \varphi^2 \tag{7}$$

Hence

$$\theta^{\mu\nu} = \theta_B^{\mu\nu} - \frac{1}{6} \sum_{\substack{\text{spin}\\ \text{zero}\\ \text{fields}}} (\partial^\mu\partial^\nu - g^{\mu\nu}\Box)\,\varphi^2 \tag{8}$$

(The unique role of spin zero has not as yet been understood.) In renormalized perturbation theory, it is $\theta^{\mu\nu}$, as given by (8), rather than $\theta_B^{\mu\nu}$ which has finite matrix elements. (For scalar particles, it is also necessary to shift the scalar field φ by an infinite constant.)

Since in the usual Einstein gravity theory $\theta_B^{\mu\nu}$ rather than $\theta^{\mu\nu}$ is the source of gravity, it is necessary to find a new theory of gravity in which $\theta^{\mu\nu}$ is the source, at least to lowest order in the gravitational interaction. This modification must be consistent with the usual tests of general relativity. It will depend only on spin zero fields, since the difference between $\theta^{\mu\nu}$ and $\theta_B^{\mu\nu}$ involves only these fields. Such a modified field theory has been found.[3] It is derived from the action

$$I - \int d^4x \sqrt{-g} \left[\frac{R}{16\pi G} - \frac{R}{12}\varphi^2 + \mathcal{L}_m\right] \tag{9}$$

In the above R is the Riemann curvature, g is the determinant of the metric tensor, G is the gravitational coupling strength, and $\mathcal{L}_m$ is the matter Lagrangian. We have assumed that only one spin zero matter field φ is present. The field equation for gravity is

$$R_{\mu\nu} - \tfrac{1}{2} g_{\mu\nu} R = - \frac{8\pi G}{1 - 4/3\,\pi G\varphi^2}\ \theta_{\mu\nu} \tag{10}$$

and to lowest order in G, our tensor is indeed the source of gravity. If $\mathcal{L}_m$ refers only to the field φ, with a mass μ and a $-\lambda\varphi^4$ self-interaction, then the matter equations are

$$\Box\varphi = -\mu^2\varphi^2 - 4\lambda\varphi^3 - \frac{1}{6} R\varphi \tag{11a}$$

R can be eliminated between (10) and (11a), so that the final matter equation is

$$\Box\varphi = -\mu^2\varphi^2 - 4(\lambda + \tfrac{1}{3}\pi G\mu^2)\varphi^3 \tag{11b}$$

Hence the only effect of our new gravitational theory is to change, in a universal way, the strength of the quartic self-interaction; this is obviously consistent with the principle of equivalence.

References

1. This work was done in collaboration with C. G. Callan, Jr. and S. Coleman; see "A New Improved Energy-Momentum Tensor," Ann. Phys. (N.Y.) 59, 42 (1970).
2. Scale and conformal transformations are discussed in detail in Ref. 1, as well as by S. Coleman and R. Jackiw in "Canonical and Noncanonical Scale Symmetry Breaking," MIT preprint. For a less general treatment and a survey of the older literature see G. Mack and A. Salam, Ann. Phys. 53, 174 (1969).
3. Alternate theories of gravity closely related to the present one have been discussed by F. Gursey, Ann. of Phys. 24, 211 (1963) and R. Penrose, Proc. Roy. Soc. (London) 284A, 204 (1965).

SU(3) × SU(3) AND DILATATION INVARIANCE
OF STRONG INTERACTIONS†

P. L. F. Haberler‡
Department of Physics
Duke University
Durham, North Carolina 27706

After many lectures on the more formal group theoretical character of the conformal and De Sitter groups it seems worthwhile to look into the dynamical consequences of the conformal group, especially its implications for hadron physics.

Stimulated by the fundamental work of Wess,[1] Kastrup,[2] Mack[3] and Wilson[4] recently much work[5] has been done on the possible application of the conformal group to strongly interacting particles. One first important result[6],[7] which has been established is that for a wide class of Lagrangian theories, which include all renormalizable interactions except φ^3 coupling, scale invariance implies invariance under conformal transformations.

The fundamental quantity in the study of scale invariance is the local energy momentum tensor, $\Theta_{\mu\nu}(x)$. With its help, one can construct the generators of translations and Lorentz rotations in the well known way. Now a dilatation, or change of scale, obviously changes coordinates by $x_\mu \to \lambda x_\mu$. The generator of the dilatation, D, can by analogy with the Lorentz generators be written as

$$D = \int d^3x \, \Theta_{o\mu} x^\mu \sim -ix^\mu \partial_\mu .$$

It generates scale transformations on fields. For a finite scale change, $x \to \lambda x$, a field transforms as

$$\varphi(x) \to \lambda^d \varphi(\lambda x)$$

†Presented at the Symposium on De Sitter and Conformal Groups, University of Colorado, Summer 1970.
‡Supported in part by the ARO(D) and N. S. F.

where d is the dimension of φ. As is well known for free fields at least, d = 1 for spin 0, d = 3/2 for spin 1/2, and so on.

Now we can define[7] a current $\mathcal{D}_\mu = \Theta_{\mu\nu} x^\nu$ such that D is the space integral of its time component. Then time independence of D or scale invariance of the theory corresponds to $\mathcal{D}$ being divergence-free. The divergence of $\mathcal{D}_\mu$ is easily seen to be just $\Theta_\mu{}^\mu$, the trace of the energy-momentum tensor. We want the limit of scale invariance to correspond to the vanishing from the Lagrangian of terms having dimensional coupling constants, since a theory with only dimensional coupling constants should be scale invariant. Therefore, $\Theta_\mu{}^\mu$ should be proportional to those terms in the Lagrangian having dimensional coupling constants. However as we heard from Professor Jackiw he and several other people[8),9),10)] have found that due to divergences which are inherent in all nontrivial field theories a formally scale invariant interaction term may give rise to terms which break scale invariance unless the expansion parameter (unrenormalized) of perturbation theory has a fixed value. We will return to this point later on.[34)] Besides this very serious problem Wilson[11)] remarked that the dimension d of a field may change due to the interaction (anomalous dimensions). Another important point which has been dealt with quite often in the literature[5)] concerns the question whether conformal invariance can be realized in the Goldstone way, i.e., whether there exists a zero mass scalar particle which couples to the vacuum through $\Theta_{\mu\nu}$. This would allow some masses to remain nonzero in the symmetry limit. This possibility has led many people[5)] to look for a combined scale and chiral invariance of strong interactions, because of similarities which I shortly sketch below.[5)]

Although a detailed theory of strong interactions does not yet exist it is reasonable to suppose that SU(3) x SU(3) is nearly a good symmetry. We write the energy density Θ_{00} in the form

$$\Theta_{00} = \Theta_{00} + u \qquad (1)$$

where Θ_{00} is invariant under SU(3) x SU(3) and u contains SU(3) singlet and octet components. From the work of Gell-Mann, Oakes and Renner[12)] it seems plausible that u has the simple form $u_0 + c\,u$ with c near $-\sqrt{2}$ (so that the pion mass is nearly zero), the u_i being scalar components of a $(3,\bar{3}) \oplus (\bar{3},3)$ representation. Under standard[5)] assumptions

$$\partial^\mu \mathfrak{A}^i_\mu = i\,[F^5_i , u] \qquad i = 1\text{-----}8 \qquad (2)$$

and the axial current divergences are dominated by a pseudoscalar octet in $(3,\bar{3}) + (\bar{3},3)$.

When u goes to zero the axial vector currents are conserved and the pseudoscalar octet becomes massless. However other particles apparently remain massive in this limit. The numerical analysis of von Hippel and Kim[13] indicates that the nucleon mass remains essentially unchanged as $u \to 0$. It also seems that the η' remains massive in this limit. Thus it is reasonable to assume that $\hat{\Theta}_{00}$ contains a scale breaking (scalar) operator δ in addition to a part of dimension d = -4, $\bar{\bar{\Theta}}_{00}$

$$\hat{\Theta}_{00} = \bar{\bar{\Theta}}_{00} + \delta$$

$$\Theta_{00} = \bar{\bar{\Theta}}_{00} + \delta + u \tag{3}$$

δ is to be SU(3) x SU(3) invariant, as is $\bar{\bar{\Theta}}_{00}$. It was Wilson[4] who first advocated the existence of δ. Mass shifts of the order of 1 GeV are to be attributed to the operator δ, while u causes shifts of a few hundred MeV.

In the limit where δ goes to zero we distinguish two types of theories. In the first, all masses vanish in the limit of scale invariance and the vacuum is normal under scale transformations. The second alternative is that of spontaneous breakdown of scale invariance, in which case the vacuum is degenerate.

Recently doubts[14] have been raised whether scale invariance can be realized through the Nambu-Goldstone way. We will come to this point later. For the moment we note that the vanishing of all bare masses and dimensional coupling constants is one necessary condition for a combined scale and chiral invariance of strong interactions.

In the following we are going to investigate the condition of vanishing of all bare dimensional quantities in the framework of Lagrangian field theory. To accomplish this we have to choose a model field theory which supposedly is capable of describing strong interactions. A promising candidate is the field theory of mesons and baryons with pseudoscalar Yukawa coupling. In this framework Drell, Levy and Yan[15] and others[16] were able to derive useful results for inelastic electron-proton scattering. Also Padé-approximant calculation based on this field theory yields results which agree very well with the data.[17],[18]

The Lagrangian of the pion-nucleon system is given by[19] (SU(2) case):

$$L = \bar{\Psi}(i\gamma\partial - M)\Psi + \tfrac{1}{2}(\partial\mu\vec{\pi})^2 - \frac{\mu^2}{2}\vec{\pi}^2 - i g_0 \bar{\Psi}\gamma\vec{\tau}\cdot\vec{\varphi}\Psi$$

$$+ \delta M \bar{\Psi}\Psi + \frac{\delta\mu^2}{2}\vec{\pi}^2$$

$$\delta M = M - M_0, \quad \delta\mu^2 = \mu^2 - \mu_0^2 \qquad (4)$$

where M and μ^2 are the physical masses of the nucleon and pion respectively, M_0 and μ_0^2 are the corresponding bare masses, g_0 is the unrenormalized π - N coupling constant and Ψ and φ are the unrenormalized nucleon and pion fields respectively.

The interaction Lagrangian in the SU(3) case is given by[20]

$$L_I = -2 i g_0 \bar{B}_{\gamma 5}[\alpha D^a + (1-\alpha)F^a] BP_a \qquad (5)$$

plus the selfmass terms which we do not explicitly write down. α is the mixing parameter which measures the relative strength of the symmetric and antisymmetric couplings (D and F types respectively). B and P_a are unrenormalized baryon and mesons fields respectively.

In the framework of this theory one can derive the following exact relation for the bare masses.[21]

Nucleons:

$$M_0 = M Z_2 + Z_2 \int_{(M+\mu)}^{\infty} a\,[r_1(a) - r_1(-a)]\,da \qquad (6)$$

Baryon octet:[22]

$$M_0^i = M^i Z_2^i + Z_2^i \int_{(M+\mu)}^{\infty} a[r_1^i(a) - r_1^i(-a)]\,da \qquad (7)$$

where Z_2 (Z_2^i) is the wavefunction renormalization constant for the nucleon (baryon octet) and $r_1(a)$ ($r_1^i(a)$) is the renormalized spectral function of the fermion propagator

Mesons:

$$\mu_0^2 = \mu^2 Z_3 + Z_3 \int_{4M^2}^{\infty} a^2 \rho(a^2)\,d a^2 \qquad (8)$$

Meson octet:[22]

$$\mu_0^{2i} = \mu^{2i} Z_3^i + Z_3^i \int_{4M^2}^{\infty} a^2 \rho^i(a^2)\,d a^2 \qquad (9)$$

where Z_3 (Z_3^{i}) and $\rho(a^2)$ ($\rho^{i}(a^2)$) are the corresponding quantities for the pion (meson octet). Now let us discuss these equations from the point of view that the bare masses have to vanish in the limit of scale invariance. We start with M_0. From Eqs. (6) and (7) we learn[23) that there are two possibilities for M_0 to vanish

$$\text{I.}\quad Z_2 \neq 0, \quad M + \int_{(M+\mu)}^{\infty} a[r_1(a) - r_1(-a)]\, da = 0 \tag{10}$$

or

$$\text{II.}\quad Z_2 = 0, \quad \left| M + \int_{(M+\mu)}^{\infty} a[r_1(a) - r_1(-a)]\, da \right| < \infty \tag{11}$$

Case I, which one could interpret as the case where the nucleon is a fundamental object, has been dealt with in the literature[24) to calculate the pion-nucleon coupling constant without much success.
Case II leads to the conclusion that when the wavefunction renormalization constant vanishes, the bare mass has to vanish. The condition $Z_2 = 0$ is known as the condition for compositeness of a particle in field theory.[25) Salam[26) conjectured that

$$\delta M = M - M_0 = 0 \tag{12}$$

when $Z_2 = 0$ and Hagen[27) later on made this statement more precise in showing if the vacuum is nondegenerate actually

$$M = M_0 = 0 \tag{13}$$

from $Z_2 = 0$ follows in a $\gamma 5$ invariant theory. Hagen also noted in this case that a spontaneous generation of symmetry emerges. There is still another possibility which has to the best of the author's knowledge not yet been reported in the literature, namely that from $Z_2 = 0$, $M_0 = 0$ follows but $M \neq 0$. In this case the vacuum has to be degenerate and we expect that the Goldstone-Nambu theorem[28) applies, in that there exists a spin zero, mass zero particle. In view of the work of Ref. 13 it is very important to know whether such a possibility exists. Therefore we have investigated this question in our chosen field theory for strong interactions.

Before I tell you the results I should point out that usually you expect[27) M_0 to be infinite when $Z_2 = 0$. Also straightforward perturbation theory gives you always $M_0 = -\infty$ and therefore we introduce a cutoff to make the integrals well defined. Later on we will see that the cutoff plays a special role.[34)

First let us see what lowest order perturbation (unrenormalized) gives. For Z_2 we find

$$Z_2 = 1 - \int_0^{\Lambda^2} dL \int_0^1 \frac{dz(1-z)\, z[M^2(1-z)^2 + Lz]}{[M^2(1-z)^2 + \mu^2 z + Lz]^2} \tag{14}$$

and for δM

$$\delta M = M \int_0^{\Lambda^2} dL \int_0^1 \frac{dz\, z(1-z)}{[M^2(1-z)^2 + \mu^2 z + Lz]} \tag{15}$$

From Eqs. (14) and (15) we learn the following

1. For $M = 0$ (physical mass of the nucleon) indeed $\delta M = M - M_0 = 0$. But letting $M \to 0$ in the expression for Z_2 we find that Z_2 diverges (infrared). Therefore it seems that Z_2 has nothing to do with δM in this case. This observation is confirmed when one looks at the rainbow approximation[29] (summing up the rung-diagrams) and in the Zachariasen[30] version of this model.[31]

2. For $M \neq 0$ one readily sees upon comparing Eq. (14) and Eq. (15) that for $\mu = 0$ (physical mass of the pion) indeed $M_0 = 0$ from $Z_2 = 0$ follows. This relation can be written in the beautiful form

$$Z_2 = 1 - \frac{\delta M}{M} \tag{16}$$

or

$$M_0 = Z_2 M$$

The validity of this relation is confirmed in the above mentioned models[29],[31] and I think that it is true in general[32] for theories where the limit boson mass → 0 exists.

When one looks at the SU(3) case these results do not change. One finds in this case[31] an octet of massless pseudoscalar mesons emerging while the masses assume the value of the nucleon mass in the symmetry limit.

These findings suggest that the condition $Z_2 = 0$ (and all Z's = 0 in general[26]) lead to a spontaneous generation of SU(2) × SU(2) (SU(3) × SU(3)). But the emergence of massless pseudoscalar Goldstonous is not yet enough to have a chiral symmetry. One has to show that the axial vector current is conserved when the wavefunction renormalization constants are zero. Following Gell-Mann and Levy[33] we perform the following chiral transformations on our Lagrangian equation (4):

$$\Psi \rightarrow (1 + i\,\vec{\tau}\cdot\vec{v}\,\gamma_5)\,\Psi$$

$$\vec{\pi} \rightarrow \vec{\pi} \tag{17}$$

The axial current is given then by[33)]

$$\mathfrak{A}^i_\mu = \bar{\Psi}\,\gamma_\mu\,\gamma_5\,\tau^i\,\Psi \tag{18}$$

and its divergence

$$\partial^\mu\,\mathfrak{A}^i_\mu = 2i\,M_o\,\bar{\Psi}\,\gamma_5\,\tau^i\,\Psi + g_o\,\bar{\Psi}\,\Psi\,\varphi^i \tag{19}$$

where all quantities are understood to be unrenormalized. We have calculated $\langle N|\,\mathfrak{A}^i_\mu\,|N\rangle$ and $\langle N|\partial^\mu\,\mathfrak{A}^i_\mu\,|N\rangle$, $|N\rangle$ fixed physical nucleon state and found that in the limit where $\mu = 0$

$$\langle N|\,\mathfrak{A}^i_\mu\,|N\rangle = M\,\gamma_\mu\,\gamma_5\,\tau^i \tag{20}$$

and

$$\langle N|\partial^\mu\,\mathfrak{A}^i_\mu\,|N\rangle = Z_2^2\,(Z_2 + Z_3^{\frac{1}{2}})\,M\,\gamma_5 \tag{21}$$

for zero momentum transfer squared.[35)] If one formally renormalizes Eq. (19) one finds[34)]

$$\partial^\mu\,\mathfrak{A}^i_{\mu\,R} = 2(M - \delta M)i\,Z_2\,\bar{\Psi}_R\,\gamma_5\,\tau^i\,\Psi_R + g_o Z_3^{\frac{1}{2}}\,Z_2\,\bar{\Psi}_R\,\Psi_R\,\varphi^i_R \tag{22}$$

which has essentially the same Z-factors as Eq. (21) when one uses the condition (16). (The additional Z_2 factor in (21) comes from the sandwiching between physical states.) Therefore we conclude that Eq. (20) and (21) are in general true. Therefore we find that if $Z_2 = 0$, indeed, the axial current is conserved provided that the pion mass is zero. Also the equation $Z_2 = 0$ does not mean that the axial current itself vanishes identically when $Z_2 = 0$. It is an easy exercise to convince oneself (in lowest order) that

$$\frac{\partial}{\partial t}\,Q_5^i = 0$$

where

$$Q_5^i = \int d^3x\,\mathfrak{A}^i_o(x) \quad . \tag{23}$$

Therefore we conclude that the condition $Z_2 = 0$ indeed leads to a spontaneous generation of chiral symmetry ($SU(2) \times SU(2)$ in this case), which at the same time turns out to be spontaneously broken. Although the connection Z's = 0 (bootstrap) and the emergence of higher symmetries has been noted a long time ago[35] the simultaneous breakdown of the emerging symmetry has to the best of the knowledge of the author not yet been reported in the literature.

Although many people may argue that this result seems to hold only in a special model, I would like to emphasize again the relevance of this field theory for what is going on in nature: these results again support the work of Ref. 13. Further, although one usually believes that the σ-model[33] is a better model for strong interactions these results suggest that the advantages of the σ model (chiral invariance) are actually hidden also in the simpler γ_5 theory.

Now let us turn to the bare mass of the mesons. At first sight it seems that again $\mu_o^2 = 0$ when $Z_3 = 0$. But one has to be very careful in this case. Hagen[27] actually claimed that the opposite is true: if $Z_3 = 0$, μ_o^2 is infinite. But other research workers in this field[36] have found the condition

$$Z_3 \; \delta\mu^2 = 0 \tag{24}$$

i.e. $\delta\mu^2$ is not infinite. This was confirmed recently by Kang and Land[37] who reexamined this problem very carefully. They showed that condition (24) is indeed necessary for a complete bootstrap. Furthermore they showed that the complete bootstrap condition imposes a restriction on the propagator, namely that the propagator requires a subtracted Lehmann representation[21] in this case.

Since (24) is compatible with $\mu_o^2 = 0$ we conclude again that from $Z_3 = 0$ a vanishing of the bare mass μ_o^2 in the symmetry limit results. Indeed calculations based on perturbation theory confirm this.[31]

Now after we have made it very plausible that the vanishing of all renormalization constants leads to the necessary conditions for chiral invariance, we still have to investigate whether these conditions also lead to scale invariance. There the big question arises whether as in the case of chiral symmetry or Goldstone-Nambu realization is possible. It was claimed in the literature[38] that the σ-model provides an example for a Goldstone solution, the σ-meson acting then as a Goldstone. But a recent careful renormalization of the σ-model[39],[40] shows that when the physical mass of the σ vanishes, all[41] other masses, including the nucleon mass, vanish also. Also on general grounds Genz and others[14] have provided indication that the Goldstone situation cannot arise in the case of scale invariance.

But let us look into meson-nucleon theory when all the wavefunctions renormalization constants vanish. Following Ref. 7, the divergence of the dilatation current is given by

$$\partial^{\mu} \mathcal{D}_{\mu} = \theta^{\mu}_{\ \mu} = M_o \bar{\Psi}\Psi + \mu_o^2 \vec{\varphi}^2 \tag{25}$$

Formally renormalizing this expression introduces Z-factors. However, taking matrix elements between physical nucleon states or physical pion states shows[31] that, only if all physical masses vanish, $\partial^{\mu}\mathcal{D}_{\mu} = 0$ follows. However we still want to be cautious because this emerges only in a lowest order calculation. Nevertheless we feel that it is a strong indication against a Nambu-Goldstone realization of scale invariance.

So let me sum up. You have seen that there is a strong connection between the bootstrap apprach and the way symmetries arise and how they are realized in the symmetry limit. We found that chiral symmetry has to be realized in the Goldstone-Nambu way, while scale invariance seems to demand that all physical masses go to zero in the limit. From this immediately follows that a quantity δ must exist, as introduced by Wilson[8] who claims that Glashow was the first who considered it. Therefore this suggests the following picture for $\theta_{00} = \bar{\bar{\theta}}_{00} + u + \delta$ in the symmetry limit:

When u goes to zero, the pseudoscalar masses go to zero and the baryon masses stay finite. In the limit $\delta \to 0$ all masses go to zero.

Furthermore, our findings strongly suggest that Abdus Salam's[26] original hypothesis is right, namely that all Z's = 0 is the field theoretic equivalent of the bootstrap hypothesis.

Acknowledgment

The author is particularly indebted to Professor Roman Jackiw for some very penetrating questions. Useful discussions with Prof. A. Barut and Prof. H. Kastrup are also gratefully acknowledged. Also he thanks Prof. S. Bludman for pointing out to him Refs. 35 and 36.

References

1. J. Wess, Nuovo Cimento 18, 1086 (1960).
2. H. A. Kastrup, Phys. Rev. 150, 1183 (1966); Nucl. Phys. B15, 179 (1970).
3. G. Mack, Nucl. Phys. B5, 499 (1968).
4. K. Wilson, Phys. Rev. 179, 1499 (1969).
5. P. Carruthers, preprint Calt-68-266 (1970). There all references can be found.
6. D. Gross and J. Wess, preprint CERN TH-1076 (1969).

7. C. Callan, S. Coleman and R. Jackiw, to be published in Ann. Phys. (N.Y.).
8. K. Wilson, preprint SLAC-PUB 741 (1970).
9. C. Callan, preprints Calt-68-2570259 (1970).
10. K. Symanzik, DESY preprint (1970).
11. K. Wilson, preprint SLAC-PUB 737 (1970).
12. S. Glashow and S. Weinberg, Phys. Rev. Letters 20, 227 (1968); M. Gell-Mann, R. J. Oakes and B. Renner, Phys. Rev. 175, 2195 (1968).
13. F. V. Hippel and J. K. Kim, Phys. Rev. Letters 22, 740 (1969); Phys. Rev. D1, 151 (1970).
14. H. Genz, Phys. Letters 31B, 146 (1970); D. Maison and H. Reeh, MPI preprints (1970); H. Kastrup, private communication.
15. S. D. Drell, D. J. Levy and T. W. Yan, I-IV Phys. Rev. Letters 22, 744 (1969); Phys. Rev. 187, 2159 (1969); Phys. Rev. D1, 1035 (1970); Phys. Rev. D1, 1617 (1970); Phys. Rev. D1, 2402 (1970).
16. Shau-Jin Chang and P. Fishbane, Phys. Rev. Letters 24, 847 (1970) and Illinois preprints (1970).
17. See D. Bessis, V. Grecchi, G. Turchetti, Phys. Letters 28B, 576 (1969) and Phys. Rev. (to be published) for a comprehensive review on the subject.
18. L. B. Redei, Phys. Rev. D2, 239 (1970).
19. J. Bjorken and S. D. Drell, Relativistic Quantum Fields (McGraw-Hill, New York, 1965).
20. P. Carruthers, "Introduction to Unitary Symmetry," Interscience Tracts on Physics and Astronomy No. 27.
21. H. Lehmann, Nuovo Cimento 11, 342 (1954). We follow the notation of W. Thirring, Principles of Quantumelectrodynamics (Academic Press, Inc., New York, 1958).
22. No sum over i implied.
23. We restrict ourselves in the following to the SU(2) case. But the same conclusions hold for the SU(3) case.
24. R. Acharya and M. Y. Han, unpublished, and C. Hagen, J. Math. Phys. 5, 900 (1964).
25. B. Jouvet, Nuovo Cimento 5, 1 (1957); M. Gell-Mann and F. Zachariasen, Phys. Rev. 123, 1065 (1961).
26. Abdus Salam, Nuovo Cimento 25, 224 (1962). See also P. E. Kaus and F. Zachariasen, Phys. Rev. 171, 1597 (1968).
27. C. Hagen, Ann. Phys. 31, 185 (1965).
28. See Y. Nambu, "Symmetry Breakdown and Small Mass Bosons," EFI 70-74 for a comprehensive review.

29. P. L. F. Haberler, Duke preprint, April 1970, unpublished.
30. F. Zachariasen, Phys. Rev. 121, 1851 (1961); W. Thirring, Phys. Rev. 126, 1209 (1962).
31. P. L. F. Haberler, to be published.
32. P. L. F. Haberler, to be published.
33. M. Gell-Mann and M. Levy, Nuovo Cimento 16, 705 (1960).
34. R means renormalized, g_0 is the unrenormalized coupling constant. Since we are taking a cutoff more seriously[15)] the wavefunction renormalization constants are finite but cutoff dependent. Therefore an explicit renormalization of g_0 is not necessary. Also in a forthcoming paper by N. Lipshutz, O. Weaver and the author it is shown that the Z = 0 condition leads directly to a fixed point condition of the renormalization group, whereby $g_0^2/4\pi = 15$ emerges. See in this connection also a preprint by K. Wilson, SLAC-PUB-733 (1970).
35. R. H. Capps, Phys. Rev. 132, 2749 (1963); Phys. Rev. Letters 10, 312 (1964); F. Zachariasen, in Seminar on High Energy Physics and Elementary Particles, Trieste (1965); C. Hagen, Ref. 27; R. Cutkosky, Phys. Rev. 131, 1888 (1963); S. A. Bludman and N. G. Deshpande, Nuovo Cimento 45, 656 (1966).
36. S. A. Bludman and N. G. Deshpande, Phys. Rev. 143, 1239 (1966).
37. K. Kang and D. J. Land, Nuovo Cimento 63A, 1953 (1969).
38. G. Mack and Abdus Salam, Ann. Phys. 53, 174 (1969).
39. B. W. Lee, Nucl. Phys. B9, 649 (1969); J. L. Gervais and B. Lee, Nucl. Phys. B12, 627 (1969); J. L. Basdevant and B. W. Lee, Phys. Rev. to be published.
40. J. A. Mignaco and E. Remiddi, Pisa preprints UGEN(CMP-3905)70, UGEN(CMP-3904)70.
41. P. L. F. Haberler, to be published.

COMMUTATION RELATIONS FOR THE BROKEN CONFORMAL INVARIANCE IN QUANTUM FIELD THEORY†‡

F. K. Cheung
Department of Physics
University of Notre Dame
Notre Dame, Indiana 46556

The interest[1] in conformal invariance has recently been greatly revived by Gell-Mann[2] and others[3] in connection with SU(3) x SU(3) as applied to hadrons.[4] Like SU(3) x SU(3), the dilatation-conformal symmetry is badly broken by the strong interaction as is evidenced by the nonvanishing masses of hadrons. There are, however, at least two basic differences between the conformal symmetry and an internal symmetry such as SU(3) x SU(3). First of all, the conformal generators do not commute with the Hamiltonian even in the limit of exact symmetry; and secondly, the conformal symmetry is a geometric symmetry induced by space-time transformation. We shall here mainly make use of the latter property of the conformal invariance, which enables us in Lagrangian model field theories to define the infinitesimal generators in terms of the field operators through the Noether's theorem independent of the manner in which the symmetry may be broken. We shall later also consider a direct geometrical definition for the dilatation-conformal generators. But we shall see that in all renormalizable Lagrangian models containing fields of spin 0, $\frac{1}{2}$, and 1, the Noether's generators and the corresponding geometrical ones are in fact equivalent. The commutation relations among these generators will be obtained. We shall demonstrate that the breaking is completely determined by the divergence of the dilatation current and that the broken algebra involving only the Poincaré-conformal generators and the divergence of the dilatation current is model independent. At the end, we shall also give a direct geometrical meaning to the breaking terms of the conformal invariance.

---

†Presented at the Symposium on De Sitter and Conformal Groups, University of Colorado, Summer 1970.

‡Research supported in part by the U. S. Atomic Energy Commission under Contract No. AT(11-1)-427.

The infinitesimal scale and special conformal transformations are defined by their action on a space-time point x_μ in the Minkowski space,

$$\epsilon S:\quad x_\mu \to x'_\mu = x_\mu + \epsilon x_\mu$$

$$\alpha_\mu A^\mu: x_\mu \to x'_\mu = x_\mu + \alpha_\mu x^2 - 2x_\mu(\alpha\cdot x)$$

where we notice that the special conformal generator A_μ acts non-linearly on x_μ.

For a given Lagrangian $\mathcal{L}(\varphi^i, \varphi^i_\mu)$ which depends on x_μ through the space-time dependence of the fields and their derivatives, S and A_μ will induce a corresponding change in these fields and hence the Lagrangian, very much like what happens under the Lorentz transformation. One demands on physical grounds that under S and A_μ the action integral

$$W = \int d^4x\, \mathcal{L}(\varphi^i(x), \varphi^i_\mu(x))$$

be invariant up to an integral of a divergence so that the field equations will remain unchanged. This gives rise to a current which becomes divergenceless when S and A_μ is a symmetry of the dynamics described by $\mathcal{L}(\varphi^i, \varphi^i_\mu)$. This is the content of the Noether's theorem, in accordance with which we may now write down, for a given Lagrangian $\mathcal{L}(\varphi^i, \varphi^i_\mu)$, the infinitesimal generators of the Poincaré-conformal generators even if the symmetry is broken,

$$P_\mu = \int d^3x\, T_{o\mu}(x)$$

$$M_{\mu\nu} = \int d^3x\, (x_\nu T_{o\mu} - x_\mu T_{o\nu})$$

$$S = \int d^3x\, \Big(x^\lambda T_{o\lambda} - \sum_i \ell_i \frac{\partial\mathcal{L}}{\partial\varphi^i_o}\varphi^i\Big)$$

$$A_\mu = \int d^3x \Big[2x_\mu x^\lambda T_{o\lambda} - x^2 T_{o\mu} - \sum_i (2\ell_i x_\mu + ix^\lambda \sigma_{\mu\lambda}) \frac{\partial\mathcal{L}}{\partial\varphi^i_o}\varphi^i$$

$$+ g_{\mu o} \sum_{\text{boson}} \ell_i (\varphi^i(x))^2 \Big] \tag{1}$$

where we have used the metric (1,-1,-1,-1), $\varphi_\mu = \frac{\partial\varphi}{\partial x^\mu}$ and

$$T_{\mu\nu} = \sum_i \frac{\partial \mathcal{L}}{\partial \varphi^i_\mu} \varphi^i_\nu - g_{\mu\nu} \mathcal{L}$$

is the canonical energy-momentum tensor.[5] The last term within the integral for A_μ is summed over boson fields only and is necessary in order to make $A_{\mu\nu}$ divergenceless when the symmetry is exact. ℓ_i is the dimension of the field φ^i such that under an infinitesimal scale transformation, a boson field of dimension ℓ, for instance, will transform as,

$$S:\quad \varphi(x) \to \varphi'(x') = \varphi(x) + \ell\epsilon\, \varphi(x)$$

But

$$\varphi'(x') \equiv \varphi'(x+\epsilon x) = \varphi'(x) + \epsilon x_\mu \frac{\partial}{\partial x_\mu} \varphi(x) + o(\epsilon^2)$$

and

$$\varphi'(x) = (1 - i\epsilon S)\,\varphi(x)\,(1 + i\epsilon S)$$

$$= \varphi(x) - i\epsilon[S, \varphi(x)] + o(\epsilon^2)$$

Hence

$$[S, \varphi(x)] = \frac{1}{i}\left[-\ell + x^\lambda \frac{\partial}{\partial x^\lambda}\right] \varphi(x) \qquad (2)$$

Similarly, under an infinitesimal special conformal transformation,

$$\left[A_\mu, \varphi(x)\right] = \frac{1}{i}\left[-2\ell x_\mu + 2x_\mu x^\lambda \frac{\partial}{\partial x^\lambda} - x^2 \frac{\partial}{\partial x^\mu}\right] \varphi(x) \qquad (3)$$

We see that the dimension ℓ of $\varphi(x)$ is not determined by these transformations, but rather by the transformation property of the canonical momentum of $\varphi(x)$, $\pi(x) = \frac{\partial \mathcal{L}}{\partial \varphi_0}$, which in turn depends on the equal-time canonical commutation relation,

$$[\varphi(x), \pi(y)]_{x_0 = y_0} = i\,\delta^3(\vec{x} - \vec{y}) \qquad (4)$$

Thus, using this equation, one gets from Eq. (1)

$$[S, \pi(x)] = \frac{1}{i}\left[x^{\lambda}\frac{\partial}{\partial x^{\lambda}} + 3 + \ell\right]\pi(x) \tag{5}$$

However, when the dilatation symmetry is exact, dS/dt = 0, we may also get the commutator by direct differentiation of Eq. (2),

$$[S, \pi(x)] = \frac{1}{i}\left[x^{\lambda}\frac{\partial}{\partial x^{\lambda}} + 1 - \ell\right]\pi(x) \tag{6}$$

Hence, to be consistent the boson field must have a scale $\ell = -1$. Similarly, the unrenormalized fermion fields are limited to the scale $\ell = -3/2$.

To derive the commutation relation for the conformal generators, one needs to make use of the equal-time canonical commutation relations such as Eq. (4) and

$$[\pi(x), \nabla_i \varphi(y)]_{x_o = y_o} = \frac{1}{i}\frac{\partial}{\partial y_i}\delta^3(\vec{x} - \vec{y})$$

However, for reasons to become clear later it is better to use, wherever possible, the Schwinger's[6] equal-time commutator among various components of the energy momentum tensor,

$$i[T_{oo}(x), T_{oo}(y)]_{x_o = y_o} = [T_{ok}(\vec{x}) + T_{ok}(\vec{y})]\frac{\partial}{\partial x_k}\delta^3(\vec{x} - \vec{y}) - \tau^{oo,oo}(\vec{x}, \vec{y})$$

$$i[T_{oo}(x), T_{oi}(y)]_{x_o = y_o} = [T_{ik}(\vec{x}) + T_{oo}(\vec{y})\delta_{ik}]\frac{\partial}{\partial x_k}\delta^3(\vec{x} - \vec{y}) - \tau^{oo,oi}(\vec{x}, \vec{y})$$

$$i[T_{oi}(x), T_{oj}(y)]_{x_o = y_o} = [T_{oj}(\vec{x})\frac{\partial}{\partial x^i} + T_{oi}(y)\frac{\partial}{\partial x^j}]\delta^3(\vec{x} - \vec{y}) - \tau^{oi,oj}(\vec{x}, \vec{y})$$

where the τ's are the Schwinger terms; they are in general q-number quantities but each of them involves at least two three-divergence such that our result will be for most part independent of Schwinger terms. We will neglect their presence in the following discussion.

Now using these commutators, it is not difficult but exceedingly tedious to verify the following relations,

$$[P_\mu, P_\nu] = 0$$

$$[M_{\mu\nu}, P_\lambda] = i\, g_{\mu\lambda} P_\nu - i\, g_{\nu\lambda} P_\mu$$

$$[M_{\mu\nu}, M_{\lambda\sigma}] = i\, g_{\mu\lambda} M_{\nu\sigma} - i\, g_{\mu\sigma} M_{\nu\lambda} + i g_{\nu\sigma} M_{\mu\lambda} - i\, g_{\nu\lambda} M_{\mu\sigma}$$

$$[S, S] = 0$$

$$[S, P_\mu] = - i\, P_\mu + i\, g_{\mu o} \int d^3x\, \partial^\lambda S_\lambda$$

$$[S, M_{\mu\nu}] = i \int d^3x\, (g_{\mu o} x_\nu - g_{\nu o} x_\mu)\, \partial^\lambda S_\lambda$$

$$[A_\mu, P_\nu] = 2i (g_{\mu\nu} S - M_{\mu\nu}) + 2i\, g_{\nu o} \int d^3x\, x_\mu \partial^\lambda S_\lambda$$

$$[A_\lambda, M_{\mu\nu}] = i\, g_{\nu\lambda} A_\mu - i g_{\mu\lambda} A_\nu + 2i \int d^3x\, (g_{\mu o} x_\nu - g_{\nu o} x_\mu)\, x_\lambda \partial^\sigma S_\sigma$$

$$[A_\mu, S] = -i\, A_\mu + i\, g_{\mu o} \int d^3x\, x^2 \partial^\lambda S_\lambda$$

$$[A_\mu, A_\nu] = 2i \int d^3x\, (g_{\mu o} x_\nu - g_{\nu o} x_\mu)\, x^2\, \partial^\lambda S_\lambda \tag{7}$$

We would like to remark that:

(a) All the breaking terms come from the divergence of the dilatation current $\partial^\lambda S_\lambda$. With the definition given by Eq. (1), we have

$$\partial^\lambda S_\lambda = -4\mathcal{L} - \sum_i \left[\ell_i \frac{\partial \mathcal{L}}{\partial \varphi^i} \varphi^i + (\ell_i - 1) \frac{\partial \mathcal{L}}{\partial \varphi^i_\mu} \partial_\mu \varphi^i \right]$$

which means that those terms in the Lagrangian having dimensionless coupling constants such as the kinetic energy terms, interaction terms like $\lambda\varphi^4$, $\bar{\psi}\gamma_\mu\psi A^\mu$, or $\bar{\psi}\gamma_5\psi\varphi$ will not contribute to the divergence of the dilatation current. Furthermore, all the breaking terms are simple moments of $\partial^\lambda S_\lambda$. The commutators $[S, M_{\mu\nu}]$ and $[A_\mu, A_\nu]$ contain only odd moments of $\partial^\lambda S_\lambda$, $[A_\mu, S]$ only even moments of $\partial^\lambda S_\lambda$, while $[A_\mu, P_\nu]$ and $[A_\lambda, M_{\mu\nu}]$ have both even and odd moments of $\partial^\lambda S_\lambda$. Hence, it is quite possible to construct a model where all the odd moments of $\partial^\lambda S_\lambda$ say vanish, so that the commutators $[S, M_{\mu\nu}]$ and $[A_\mu, A_\nu]$ are the same as in the exact symmetry limit. In any case, the breaking of the symmetry is completely determined by $\partial^\lambda S_\lambda$.

In specifying $\partial^\lambda S_\lambda$, we have also specified the model of breaking. In particular, in the limit of exact symmetry $\partial^\lambda S_\lambda = 0$, the above commutation relations reduce to the closed algebra of the conformal group. We notice once again that S and A_μ do not commute with the Hamiltonian P_o even in the limit of exact symmetry.

(b) All the breaking terms in the above relations carry a factor $g_{\alpha o}$ with one zero-component subscript. This means, for instance, that only the zero-component of P_μ and $M_{\mu\nu}$ induce an explicit time dependence on the generators S and A_λ in the commutators $[S, P_\mu]$, $[S, M_{\mu\nu}]$, $[A_\mu, P_\nu]$ and $[A_\mu\ M_{\lambda\sigma}]$. Similarly, only the zero-component of A_μ can induce a time dependence on S and A_μ. This implies in particular that the commutators $[S, P_i]$, $[S, M_{ij}]$, $[A_\mu, P_i]$, $[A_\mu\ M_{ij}]$, $[A_i, S]$ and $[A_i, A_j]$ with $i, j = 1,2,3$ are not changed by any conformal breaking.

(c) Inside a three-dimensional integral in the above equations, a term like $x_\mu \partial^\lambda S_\lambda$ may be replaced by $-\frac{1}{2}\partial^\lambda A_{\lambda\mu}$, since these two expressions can differ at most by a total divergence.[7] In this sense, the special conformal generator A_μ is not independent of the dilatation generator S. In other words, knowing S_μ, A_μ may be obtained by taking moments of the dilatation current S_μ. This will become even clearer when we consider the geometric definitions of S and A_μ.

(d) In deriving the above equations, partial integrations are quite often performed and what have been integrated out are dropped. This means that we have assumed the fields and the energy-momentum tensor to be so localized spatially that for instance,

$$\lim_{\vec{x}\to\infty} x_i\, T_{\mu\nu}(\vec{x}, x_o) = 0$$

(e) There are terms of the form $\int d^3x\, \ell(\ell+1)\, \varphi^2(x)$ on the right-hand side of the last two of the above equations. The simple form given holds only if we have assumed that the boson fields have the dimension $\ell = -1$.

(f) The commutation relations for the exact conformal symmetry are invariant under the following transformation,

$$A_\mu \to A_\mu$$

$$P_\mu \to \frac{1}{c} P_\mu$$

$$S \to S$$

and

$$M_{\mu\nu} \to M_{\mu\nu}$$

for any constant c.

This is no longer true for the broken algebra; we may say that the above symmetry transformation is also broken by the existence of the divergence of the dilatation current.

The above considerations based on Lagrangian models suffer from one serious drawback, namely, the canonical equal-time commutators used to derive these relations such as Eq. (4) are for bare fields and their canonical conjugate momenta only. Even for renormalizable Lagrangians, the canonical equal-time commutators for the renormalized fields and their conjugate momenta will be multiplied by some cut-off dependent and hence scale dependent renormalization constant. Quite possibly, the renormalized fields will no longer transform under S with a definite dimension.[8] This prompts us to consider the new definition of S and A_μ given by Gell-Mann,[2]

$$\bar{S} = \int d^3 x^\lambda \, \theta_{o\lambda}$$

$$\bar{A}_\mu = \int d^3 \, [2x^\lambda x_\mu \, \theta_{o\lambda} - x^2 \theta_{o\mu}]$$

$$\bar{P}_\mu = \int d^3 \, \theta_{o\mu}$$

$$\bar{M}_{\mu\nu} = \int d^3 \, [x_\nu \, \theta_{o\mu} - x_\mu \, \theta_{o\nu}] \tag{8}$$

These definitions coincide with our geometrical intuition on these generators; they depend only on the physical quantities $\theta_{\mu\nu}$ but not on the Lagrangian, and hence are model-independent. The divergence of the dilatation current $\bar{S}_\mu = x^\lambda \theta_{\lambda\mu}$ is now simply given by the trace of the symmetry energy momentum tensor $\theta^\mu{}_\mu$. It follows that unlike the canonical $T_{\mu\nu}$, $\theta_{\mu\nu}$ is traceless in the limit of exact scale invariance. For renormalizable Lagrangians containing fields of spin 0, $\frac{1}{2}$, and 1, Coleman et al[9] had shown that $\theta_{\mu\nu}$ differs from the symmetrized canonical $T_{\mu\nu}$ only by the Huggins term[10] $\frac{-1}{6}(\partial_\mu\partial_\nu - g_{\mu\nu}\Box)\,\varphi^2$ for each scalar or pseudoscalar field φ. It is then easy to check that for all the renormalizable Lagrangians considered in Ref. 9,

$$\partial^\mu \bar{S}_\mu = \theta^\mu{}_\mu$$

$$\partial^\mu S_\mu = T^\mu{}_\mu - \sum_i \ell_i \left[\frac{\partial \mathcal{L}}{\partial \varphi^i} \varphi^i + \varphi^i_\mu \, \varphi^{i\mu} \right]$$

so that

$$\partial^\mu S_\mu - \partial^\mu \bar{S}_\mu = - \sum_i \left[\varphi^i \Box \varphi^i + \ell_i \varphi^i \partial_\mu \frac{\partial \mathcal{L}}{\partial \varphi^i_\mu} \right]$$

from which we immediately obtain

$$\partial^\mu S_\mu = \partial^\mu \bar{S}_\mu \tag{9}$$

and hence

$$\partial^\mu A_{\mu\nu} = \partial^\mu \bar{A}_{\mu\nu} = - 2x_\nu \partial^\mu \bar{S}_\mu \tag{10}$$

provided that

$$\partial_\mu \varphi^i = - \ell_i \frac{\partial \mathcal{L}}{\partial \varphi^i_\mu} \tag{11}$$

which is true if there is no derivative coupling in $\mathcal{L}$ and the dimension of spin zero bosons is -1. Note that if we only want to preserve the Poincaré generators, the Huggins term is far from unique. For instance, multiplication with any polynomial in the Klein-Gordon operator, $P(\Box)$, to the Huggins term will not affect P_μ and $M_{\mu\nu}$. The coefficient -1/6 is again required by the canonical dimension of the bosons. After renormalization, the dimension of the boson field will in general change; this coefficient must then be changed accordingly in order to preserve Eq. (9) and Eq. (10).

It now follows from Eqs. (9) and (10) that

$$S = \bar{S}$$

$$A_\mu = \bar{A}_\mu \tag{12}$$

which together with P_μ and $\bar{P}_\mu$ and $M_{\mu\nu} = \bar{M}_{\mu\nu}$ show that the geometrical generators defined by Eq. (8) are in fact identical with those defined through Noether's theorem as given by Eq. (7).

Having demonstrated their equivalence in Lagrangian models, we may now discard the Noether's definition in favor of the geometrical one. Since the geometrical generators of the conformal group depend only on the energy-momentum tensor $\theta_{\mu\nu}$, all we need to obtain the commutation relations for these generators are the equal-time commutators among components of $\theta_{\mu\nu}$. Now since the Schwinger's equal-time commutators among various components of the energy-momentum tensor follow from locality and Lorentz covariance alone, they are the same for $\theta_{\mu\nu}$ and $T_{\mu\nu}$; furthermore, they are not affected by renormalization.[11] With this reasoning, it is straightforward to show that the generators $\bar{S}$, $\bar{A}_\mu$, $\bar{P}_\mu$, and $\bar{M}_{\mu\nu}$ again satisfy the commutation relations of Eq. (7). We conclude that the broken algebra involving only the Poincaré-conformal generators and the divergence of the dilatation current is model-independent.

In order to give a direct geometrical meaning to the breaking of the conformal algebra, let us consider the well known representation of the Poincaré-conformal generators by differential operators acting on a Hilbert space of $L(p^2)$

$$\begin{aligned}
\bar{P}_\mu &= i\frac{\partial}{\partial x_\mu} \\
\bar{M}_{\mu\nu} &= i\left(x_\nu \frac{\partial}{\partial x^\mu} - x_\mu \frac{\partial}{\partial x^\nu}\right) \\
\bar{S} &= i\,x^\lambda \frac{\partial}{\partial x^\lambda} \\
\bar{A} &= i\left(x^2 \frac{\partial}{\partial x^\mu} - 2x_\mu x^\lambda \frac{\partial}{\partial x^\lambda}\right)
\end{aligned} \tag{13}$$

where we have omitted some constants in S and A_μ which are necessary to make them hermitian differential operators. Such omissions are justified since they play no role at all in the commutation relations. The form of the differential operators follows directly from the action of the infinitesimal generators on a space-time point x_μ. Thus, when the infinitesimal scale generator operates on a squared integrable function, we have

$$\begin{aligned}
\bar{S}[f(x)] &= f(x + \epsilon x) \\
&= f(x) - i\epsilon\left(i\,x^\lambda \frac{\partial}{\partial x^\lambda}\right) f(x) + 0(\epsilon^2)
\end{aligned}$$

hence

$$\bar{S} = i\,x^\lambda \frac{\partial}{\partial x^\lambda}$$

Similarly, under A_μ, we have

$$\bar{A}_\mu [f(x)] = f(x_\mu + \alpha_\mu x^2 - 2x_\mu \alpha \cdot x)$$

$$= \left[1 - i\alpha_\mu \, i\left(x^2 \frac{\partial}{\partial x^\mu} - 2x_\mu x^\lambda \frac{\partial}{\partial x^\lambda} \right) + 0(\alpha^2) \right] f(x)$$

which gives

$$\bar{A}_\mu = i\left(x^2 \frac{\partial}{\partial x^\mu} - 2x_\mu x^\lambda \frac{\partial}{\partial x^\lambda} \right)$$

If we form commutation relations for the generators given by Eq. (13), we get just the Lie algebra of the conformal group. Now if the symmetry is broken, the generators S and A_μ will be explicitly time-dependent. To obtain the broken algebra, we first assume the time-dependent Heisenberg equations of motion for S and A_μ,

$$[\bar{S}, \bar{P}_\mu] = -i\bar{P}_\mu + ig_{\mu o} \frac{d\bar{S}}{dt} \tag{14}$$

$$[\bar{A}_\mu, \bar{P}_\nu] = 2i\,(g_{\mu\nu} \bar{S} - \overline{M}_{\mu\nu}) + ig_{\nu o} \frac{d\bar{A}_\mu}{dt} \tag{15}$$

We next postulate the commutator,

$$i\,[x_\mu, S(x_o)] = x_\mu$$

We then have

$$[\bar{S}, M_{\mu\nu}] = i\,(g_{\mu o} x_\nu - g_{\nu o} x_\mu) \frac{d\bar{S}}{dt} \tag{16}$$

and

$$[\bar{S}, \bar{S}] = 0 \tag{17}$$

If we now also make use of the fact that for the representation given we have the following simple relation among $\bar{A}_\mu$, $\bar{P}_\mu$, and $\bar{S}$,

$$\bar{A}_\mu = x^2 \bar{P}_\mu - 2x_\mu \bar{S} \tag{18}$$

one can then verify the following relations,

$$[\bar{A}_\lambda, \bar{M}_{\mu\nu}] = i(g_{\nu\lambda}\bar{A}_\mu - g_{\mu\lambda}\bar{A}_\nu) + i(g_{\mu o}x_\nu - g_{\nu o}x_\mu)\frac{d\bar{A}_\lambda}{dt}$$

$$[\bar{A}_\mu, S] = -i\bar{A}_\mu + i x^2 g_{\mu o}\frac{d\bar{S}}{dt}$$

$$[\bar{A}_\mu, \bar{A}_\nu] = 2i x^2 (x_\nu g_{\mu o} - x_\mu g_{\nu o})\frac{d\bar{S}}{dt} \tag{19}$$

plus the usual Poincaré algebra which is not affected by the conformal breaking. Taking into account the relation that $\partial^\mu\bar{A}_{\mu\nu} = -2x_\nu\,\partial^\mu\bar{S}_\mu$ there is a one-to-one correspondence between these relations and those given by Eq. (7). Indeed, it is easy to show for instance,

$$\begin{aligned}\frac{d\bar{S}}{dt} &= \frac{d}{dt}\int d^3x\,(x^o\,\theta_{oo} - x^i\,\theta_{oi}) \\ &= \int d^3x\,\theta_{oo} - \int d^3x\,x^i\frac{\partial}{\partial x_j}\,\theta_{ij} \\ &= \int d^3x\,\theta^\mu{}_\mu = \int d^3x\,\partial^\lambda S_\lambda\end{aligned} \tag{20}$$

We see that the breaking of the conformal invariance is directly given by the time variation of the conformal generators $d\bar{S}/dt$ and $d\bar{A}_\mu/dt$.

The above discussion strongly suggests that the broken conformal algebra we have obtained is not only model-independent but also more general than the particular representation we have chosen. We think that quite generally, for any satisfactory definition one may give to the conformal generators S and A_μ, their commutation relations will take the form given by Eq. (7).

Acknowledgments

We would like to thank Drs. W. D. McGlinn, S. K. Bose, and P. C. DeCelles for helpful discussions.

References

1. H. Bateman, Proc. London Math. Soc. 8, 223 (1910); P. A. M. Dirac, Ann. Math. 37, 429 (1936); H. A. Kastrup, Ann. Phys. 7, 388 (1962); J. Wess, Nuovo Cimento 18, 1086 (1960); F. Gursey, Ann. Phys. 24, 211 (1963); T. Fulton, R. Rohrlich, and L. Witten, Rev. Mod. Phys. 34, 442 (1962) and references cited therein.

2. M. Gell-Mann, "Symmetry Violation in Hadron Physics," CALT Preprint 08-2441 (1970).
3. For example, see H. A. Kastrup, Phys. Rev. 142, 1060 (1966); 143, 1041 (1966); Nucl. Phys. 58, 5611 (1964); B15, 179 (1970); G. Mack, Nucl. Phys. 135, 499 (1968); K. Wilson, Phys. Rev. 179, 1499 (1969).
4. M. Gell-Mann, R. Oakes, and B. Renner, Phys. Rev. 175, 2195 (1968).
5. F. J. Belinfante, Physica 7, 449 (1940).
6. J. Schwinger, Phys. Rev. 127, 324 (1962); 130, 406, 800 (1963); D. G. Boulware and S. Deser, J. Math. Phys. 8, 1468 (1967).
7. G. Mack and A. Salam, Ann. Phys. 53, 174 (1969); D. J. Gross and J. Wess, CERN Preprint TH-1076 (1970).
8. K. Wilson, Phys. Rev. 179, 1499 (1969).
9. C. G. Callan, Jr., S. Coleman, and R. Jackiw, Ann. Phys. (to be published).
10. E. Huggins, Ph.D. Thesis, California Institute of Technology (1962).
11. D. G. Boulware and S. Deser, J. Math. Phys. 8, 1648 (1967).

Section B: Dynamical Groups and Current Algebra

CLOSED ORBITS AND SO(4,2) SYMMETRY IN RELATIVISTIC TWO-BODY THEORY†

C. M. Andersen and Hans C. von Baeyer
Department of Physics
College of William and Mary
Williamsburg, Virginia 23185

Many of us have been fascinated for some time by the special symmetries associated with the nonrelativistic spinless hydrogen atom, or, as it is often called, the Kepler problem. In this problem we have one particle bound to a second particle, treated as infinitely massive, by a $1/r$ potential. From the classical point of view, this problem is very special because all the bound orbits are closed: in fact they are ellipses. From the quantum point of view the special feature is the presence of the so-called "accidental degeneracies." In either case we have an SO(4) group of transformations (an invariance group) which may be applied to the trajectories or to the states resulting in a family of trajectories or states with the same energy. The six generators of this group are the three components of the conserved angular momentum and the three components of the conserved Runge-Lenz vector.[1)]

We may even extend this group to the noninvariance group SO(4,2) which in the quantum mechanical case has a single irreducible representation whose basis vectors are in one-to-one correspondence with the entire set of bound states and wherein the several basis vectors associated with any one of the various representations of the SO(4) subgroup correspond to states of the same energy.

The problem just described is a nonrelativistic one-body problem. We now ask how to formulate a relativistic one-body problem with the same symmetry. We impose three conditions.

(1) It should reduce to the Kepler problem in the nonrelativistic limit.

(2) It should have the relativistic energy-momentum relation in the free particle limit, i.e. as the potential goes to zero, the energy-momentum relation should become $E^2 = p^2 + m^2$.

---

†Presented by C. M. Andersen at the Symposium on De Sitter and Conformal Groups, University of Colorado, Summer 1970.

Examples of theories satisfying these first two conditions are given by the following sets of equations:

(a) $$d\vec{p}/dt = -\vec{\nabla}V(r), \quad V = -k/r$$

$$\vec{p} = m\gamma\vec{v}$$

which implies

$$E \equiv m\gamma + V = \text{const}$$

$$(E - V)^2 = p^2 + m^2 .$$

This is a vector theory in the sense that V is treated as the fourth component of a vector. It is familiar to us because it properly describes the Coulomb electromagnetic interaction. The space-like components of the 4-vector A_μ are set equal to zero and V is the time-like component A_0.

(b) $$d\vec{p}/dt = -\gamma^{-1} \vec{\nabla}V(r); \quad V = -k/r$$

$$\vec{p} = E\vec{v}; \quad E \equiv (m + V)\gamma$$

which implies

$$E = \text{const}$$

$$E^2 = p^2 + (m + V)^2 .$$

This is a scalar theory, and the potential V is now more closely associated with the mass m than with the energy E. In fact the "effective rest mass" is now (m + V) and, for an attractive potential, may go through zero and become negative if the particle comes sufficiently close to the origin.

(c) $$d\vec{p}/dt = -(m/E) \vec{\nabla}V(r); \quad V = -k/r$$

$$\vec{p} = E\vec{v}; \quad E \equiv m\gamma \sqrt{1 + 2V/m}$$

which implies

$$E = \text{const}$$

$$E^2 = p^2 + m^2 + 2mV.$$

This again is a scalar theory with an effective rest mass which depends on the position coordinates.

In addition to these three examples, there are numerous theories in which V(r) is a component of a tensor of rank two or higher.

Now we state the remaining condition.

(3) There should exist closed orbits in the classical theory or "accidental degeneracies" in the corresponding quantum theory.

All three conditions are met in Example (c) wherein the bound orbits not only are closed but are ellipses as in the nonrelativistic problem.

Note that in Example (c) the equations of motion

$$\frac{d\vec{p}}{dt} = -\left(\frac{m}{E}\right)\vec{\nabla}V$$

$$\frac{d\vec{x}'}{dt} = \left(\frac{m}{E}\right)\frac{\vec{p}}{m} \quad ,$$

apart from the presence of the quantity m/E, which is a constant of the motion, are the same as for the nonrelativistic Kepler problem. On the other hand in Examples (a) and (b) factors of γ are brought into the equations of motion and γ is not a constant of the motion. Because of the similarity of Example (c) to the Kepler problem we may expect a conserved Runge-Lenz vector to exist there. It is given by

$$\vec{A} = \frac{\vec{r}}{r} - \frac{\vec{p} \times \vec{J}}{mk}$$

with $\vec{J} = \vec{r} \times \vec{p}$ and has exactly the same form as in the Kepler problem.

For bounded trajectories, $\vec{A}$ is closely associated with the eccentricity of the ellipse of the orbit. The distance from the center of the ellipse to one of the foci is given by

$$\vec{e} = \frac{mk}{m^2 - E^2}\vec{A} \ .$$

For unbounded trajectories, that is, in the case of scattering, the existence of the conserved quantities energy, angular momentum, and Runge-Lenz vector allows us by means of a three-line calculation involving no integration or differentiation to determine the scattering angle θ:

$$\tan\frac{\theta}{2} = \frac{mk}{p^2 b} = \frac{k(1 - v^2)}{mv^2 b} \quad ,$$

where b is the impact parameter and v is the velocity of particle long before and long after the scattering. Note that as $v \to 0$ this becomes the Rutherford formula.

The relation between (c) and a classical theory of point particles with interaction mediated by a scalar field has been discussed by the authors elsewhere.[2)]

For the corresponding Schrödinger theory, i.e. in a wave equation in which E and $\vec{p}$ are treated as operators and the Hamiltonian is $H = p^2 + m^2 + 2mV$, the energy spectrum is just that for the Dirac hydrogen atom with quantum number k set equal to the principal quantum number n, i.e.

$$E_n = m\sqrt{1 - (k/n)^2} \quad ,$$

where the principal quantum number n takes on the values 1,2,... . Of course the degeneracies are the same as for the nonrelativistic hydrogen atom. E_n is independent of the other quantum numbers ℓ and m satisfying $\ell = 0, 1, \ldots, n-1$ and $-\ell \leq m \leq \ell$.

Now we want to look for a relativistic two-body theory. Dirac[3)] tells us that we may try to do this using a "point" form of dynamics, i.e. using 4-vectors and manifestly covariant equations or by using an "instant" form of dynamics in which we work with 3-vectors. However, even using the instant form we can have Poincaré invariance. The particular instant form we have turned to is due to Bakamjian and Thomas.[4)] For the two-body problem one makes a contact transformation from the coordinates $\vec{x}_1 \vec{x}_2$ and momenta $\vec{p}_1 \vec{p}_2$ of the particles to new coordinates $\vec{R}$, $\vec{r}$ and new momenta $\vec{P}$, $\vec{p}$. The total momentum of the system is $\vec{P} = \vec{p}_1 + \vec{p}_2$, and $\vec{R}$ may be thought of as the position of the center of mass. The upper case variables, $\vec{R}$ and $\vec{P}$, are referred to as "external variables." $\vec{r}$ is related to the difference in positions of the two particles, and $\vec{p}$ is related to the difference in their momenta. The lower case variables, $\vec{r}$ and $\vec{p}$, are referred to as "internal variables." The center of mass energy M is taken to be a scalar function of the internal variables, i.e. we have $M = M(\vec{r}^2, \vec{r}\cdot\vec{p}, \vec{p}^2)$. The total Hamiltonian is given by

$$H_{total} = \{ |\vec{P}|^2 + M^2(\vec{r}^2, \vec{r}\cdot\vec{p}, \vec{p}^2) \}^{\frac{1}{2}}$$

and is independent of $\vec{R}$. Our freedom of choice consists in selecting the functional form of the center of mass energy M. We need to apply three conditions

(i) We recover the Kepler problem in the nonrelativistic limit.

(ii) The energy-momentum relation should reduce to

$$E = \sqrt{p_1{}^2 + m_2{}^2} + \sqrt{p_2{}^2 + m_2{}^2}$$

in the free particle limit. Examples of the choices open to us at this stage are the forms:

$$\text{(A)} \qquad M = \sqrt{p^2 + m_1{}^2} + \sqrt{p^2 + m_2{}^2} + V(r)$$

This form was given by Bakamjian and Thomas. With the proper choice of $V(r)$ one gets Hamiltonians invented by Breit and by Darwin which correctly give the first few relativistic correction terms for the hydrogen atom problem.

$$\text{(B)} \qquad M = \sqrt{p^2 + (m_1 + V(r))^2} + \sqrt{p^2 + (m_2 + V(r))^2}$$

which is patterned after the first scalar interaction.

$$\text{(C)} \quad M = \sqrt{p^2 + m_1{}^2 + 2m_r V(r)} + \sqrt{p^2 + m_2{}^2 + 2m_r V(r)}$$

where m_r is the reduced mass $m_1 m_2/(m_1 + m_2)$. This is patterned after the second scalar interaction.

Condition (iii) is the existence of a conserved Runge-Lenz vector. This we can accommodate as well. We use form (C) with $V = -k/r$. The Runge-Lenz vector is a function of the internal variables $\vec{r}$ and $\vec{p}$ and has the same form as before

$$A(\vec{r}, \vec{p}) = \frac{\vec{r}}{r} - \frac{\vec{p} \times \vec{J}^{\,\text{int.}}}{m_r k}$$

with $\vec{J}^{\,\text{int.}} = \vec{r} \times \vec{p}$.

Let us explore a few of the properties of this theory. We define a "little" Hamiltonian

$$h(\vec{r}, \vec{p}) = p^2/2m_r - k/r,$$

and "imbed" this little Hamiltonian into the full Hamiltonian in the following way

$$E = H(\vec{R}, \vec{P}, r, p) - \{ |\vec{P}|^2 + (\sqrt{m_1{}^2 + 2m_r h(\vec{r}, \vec{p})} + \sqrt{m_2{}^2 + 2m_r h(\vec{r}, \vec{p})}\,)^2 \}^{\frac{1}{2}}$$

From this relation, we obtain the Hamiltonian equations of motion

$$\frac{d\vec{P}}{dt} = 0, \qquad \frac{d\vec{r}}{dt} = \frac{\vec{p}}{\mathcal{E}} ,$$

$$\frac{d\vec{R}}{dt} = \frac{\vec{P}}{E}, \qquad \frac{d\vec{p}}{dt} = -\frac{m_r k\vec{r}}{\mathcal{E}r^3} ,$$

where in the center of mass frame the reduced energy $\mathcal{E}$ has the form

$$\mathcal{E} = \frac{\sqrt{m_1{}^2 + 2m_r h} \cdot \sqrt{m_2{}^2 + 2m_r h}}{\sqrt{m_1{}^2 + 2m_r h} + \sqrt{m_2{}^2 + 2m_r h}} .$$

Note that the equations of motion for the internal variables have just the same form as in the relativistic one-body problem except that we must distinguish the reduced energy $\mathcal{E}$ from the total energy E.

The proof that $\vec{A}(\vec{r},\vec{p})$ is conserved is simple, namely $\vec{A}$ "commutes" with P and with $h(\vec{r},\vec{p})$ and thus with H.

So far we have not related the canonical variables $\vec{r},\vec{p},\vec{R},\vec{P}$ to the coordinates $\vec{x}_1$, $\vec{x}_2$ and momenta $\vec{p}_1$, $\vec{p}_2$, other than to say that this relation is a contact transformation, i.e. that the Poisson brackets $\{\ ,\}_{\vec{x}_1\vec{x}_2\vec{p}_1\vec{p}_2}$ and $\{\ ,\}_{\vec{R}\vec{P}\vec{r}\vec{p}}$ are equal. The relation has been worked out in detail by Bakamjian and Thomas[4)] in such a way as to explicitly give Poincaré invariance. These relations are quite complicated so we give them here only for the special case of total momentum equal to zero. They look as follows:

$$\vec{R} = \frac{\sqrt{m_1{}^2 + 2m_r h}\ \vec{x}_1 + \sqrt{m_2{}^2 + 2m_r h}\ \vec{x}_2}{\sqrt{m_1{}^2 = 2m_r h} + \sqrt{m_2{}^2 + 2m_r h}}$$

$$\vec{P} = \vec{p}_1 + \vec{p}_2 = 0 ,$$

$$\vec{r} = \vec{x}_1 - \vec{x}_2 ,$$

$$p = \frac{\sqrt{m_2{}^2 + 2m_r h}\ \vec{p}_1 - \sqrt{m_1{}^2 + 2m_r h}\ \vec{p}_2}{\sqrt{m_1{}^2 + 2m_r h} + \sqrt{m_2{}^2 + 2m_r h}}$$

The total angular momentum $\vec{J}^{\text{total}} = \vec{R} \times \vec{P} + \vec{r} \times \vec{p}$ is a conserved quantity. Using the generalizations of the above relations for the case $\vec{P} \neq 0$, one obtains a Hamiltonian

$$H = H(\vec{x}_1, \vec{p}_1, \vec{x}_2, \vec{p}_2)$$

with the desired symmetry properties. In the case that one particle is infinitely massive we recover the relativistic one-body problem that we discussed earlier as Example (c).

The discussion of the SO(4,2) noninvariance group[1] goes through unscathed by the imbedding of $h(\vec{r},\vec{p})$ into the two-body Hamiltonian. The SO(4,2) Poisson bracket relations are among quantities formed only from the internal variables $\vec{r}$ and $\vec{p}$.

The existence of a conserved Runge-Lenz vector allows us to find the scattering angle. We find

$$\tan\frac{\theta}{2} = \frac{mk}{p^2 b} = \frac{mk}{bv^2 \mathcal{E}^2} \quad .$$

However, $\mathcal{E}^2$ is related to the initial velocity v by an algebraic equation which is quartic in $\mathcal{E}^2$ and in v^2.

Summary

First we have displayed a one-body theory with relativistic kinematics and SO(4,2) symmetry. This theory can be made quantum mechanical simply by interpreting the energy and momentum variables as operators. The Hamiltonian is $H = (p^2 + m^2 - 2mk/r)^{\frac{1}{2}}$.

Secondly we have generalized this theory to a Poincaré invariant two-body theory with the same symmetry. The device we used was Bakamjian-Thomas theory and this can also be quantized easily.

Fronsdal,[5] starting from an approximation to the Bethe-Salpeter equation, has obtained a relativistic classical mechanics which also has closed ellipses. However, he has put the particle with mass m_1 on the mass shell and kept the other one off, whereas in our theory the particles are treated symmetrically. In the potential theory limit ($m_2/m_1 \to 0$) his mechanics coincides with ours.[2] Our hope is that the quantum version of our theory can also be shown to be a well-defined approximation to a Bethe-Salpeter equation.[6]

Acknowledgment

We would like to thank Professor E. Remler for drawing our attention to Bakamjian-Thomas theory and for several illuminating conversations.

References

1. G. Györgyi, Nuovo Cimento 53A, 717 (1968).
2. C. M. Andersen and Hans C. von Baeyer, "On Classical Scalar Field Theories and the Relativistic Kepler Problem," to be published.

3. P. A. M. Dirac, Rev. Mod. Phys. 21, 392 (1949).
4. B. Bakamjian and L. H. Thomas, Phys. Rev. 92, 1300 (1953).
5. C. Fronsdal, "The Relativistic Kepler Problem," UCLA preprint, March 1970.
6. E. A. Remler has pointed out in an unpublished paper (Princeton, 1966) that form (C) of the Hamiltonian is precisely of the type which can be related to the Blankenbecler-Sugar equation.

REPRESENTATIONS OF THE DYNAMICAL GROUP O(4,2) REALIZED IN THE DYONIUM ATOM†‡

A. O. Barut
Institute for Theoretical Physics
University of Colorado
Boulder, Colorado 80302

I. Dyonium and Its Energy Spectrum

We consider the bound and scattering states of two massless particles having both electric and magnetic charges. The usual H-atom is a special case of this new class of atoms, called dyonium. Let m_1 and m_2 denote the masses of the particles and $q_1 = (e_1, g_1)$ and $q_2 = (e_2, g_2)$ be the charges, where e_i represent the electric and g_i the magnetic charges. Let the particle 2 be at the origin of the coordinate system, and particle 1 to move in the field produced by particle 2. The electromagnetic field of the particle 2 can be described by two vector potentials

$$A_\mu = (A_o, -\underset{\sim}{A}) \quad \text{and} \quad \tilde{A}_\mu = (\tilde{A}_o, -\underset{\sim}{\tilde{A}}) \quad , \tag{1}$$

where, in our coordinate system,

$$A_o = \frac{e_2}{r}, \quad \tilde{A}_o = \frac{g_2}{r}, \quad \underset{\sim}{A} = g_2 \, \underset{\sim}{D}(\underset{\sim}{r}), \quad \underset{\sim}{\tilde{A}} = - e_2 \, \underset{\sim}{D}(\underset{\sim}{r}) \tag{2}$$

with

$$\underset{\sim}{D}(\underset{\sim}{r}) = \underset{\sim}{r} \times \underset{\sim}{\hat{n}}(\underset{\sim}{r}\cdot\underset{\sim}{\hat{n}})/[r(r^2 - (\underset{\sim}{r}\cdot\underset{\sim}{\hat{n}})^2]^{-1} \quad , \tag{3}$$

where $\underset{\sim}{\hat{n}}$ is an arbitrary unit vector.

Note that A_μ is a vector under parity, and $\tilde{A}_\mu$ an axial vector (under parity: $\underset{\sim}{E} = -\underset{\sim}{E}$, $\underset{\sim}{B} \to \underset{\sim}{B}$, $e \to e$, $g \to -g$; $\underset{\sim}{j}_e \to -\underset{\sim}{j}_e$; $\underset{\sim}{j}_m \to \underset{\sim}{j}_m$). Note also the singularity line in the vector potential $\underset{\sim}{D}(\underset{\sim}{r})$ in (3) at $r^2 = (\underset{\sim}{r}\cdot\underset{\sim}{\hat{n}})^2$, i.e. along $\hat{n}$. Nevertheless we have $\nabla \times \underset{\sim}{D}(\underset{\sim}{r}) = \underset{\sim}{r}/r^3$, independent of $\hat{n}$.

---

†Presented at the Symposium on De Sitter and Conformal Groups, University of Colorado, Summer 1970.

‡Supported in part by the Air Force Office of Scientific Research under Grant AF-AFOSR-30-67.

The relativistic Lagrangian for particle 1 in the field (1) is given by

$$L = mc\sqrt{u_1^2} + \frac{e_1}{c} A_\mu u_1^\mu + \frac{g_1}{c} \tilde{A}_\mu u_1^\mu , \tag{4}$$

which leads to the canonical momentum

$$P_\mu = mc\, u_{1\mu} + \frac{e_1}{c} A_\mu + \frac{g_1}{c} \tilde{A}_\mu \tag{5}$$

and the Minkowski force

$$K_\mu = \left(\frac{e_1}{c} F_{\mu\nu} + \frac{g_1}{c} \tilde{F}_{\mu\nu}\right) u_1^\nu , \tag{6}$$

where†

$$F_{\mu\nu} = A_{\nu,\mu} - A_{\mu,\nu} \quad \text{and} \quad \tilde{F}_{\mu\nu} = \tilde{A}_{\nu,\mu} - \tilde{A}_{\mu,\nu} = \tfrac{1}{2}\epsilon_{\mu\nu\lambda\rho} F^{\lambda\rho} \tag{7}$$

The Eqs. (4)-(7), as well as the Maxwell-Dirac equations (Gaussian units)

$$F_{\mu\nu}{}^{,\nu} = \frac{4\pi}{c} j_\mu^e , \quad \tilde{F}_{\mu\nu}{}^{,\nu} = \frac{4\pi}{c} j_\mu^m , \tag{8}$$

and all equations that will follow, are invariant under the two-dimensional chiral rotation by any angle θ in the e-g plane (or j^e-j^m plane), and simultaneously in $\underset{\sim}{E}$-$\underset{\sim}{B}$ plane (or $F_{\mu\nu}$ - $\tilde{F}_{\mu\nu}$ plane).

From Eq. (5), because $u_\mu u^\mu = 1$, we have

$$\left(p_\mu - \frac{e_1}{c} A_\mu - \frac{g_1}{c}\tilde{A}_\mu\right)^2 = m^2 c^2 \tag{9}$$

Hence the Klein-Gordon Hamiltonian is given by

$$H^{(KG)} \equiv cp_o = e_1 A_o + g_1 \tilde{A}_o + [m^2 c^4 + (c\underset{\sim}{p} - e_1 \underset{\sim}{A} - g_1 \underset{\sim}{\tilde{A}})^2]^{\frac{1}{2}} \tag{10}$$

If we expand formally the square root and subtract the rest energy we also obtain the Hamiltonian in Schrödinger form

---

†
$$F_{\mu\nu} = \begin{pmatrix} 0 & E_1 & E_2 & E_3 \\ & 0 & -B_3 & B_2 \\ & & 0 & -B_1 \\ & & & 0 \end{pmatrix} , \quad \epsilon^{0123} = +1$$

$$H^{(S)} = eA_o + g\tilde{A}_o + \frac{1}{2m}\left(\underset{\sim}{p} - \frac{e}{c}\underset{\sim}{A} - \frac{g}{c}\underset{\sim}{\tilde{A}}\right)^2 \tag{11}$$

The last step is, in principle, no longer meaningful if the field is very large. If we insert (2) into (10) and (11) we get

$$H^{(KG)} = \frac{e_1 e_2 + g_1 g_2}{r} + \left[m^2 c^4 + [c\underset{\sim}{p} - (e_1 g_2 - e_2 g_1)\, \underset{\sim}{D}(\underset{\sim}{r})]^2\right]^{\frac{1}{2}} \tag{10'}$$

Here the dependence of the Hamiltonian on charges is through the two chiral invariant combinations

$$\alpha = -(e_1 e_2 + g_1 g_2) = -\vec{q}_1 \cdot \vec{q}_2 \tag{12}$$

$$\mu = (e_1 g_2 - e_2 g_1) = q_1 \times q_2 \quad , \tag{13}$$

only, as it should be. We shall see presently that μ is of the order of one (in units of $\hbar c$). Thus, the passage to the Schrödinger form does not break down because g_1 is large (see below), but for small r because of the factor $1/r$ in $\underset{\sim}{D}(\underset{\sim}{r})$.

Finally, introducing the new momentum

$$\underset{\sim}{\pi} = \underset{\sim}{p} - \mu\, \underset{\sim}{D}(\underset{\sim}{r}) \quad , \tag{14}$$

we have the Hamiltonian

$$H^{(KG)} = -\frac{\alpha}{r} + [\pi^2 + m^2 c^4]^{\frac{1}{2}} \tag{10''}$$

with its formal Schrödinger counterpart

$$H^{(S)} = \frac{\pi^2}{2m} - \frac{\alpha}{r} \quad . \tag{11''}$$

These Hamiltonians are characterized by a <u>new invariant parameter</u> μ, Eq. (13). For $\mu = 0$, we get back the same Hamiltonians as that of the ordinary atom but with a different α in general ($\mu = 0$ does not necessarily mean $g_1 = g_2 = 0$, but $\frac{g_1}{g_2} = \frac{e_1}{e_2}$). <u>Each value of μ characterizes a new system</u>. The possible values of μ can be obtained from the quantization of the angular momentum.

The conserved total angular momentum satisfying the commutation relations is given by

$$\underset{\sim}{J} = \underset{\sim}{r} \times \underset{\sim}{\pi} - \mu\, \hat{\underset{\sim}{r}} \quad . \tag{15}$$

We have

$$[H, \underset{\sim}{J}] = 0, \quad [J_i, J_j] = i\,\epsilon_{ijk}\,J_k \quad . \tag{16}$$

. The component of $\underset{\sim}{J}$ along $\hat{\underset{\sim}{r}}$ is equal to $-\mu$, which can take only integer and half-integer values. Hence

$$\mu = 0, \pm\tfrac{1}{2}, \pm 1, \pm\tfrac{3}{2}, \ldots \quad . \tag{17}$$

For a fixed μ, the values of j range as follows:

$$j = |\mu|, \; |\mu|+1, \; |\mu|+2, \ldots \tag{18}$$

We shall see these results also in another way.

In order to solve for the spectrum of (10″) or (11″), we first notice that the following 15-operators:

$$\underset{\sim}{J} = \underset{\sim}{r} \times \underset{\sim}{\pi} - \mu\hat{\underset{\sim}{r}},$$

$$\underset{\sim}{A} = \tfrac{1}{2}\,\underset{\sim}{r}\pi^2 - \underset{\sim}{\pi}(\underset{\sim}{r}\cdot\underset{\sim}{\pi}) + \frac{\mu}{r}\underset{\sim}{J} + \frac{\mu^2}{2r^2}\underset{\sim}{r} - \tfrac{1}{2}\underset{\sim}{r} \;,$$

$$\underset{\sim}{M} = \tfrac{1}{2}\,\underset{\sim}{r}\pi^2 - \underset{\sim}{\pi}(\underset{\sim}{r}\cdot\underset{\sim}{\pi}) + \frac{\mu}{r}\underset{\sim}{J} + \frac{\mu^2}{2r^2}\underset{\sim}{r} + \tfrac{1}{2}\underset{\sim}{r},$$

$$\underset{\sim}{\Gamma} = r\underset{\sim}{\pi},$$

$$\Gamma_0 = \tfrac{1}{2}\left(r\pi^2 + r + \frac{\mu^2}{r}\right) \quad ,$$

$$\Gamma_4 = \tfrac{1}{2}\left(r\pi^2 - r + \frac{\mu^2}{r}\right) \quad ,$$

$$T = \underset{\sim}{r}\cdot\underset{\sim}{\pi} - i \quad . \tag{18}$$

satisfy the commutation relations of the Lie algebra of SO(4,2). The Casimir operators are calculated to be

$$Q_2 = L_{ab}L^{ab} = 3(\mu^2 - 1)$$

$$Q_3 = \epsilon_{abcdef}L^{ab}L^{cd}L^{ef} = 0$$

$$Q_4 = L_{ab}L^{bc}L_{cd}L^{da} = \text{const.} \tag{19}$$

From (19) we see that the operators (18) realize a particular class of most degenerate oscillator-type representations of SO(4,2): The parameter $|\mu|$ has the meaning of lowest spin, and for each discrete eigenvalue n of Γ_o the spectrum of J^2 is indeed

$$j = |\mu|,\ |\mu|+1,\ |\mu|+2, \ldots n-1 \ . \tag{20}$$

This is because the operators Γ_o, Γ_4, T form an O(2,1)-Lie algebra commuting with $\underset{\sim}{J}$:

$$[\Gamma_o, \Gamma_4] = iT;\ [\Gamma_4, T] = -i\Gamma_o;\ [T, \Gamma_o] = i\Gamma_4 \tag{21}$$

$$[\Gamma_o, \underset{\sim}{J}] = [\Gamma_4, \underset{\sim}{J}] = [T, \underset{\sim}{J}] = 0 \tag{22}$$

Hence Γ_o has a discrete spectrum, and, because

$$Q^2 = \Gamma_o^2 - \Gamma_4^2 - T^2 = J^2 \ , \tag{23}$$

we have a discrete representation of O(2,1), $D_+(-j-1)$ and the spectrum n of Γ_o has the range

$$n = j+1,\ j+2,\ \ldots \ . \tag{24}$$

In the other discrete representation $D_-(-j-1)$:

$$n = -(j+1),\ -j-2,\ -j-3,\ \ldots \tag{25}$$

We shall see that n is precisely the principal quantum number.

Now we discuss the spectrum of H for various cases:

(A) $\underline{\mu = 0}$, $\underline{H^{(S)} = \frac{1}{2m} p^2 - \frac{\alpha}{r}}$: we introduce the following operator,

$$\Theta \equiv r(H^{(S)} - E) = \frac{1}{2m} rp^2 - Er - \alpha ;$$

then from (18):

$$\Theta = \frac{1}{2m}(\Gamma_o + \Gamma_4) - E(\Gamma_o - \Gamma_4) - \alpha \ , \tag{26}$$

which is thus a simple linear combination of the group generators in (18). The problem of finding the spectrum of $H^{(S)}$ is equivalent to solving the equation

$$\Theta\tilde{\Phi} = \left(\frac{1}{2m}(\Gamma_o + \Gamma_4) - E(\Gamma_o - \Gamma_4) - \alpha\right)\tilde{\Phi} = 0 \tag{27}$$

By the so-called tilt operation $\tilde{\Phi} = e^{i\theta T}\Phi$ we can then diagonalize either Γ_o or Γ_4, and we get, in the standard fashion, the bound states as the discrete eigenstates of Γ_0, with eigenvalues n and the scattering states as the continuous eigenstates of Γ_4 with eigenvalues λ. This gives the well-known H-atom solutions:

$$E_n = -\tfrac{1}{2}\frac{m\alpha^2}{n^2}, \quad \text{and} \quad E_\lambda = \tfrac{1}{2}\frac{m\alpha^2}{\lambda^2} \tag{28}$$

(B) $\mu \neq 0$, $H^{(S)} = \frac{1}{2m}\pi^2 - \frac{\alpha}{r}$: In this case we introduce three new operators

$$\Gamma_0' = \tfrac{1}{2}(r\pi^2 + r), \quad \Gamma_4' = \tfrac{1}{2}(r\pi^2 - r), \quad T' = T \tag{29}$$

which also satisfy exactly the equations (21) and (22). But instead of Eq. (23) we now have:

$$Q'^2 = J^2 - \mu^2 \quad . \tag{30}$$

Consequently, the eigenvalues of Γ_0' are

$$n' = -\varphi', \; -\varphi' + 1, \; -\varphi' + 2, \ldots$$

$$\text{with} \quad \varphi' = -\tfrac{1}{2} - \sqrt{(j+\tfrac{1}{2})^2 - \mu^2} \quad . \tag{31}$$

Hence, one finds[1)]

$$E_s = -\tfrac{1}{2}m\alpha^2 \left[s + \tfrac{1}{2} + \sqrt{(j+\tfrac{1}{2})^2 - \mu^2} \right]^{-2}$$

$$s = 0, 1, 2, 3, \ldots \tag{32}$$

(C) $\mu \neq 0$, $H^{(KG)} = [\pi^2 + m^2]^{\frac{1}{2}} - \frac{\alpha}{r}$: In this case, we define the new operators

$$\Gamma_0'' = \tfrac{1}{2}(r\pi^2 + r - \frac{\alpha^2}{r})$$

$$\Gamma_4'' = \tfrac{1}{2}(r\pi^2 - r - \frac{\alpha^2}{r})$$

$$T'' = T \tag{33}$$

which again satisfy Eqs. (21) and (22) with

$$Q''^2 = J^2 - \mu^2 - \alpha^2 \tag{34}$$

For positive values of Q'', we have then the same results as before, except for the change

$$\varphi' \to \varphi'' = -\tfrac{1}{2} - \sqrt{(j+\tfrac{1}{2})^2 - \mu^2 - \alpha^2} \ . \tag{35}$$

For the treatment of

(a) the Dirac equation,

(b) the case of large coupling constant α

(c) the O(4)-symmetry and its breaking,

we refer to other work.[1)]

The theory presented here is parity and time-reversal invariant. But because under P and T: $\mu \to -\mu$, we solve the dyonium problem for $+\mu$ and $-\mu$, and then construct parity eigenstates of the form $|\mu\rangle \pm |-\mu\rangle$.[5),6)]

Conclusions

The quantum states of the dyonium atom for a given μ (H-atom: $\mu = 0$) are in one-to-one correspondence with the basis of an irreducible representation of the dynamical group SO(4,2) with the value of a particular invariant equal to μ. (See next section.) Both bound and scattering states can be obtained from the group states by the tilting operation. The Hamiltonian in Schrödinger, Klein-Gordon and Dirac forms are exactly soluble, even for large coupling constant. The dynamical group SO(4,2) solves the problem even though we have a broken O(4)-symmetry (except $\mu = 0$, and nonrelativistic case).

II. SO(4,2)-Representations Characterized by μ

The representations of SO(4,2)-algebra given by Eq. (18) for each value of μ form a special class of representations, the so-called "oscillator-representations." We now list some of the properties of these representations:

(1) They are characterized completely by a single representation relation:[2)]

$$\{L_{AB}, L^{A}{}_{C}\} = -2(1-\mu^2)\, g_{BC} \ , \tag{36}$$

where $L_{AB} = -L_{AB}$ are the 15 generators of SO(4,2).

(2) The generators (18) can be written in terms of the boson creation and annihilation operators

$$J_K = \tfrac{1}{2}\,(a^{+}\sigma_k a + b^{+}\sigma_k b) \quad , \quad k = 1,2,3$$

$$A_i = -\tfrac{1}{2}\,(a^{+}\sigma_i a - b^{+}\sigma_i b)$$

$$M_i = -\tfrac{1}{2}\,(a^{+}\sigma_i C b^{+} - a C \sigma_i b)$$

(equation continued)

$$\Gamma_i = \frac{1}{2i}\,(a^+\sigma_i C b^+ + aC\sigma_i b)$$

$$T = \tfrac{1}{2}\,(a^+Cb^+ + aCb)$$

$$\Gamma_4 = \frac{1}{2i}\,(a^+Cb^+ - aCb)$$

$$\Gamma_o = \tfrac{1}{2}\,(a^+a + b^+b + 2) \qquad (37)$$

where σ_i are the Pauli matrices and C the antisymmetric matrix $C = \begin{pmatrix} 0 & 1 \\ -1 & 0 \end{pmatrix}$. These boson operators act on the states

$$|j_1 m_1 j_2 m_2\rangle\ N\, a_1^{+\,j_1+m_1}\ a_2^{+\,j_1-m_1}\ b_1^{+\,j_2+m_2}\ b_2^{+\,j_2-m_2}\ |0\rangle$$

$$N^{-2} = (j_1+m_1)!\ (j_1-m_1)!\ (j_2+m_2)!\ (j_2-m_2)! \qquad (38)$$

(3) We can indeed explicitly give the invariant operator with the eigenvalue μ:

$$K = \tfrac{1}{2}(a^+a - b^+b) \quad , \qquad (39)$$

which commutes with all the 15 generators. Under parity

$$P: a^+ \to b^+,\ b^+ \to -a^+ \qquad (40)$$

so that μ changes sign.

(4) The multiplicity diagram for a general representation is shown in Fig. 1.

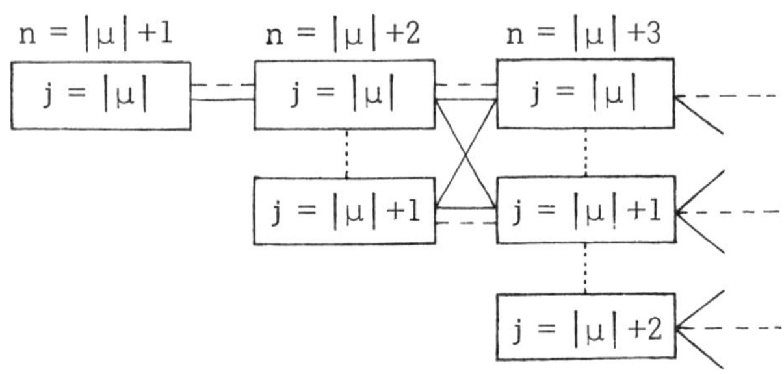

Fig. 1. Multiplicity pattern of SO(4,2)-representation. Each box is an SO(3)-representation (one spin state). Each vertical column is an SO(4)-representation. Solid lines between boxes indicate nonzero matrix elements of $\underset{\sim}{\Gamma}$ and $\underset{\sim}{M}$; broken lines the nonzero matrix elements of Γ_4 and T (Γ_o is diagonal), and dotted lines the nonzero matrix elements of $\underset{\sim}{A}$.

(5) There are two 4-vector operators (with respect to the homogeneous Lorentz group) $\Gamma_\mu^{(1)} = (\Gamma_0, \underset{\sim}{\Gamma})$ and $\Gamma_\mu^{(2)} = (\underset{\sim}{A}, T)$ satisfying

$$\{\Gamma_\mu^{(1)}, \Gamma_\mu^{(2)}\} = 0 \tag{41}$$

These representations remain irreducible when restricted to the subgroup SO(4,1) and to the subgroup SO(3,2), except for the case $\mu = 0$, which when restricted to SO(3,2) splits into two irreducible representations.[2]

(6) These discrete series of representations of SO(4,2) remain also irreducible with respect to the Poincaré subgroup. (The Poincaré subalgebra consists of the generators $\underset{\sim}{J}$, $\underset{\sim}{M}$, $P_\mu = \Gamma_\mu^{(2)} - \Gamma_\mu^{(1)}$.)

Note however that the conformal group O(4,2) has been used here as a dynamical group to describe all the rest frame states of the dyonium atom; it does not contain the physical Poincaré group as a subgroup. The momenta P_μ are outside the algebra of SO(4,2) as interpreted in this application. It is however possible to introduce P_μ additionally and to construct states of the atom with the total momentum P_μ.[3]

III. Dyonium Model In Strong Interactions

The representations of the dynamical group SO(4,2) have been used in the past four years to describe the rest frame states of hadrons: $\mu = 0$ representations for mesons, and $\mu = \frac{1}{2}$ representations for baryons. The main reasons were:

(1) the existence of more than one $j = \frac{1}{2}^+$-baryon states with the same internal quantum numbers as nucleon (e.g. N*(1470)), indicating a new quantum number n.[3]

(2) the dipole electric and magnetic form factors for the proton valid up to $t = 25\ (\text{GeV}/c)^{-2}$.

These properties follow from the $\mu = \frac{1}{2}$ representation of SO(4,2). It was concluded that some strong long-range forces inside the proton must be responsible for the excited states of the proton, without knowing what these forces could be. The model is more tractable and simple than the 3-quark model of the proton, for example, and so far agrees with experiment.

It was later discovered that the representation $\mu = \frac{1}{2}$ used was precisely identical with the dyonium system with $\mu = \frac{1}{2}$ discussed in Sec. I.[5] This remarkable correspondence cannot be accidental for the following reasons:

(1) Proton is identified with the dyonium-system $\mu = \frac{1}{2}$, $g_1 = -g_2$ (hence total $g = 0$), thus $\alpha \approx 137/4$. At large distances there is only electric Coulomb force between two protons, because $g_{tot} = 0$.

At small distances there are van der Waals forces which are short-ranged and strong and are identified with the nuclear forces. Inside the proton we have superstrong Coulomb forces, plus the vector potential $\mu\vec{D}(\vec{r})$. Thus the strength and range of strong interactions comes out correct.

(2) Spin $j = |\mu| = \frac{1}{2}$ of the ground state does not belong to one of the constituents of the proton, but to the system as a whole. This is very important for the correct dipole-form factor of the proton. The bound state of a (spin $\frac{1}{2}$ and spin 0) system with orbital angular momentum $\ell = 0$ does not give a dipole form factor.

(3) The fact that one can construct a spin $\frac{1}{2}$ state out of two spin zero particles with magnetic charges has been overlooked in the past. There is no contradiction here. The wave function is double-valued and not an eigenstate of parity. Under parity $\mu \to -\mu$, and parity eigenstates are constructed as superpositions $|\mu\rangle \pm |-\mu\rangle$.

(4) Because magnetic charge μ is an axial charge, the expectation value of it in parity eigenstates vanishes. There may not be a superselection rule for magnetic charge which may explain why it is not detected readily.[8)]

For further details we refer to the extensive published literature.[3)-6)]

IV. Infinite Component Wave Equations on the Representation-Spaces of O(4,2)

The states (38) are eigenstates of Γ_o. In fact from (37) and (38) we have

$$\Gamma_o |j_1 m_1 j_2 m_2\rangle = (j_1 + j_2 + 1) \; |j_1 m_1 j_2 m_2\rangle \tag{42}$$

Thus $(j_1 + j_2 + 1)$ is equal to n for the O(2,1)-algebra (21), and equal to n′ for the algebra (29). Because Γ_o is the component of a four vector with respect to the Lorentz group generated by J_k and M_i in Eq. (37), we have

$$e^{+i\underset{\sim}{\xi}\cdot\underset{\sim}{M}} \; \Gamma_o P^o \; e^{-i\underset{\sim}{\xi}\cdot\underset{\sim}{M}} = \Gamma_\mu P^\mu \tag{43}$$

Consequently from (42) and (43) we see that the states (38) also satisfy the following covariant wave equation

$$(\Gamma^\mu P_\mu - M(j_1 + j_2 + 1) \; e^{i\underset{\sim}{\xi}\cdot\underset{\sim}{M}} \; |j_1 m_1 j_2 m_2\rangle = 0 \tag{44}$$

If we let

$$M(j_1 + j_2 + 1) = Mn \equiv \kappa \tag{45}$$

and assume κ to be a constant, in the simplest case, we obtain

$$(\Gamma^{\mu} P_{\mu} - \kappa)\psi = 0 \ . \tag{46}$$

This is a Majorana-type equation, written now on the representation space of O(4,2), rather than O(3,2) as in the case of the original Majorana equation. The mass spectrum derived from (45) is clearly $M = \kappa/n$, by construction.

If, instead of (45), we put

$$Mn = \tfrac{1}{2}(M^2 - 6) \tag{47}$$

we obtain the mass spectrum $M = n \pm \sqrt{2b + n^2}$ which increases with increasing n. The corresponding wave equation can be written as

$$(\Gamma^{\mu} P_{\mu} - \tfrac{1}{2}P_{\mu}P^{\mu} + b)\,\psi = 0 \tag{48}$$

Next we consider the states $e^{i\theta T}|j_1 m_1 j_2 m_2\rangle$. Then from (21),

$$e^{i\theta T}\Gamma_o e^{-i\theta T} = \Gamma_o \cosh\theta + \Gamma_4 \sinh\theta \ , \tag{49}$$

and, because Γ_4 is a Lorentz-scalar, we obtain again from (42) and (43), the more general covariant equation

$$(\Gamma^{\mu} P_{\mu} \cosh\theta + \Gamma_4 M \sinh\theta - M(j_1 + j_2 + 1))e^{i\underset{\sim}{\xi}\cdot\underset{\sim}{M}} e^{i\theta T}|j_1 m_1 j_2 m_2\rangle = 0. \tag{50}$$

In this way, a class of general infinite-component wave equations can be constructed.[9] These equations generalize the H-atom and dyonium equations to the relativistic case and include the recoil effects.[10),5),6]

Bibliographical Notes

Section I.

1. A. O. Barut and G. Bornzin, "SO(4,2)-Formulation of the Symmetry Breaking in Relativistic Kepler Problems With or Without Magnetic Charges," Jour. Math. Phys. (May 1971 issue). References to earlier papers can be found here.

Section II.

2. A. O. Barut and A. Böhm, "Reduction of a Class of O(4,2)-Representations With Respect to SO(4,1) and SO(3,2)," Jour. Math. Phys. 11, 2938 (1970).
References to earlier papers can be found here.

Section III.

The O(4,2)-Models of Hadrons: See the reviews

3. A. O. Barut, "Applications of the Dynamical Group Theory to the Structure of Hadrons," in Lectures in Theoretical Physics, Vol. XB (Gordon and Breach, New York, 1968), p. 377 and Springer Tracts in Modern Physics, Vol. 50 (1969), p. 1.
4. D. Corrigan, B. Hamprecht and H. Kleinert, in Lectures in Theoretical Physics, Vol. XIA, Part I (Gordon and Breach, New York, 1969), p. 299 and Springer Tracts in Modern Physics, Vol. 49, p. 90.

For Dyonium Model see:

5. A. O. Barut, "Dyonium Atom and Molecules, Electromagnetic Origin of Strong Interactions, Proceedings Coral Gables Conference on Fundamental Interactions, Vol. II (Gordon and Breach, New York, 1970), p. 199.
6. A. O. Barut, "Atoms With Magnetic Charges as Models of Hadrons Hadrons," in Topics in Modern Physics, Essays dedicated to E. U. Condon (University of Colorado Press, Boulder, 1970).
7. See also: J. Schwinger, Science 165, 757 (1969).
8. A. O. Barut, "Superselection Rule for Magnetic Charges and the Search for Magnetic Monopoles," Phys. Rev.

Section IV.

9. E. Majorana, Nuovo Cimento 9, 335 (1932); Y. Nambu, Phys. Rev. 160, 1171 (1967); C. Fronsdal, Phys. Rev. 156, 1665 (1967); A. O. Barut and H. Kleinert, Phys. Rev. 157, 1180 (1967).
10. A. O. Barut and A. Baiquni, Phys. Rev. 184, 1342 (1969), Phys. Letters 30A, 352 (1969). A. O. Barut, Proton Form Factor, Magnetic Charges and Dyonium, Phys. Rev.

GROUP THEORETICAL BASIS FOR THE DIRAC EQUATION†

G. B. Mainland
Center for Particle Theory
Department of Physics
University of Texas
Austin, Texas 78712

The Dirac spinors, which are generally thought of as solutions to the Dirac differential equation, are developed from a group theoretical point of view. The motivation for this approach is threefold: First, by obtaining the Dirac formalism from group theoretical methods, the importance of symmetries is stressed. Second, because of the recent successes of group theoretical techniques in physics, it is significant that the Dirac formalism can be obtained from group theory as well as from a differential equation. Finally, and most significantly, the formalism which is developed here can readily be generalized so that either all half-integer spins or all integer spins are allowed. It seems very probable that such mathematical structures will be useful in describing baryons and mesons.

This work is based upon the infinite component wave equations which were originally introduced by Majorana[1] in 1932 and recently revived by Nambu.[2] The use of infinite component wave equations is intimately connected with the increasing employment of unitary representations of non-compact groups in particle physics. An example is Fronsdal's "relativistic symmetries."[3] By restricting Fronsdal's relativistic symmetries with the Dirac representation relation, one obtains the Dirac representation of the relativistic symmetries which is a non-unitary representation of a non-compact group.

A group theoretical description of leptons has also been developed by Barut and collaborators.[4),5)] In that work, the four dimensional irreducible representation of the group O(4,2) has been used in the description of leptons. While Barut's work is more concerned with lepton interactions, the major objective of this talk is to fully develop the Dirac formalism. That is, to construct the Dirac

†Presented at the Symposium on de Sitter and Conformal Groups, University of Colorado, Summer 1970.

spinors, the Dirac gamma matrices and so forth directly from the group theory.

The talk is divided into four sections: In Section I the commutation relations of the relativistic symmetries and the Dirac representation relation are given. In Section II the representation of the "intrinsic part" of the relativistic symmetries is briefly sketched. Section III is devoted to finding the representation of the whole relativistic symmetry. The spinor basis and the canonical basis are defined and their properties under a Lorentz transformation are explored. In addition, the canonical basis is expanded in terms of the spinor basis. The Dirac formalism is developed in the final part of the talk. The Dirac spinors are shown to be the transition coefficients connecting the canonical basis and the spinor basis.

Section I.

The restricted relativistic symmetry is essentially the enveloping algebra of the Poincaré group $\mathcal{E}(P)$ in certain representations adjoint by a Lorentz-vector operator Γ_μ. The relativistic symmetry is an associative algebra generated by the operators

$$P_\mu,\ M = (P^\mu P_\mu)^{\frac{1}{2}},\ L_{\mu\nu} = M_{\mu\nu} + S_{\mu\nu},\ S_{\mu\nu},\ \Gamma_\mu \qquad (1)$$

where Greek indices range from 0 through 3. The metric tensor $g_{\mu\nu}$ is given by the expression

$$g_{\mu\nu} = \begin{bmatrix} g_{oo} = 1 & & & O \\ & g_{11} = -1 & & \\ & & g_{22} = -1 & \\ O & & & g_{33} = -1 \end{bmatrix} \qquad (2)$$

The defining commutation relations of the relativistic symmetry are as follows:

$$[P_\mu, P_\nu] = 0 \qquad (3)$$

$$[L_{\mu\nu}, P_\rho] = i(g_{\nu\rho} P_\mu - g_{\mu\rho} P_\nu) \qquad (4)$$

$$[L_{\mu\nu}, L_{\rho\sigma}] = i(g_{\mu\rho} L_{\nu\sigma} + g_{\nu\sigma} L_{\mu\rho} - g_{\mu\sigma} L_{\nu\rho} - g_{\nu\rho} L_{\mu\sigma}) \qquad (5)$$

$$[M_{\mu\nu}, S_{\rho\sigma}] = 0 \tag{6}$$

$$\tfrac{1}{2}\epsilon_{\mu\nu\rho\sigma} P^{\nu} M^{\rho\sigma} = 0 \tag{7}$$

$$[P_{\mu}, S_{\rho\sigma}] = 0 \tag{8}$$

$$[P_{\mu}, \Gamma_{\nu}] = 0 \tag{9}$$

$$[S_{\mu\nu}, S_{\rho\sigma}] = -i(g_{\mu\rho} S_{\nu\sigma} + g_{\nu\sigma} S_{\mu\rho} - g_{\mu\sigma} S_{\nu\rho} - g_{\nu\rho} S_{\mu\sigma}) \tag{10}$$

$$[L_{\rho\sigma}, \Gamma_{\mu}] = [S_{\rho\sigma}, \Gamma_{\mu}] = i(g_{\sigma\mu}\Gamma_{\rho} - g_{\rho\mu}\Gamma_{\sigma}) \tag{11}$$

$$[\Gamma_{\rho}, \Gamma_{\sigma}] = -iS_{\rho\sigma} \tag{12}$$

Equations (3)-(12) define the mathematical structure which is thought to be applicable to particle physics. As a first step toward exploring the consequences of the above relations, we restrict them with the Dirac representation relation. As will be shown, the complete Dirac formalism can then be developed from the restricted structure. On the basis of this success, the Dirac representation relation is dropped and the complete structure is used to describe hadrons in the talk by A. Böhm.

The Dirac representation relation is

$$\{\Gamma_{\rho}, \Gamma_{\sigma}\} = \tfrac{1}{3} g_{\rho\sigma} \tag{13}$$

P_{μ} and $L_{\mu\nu}$ are the generators of the Poincaré group; consequently, they represent the usual physical observables momentum and angular momentum. The splitting $L_{\mu\nu} = M_{\mu\nu} + S_{\mu\nu}$ is the familiar splitting of the total angular momentum into the orbital angular momentum and the intrinsic or spin angular momentum.

Section II.

Since the representation of the intrinsic part of the relativistic symmetry is well known, it will only be sketched briefly. The second Casimir operator of $SO(3,2)_{\Gamma_{\mu}, S_{\mu\nu}}$ is

$$R = -\Gamma_{\mu}\Gamma^{\mu} - \tfrac{1}{2} S_{\mu\nu} S^{\mu\nu} \tag{14}$$

where all repeated indices are summed over the range of the index. The second order Casimir operator of $SO(3,1)_{S_{\mu\nu}}$ is

$$Q = -\tfrac{1}{2}S_{\mu\nu}S^{\mu\nu} = 1 - c^2 - k_o^2 \tag{15}$$

and the second order Casimir operator of $SO(3,1)_{\Gamma_i, S_{ij}}$ is

$$\Gamma_i\Gamma_i - S_iS_i = 1 - c^2 - k_o^2 \tag{16}$$

where

$$S_i = \tfrac{1}{2}\epsilon_{ijk}S_{jk} \tag{17}$$

and $i,j,k = 1,2,3$. The two numbers k_o and c characterize the $SO(3,1)$ representations. Since $SO(3,1)_{\Gamma_i, S_i}$ and $SO(3,1)_{S_{\mu\nu}}$ are algebraically equivalent $SO(3,1)$ subgroups, the reduction of an irreducible representation of $SO(3,2)_{\Gamma_\mu, S_{\mu\nu}}$ with respect to either of these subgroups must be the same. As a consequence, the possible values of k_o and c on the right hand side of (15) and (16) are the same. From the Dirac representation relation (13) it follows that

$$\text{(a)}\ \Gamma^\mu\Gamma_\mu = 1 \qquad \text{(b)}\ \Gamma_i\Gamma_i = -\tfrac{3}{4} \qquad \text{(c)}\ \Gamma_o^{\,2} = \tfrac{1}{4} \qquad \text{(d)}\ (\Gamma_i)^2 = -\tfrac{1}{4} \tag{18}$$

Using (18a) and (15) in (14) we obtain

$$R = -(c^2 + k_o^2) \tag{19}$$

Since R is an invariant of $SO(3,2)_{\Gamma_\mu, S_{\mu\nu}}$, the number $c^2 + k_o^2$ is a constant in an irreducible representation of $SO(3,2)$. Using (18b) in (16) yields

$$S_iS_i = \vec{S}^2 = -\tfrac{7}{4} + c^2 + k_o^2 \tag{20}$$

Because $c^2 + k_o^2$ is a constant in the Dirac representation of $SO(3,2)$, we conclude from (20) that the Dirac representation contains only one irreducible representation of $SO(3)_{S_i}$, R^s. That is, the Dirac representation contains only one spin s. Since k_o is the smallest spin

$$k_o = s \tag{21}$$

From (20)

$$s(s+1) = k_o(k_o+1) = -\tfrac{7}{4} + c^2 + k_o^{\,2} \tag{22}$$

or

$$c^2 = k_o + \tfrac{7}{4} \tag{23}$$

To find the values of k_o and c, we need to use the fact that a finite dimensional representation of SO(3,1) reduces with respect to SO(3) according to

$$\mathcal{H}(k_o, c) \longrightarrow \sum_{s=k_o}^{s=k_o+n-1} \oplus\ R^s \tag{24}$$

where n is an integer and

$$c^2 = (k_o+n)^2 \tag{25}$$

From the fact that the representation contains only one spin, we conclude that n = 1. From (25) we then have

$$c^2 = (k_o+1)^2 \tag{26}$$

Solving (26) and (23) for k_o, we find two solutions: $k_o = \tfrac{1}{2}$ and $k_o = -\tfrac{3}{2}$. The latter value of k_o is excluded because of the restriction $k_o = s \geq 0$. From (26) we then obtain

$$c = \pm\tfrac{3}{2} \tag{27}$$

Consequently

$$\overset{\text{Dirac}}{\mathcal{H}}(SO(3,2)) \underset{SO(3,1)}{\longrightarrow} \mathcal{H}(k_o=\tfrac{1}{2},\ c=\tfrac{3}{2}) \oplus \mathcal{H}(k_o=\tfrac{1}{2},\ c=-\tfrac{3}{2}) \underset{SO(3)}{\longrightarrow} R^{s=\frac{1}{2}} \oplus R^{s=\frac{1}{2}} \tag{28}$$

The basis vector

$$| f_{j_3}^{j=\frac{1}{2}\ c} \rangle \tag{29a}$$

is introduced with respect to the reduction

$$SO(3,2)_{S_{\mu\nu},\Gamma_\mu} \supset SO(3,1)_{S_{\mu\nu}} \supset SO(3)_{S_i} \supset SO(2)_{S_3} \tag{29b}$$

As a consequence of the manner in which the basis vector $|f_{j_3}^{j=\frac{1}{2}\ c}\rangle$ is defined, we have

$$S_{io}S_i|f_{j_3}^{j=\frac{1}{2}\ c}\rangle = ik_o c|f_{j_3}^{j=\frac{1}{2}\ c}\rangle \tag{30}$$

Similarly the basis vector

$$|s = \tfrac{1}{2}, s_3, c\rangle \tag{31a}$$

is introduced with respect to the reduction

$$SO(3,2)_{S_{\mu\nu},\Gamma_\mu} \supset SO(3,1)_{\Gamma_i, S_{ij}} \supset SO(3)_{S_i} \supset SO(2)_{S_3} \tag{31b}$$

and

$$\Gamma_i S_i|s = \tfrac{1}{2}, s_3, c\rangle = ik_o c|s = \tfrac{1}{2}, s_3, c\rangle \tag{32}$$

For physically motivated reasons which will become apparent later, we would like to introduce a basis system $|s = \frac{1}{2}, s_3, \mu\rangle$ in which Γ_o and $SO(3)_{S_i}$ are diagonal.

$$\Gamma_o|s = \tfrac{1}{2}, s_3, \mu\rangle = \mu|s = \tfrac{1}{2}, s_3, \mu\rangle \tag{33a}$$

From the Dirac representation relation $\Gamma_o^{\ 2} = \frac{1}{4}$. Therefore

$$\Gamma_o^{\ 2}|s = \tfrac{1}{2}, s_3, \mu\rangle = \mu^2|s = \tfrac{1}{2}, s_3, \mu\rangle = \tfrac{1}{4}|s = \tfrac{1}{2}, s_3, \mu\rangle$$

Consequently we may conclude that

$$\mu = \pm\tfrac{1}{2} \tag{33b}$$

Section III.

We now consider the whole relativistic symmetry. An irreducible representation is characterized by the eigenvalue of the mass operator $M^2 = P^\mu P_\mu$, by ϵ = sign P_o, and by the irreducible representation of $SO(3,2)_{S_{\mu\nu},\Gamma_\mu}$ as restricted by the Dirac representation relation. We introduce two basis systems into the irreducible representation space: The spinor basis is the basis in which the Lorentz transformation generated by $L_{\mu\nu}$ is explicitly split into an orbital part generated by $M_{\mu\nu}$ and a spin part generated by $S_{\mu\nu}$. The spinor basis is then the corresponding product of the two basis systems

$$| {}^{\epsilon} f_{j_3}^{j=\frac{1}{2}, c} (p) \rangle = | f^{\epsilon} (p) \rangle \times | f_{j_3}^{jc} \rangle \tag{34}$$

where $| f^{\epsilon} (p) \rangle$ is a generalized eigenvector of the "orbital" Poincaré group generated by P^{μ} and $M_{\mu\nu}$ and $| f_{j_3}^{jc} \rangle$ is a basis vector (29) of SO(3,2) as restricted by the Dirac representation relation. The canonical basis is defined by

$$| p s s_3 \mu \epsilon \rangle = U^{-1}(L(p)) \left[| \varphi^{\epsilon} (\vec{p}=0) \rangle \times | s s_3 \mu \rangle \right] \tag{35}$$

Here $L^{-1}(p)$ is a "boost" and $U^{-1}(L(p))$ changes the state from one with zero three momentum to a state with three momentum $\vec{P}$. $\varphi^{\epsilon}(p)$ is a generalized eigenvector of the "orbital" Poincaré group and $| s s_3 \mu \rangle$ is the basis vector (33) of SO(3,2) as restricted by the Dirac representation relation.

The spinor basis $| {}^{\epsilon} f_{j_3}^{j\ c} (p) \rangle$ is an eigenstate of the complete set of commuting operators

$$P^{\mu}, \quad M^2 = P^{\mu} P_{\mu}, \quad S_3 = S_{12}, \quad S_i S_i = \vec{S}^2 = \tfrac{1}{2} S_{ij} S^{ij}, \quad S_{io} S_i \tag{36}$$

Under a Lorentz transformation Λ, $| {}^{\epsilon} f_{j_3}^{j=\frac{1}{2}\ c} (p) \rangle$ transforms as follows:

$$U(\Lambda) | {}^{\epsilon} f_{j_3}^{j=\frac{1}{2}\ c} (p) \rangle = \sum_{j_3'} | f_{j_3'}^{j=\frac{1}{2}\ c} (\Lambda p) \rangle \mathcal{D}_{j_3' j_3}^{j=\frac{1}{2}\ c} (\Lambda) \tag{37}$$

where $\mathcal{D}_{j_3' j_3}^{j=\frac{1}{2}\ c} (\Lambda)$ is the representation matrix of Λ in the representation $(k_o = \frac{1}{2}, c)$.

The canonical basis $| p s s_3 \mu \epsilon \rangle$ is an eigenstate of the complete set of commuting operators

$$P^{\mu}, \; M^2 = P^{\mu} P_{\mu}, \quad W^{\mu} = \tfrac{1}{2} \epsilon^{\mu\nu\rho\sigma} P_{\nu} L_{\rho\sigma}, \; W = -W_{\mu} W^{\mu}, \; \frac{P_{\mu} \Gamma^{\mu}}{M} \tag{38}$$

Since $P_{\mu} \Gamma^{\mu}$ is a Lorentz invariant

$$\frac{2P^{\mu}\Gamma_{\mu}}{m} |pss_3\mu\ \epsilon\rangle = U(L^{-1}(p))\frac{2P^{\mu}\Gamma_{\mu}}{M} U(L(p))\ |pss_3\mu\ \epsilon\rangle$$

$$= U(L^{-1}(p))\ \frac{2P^{\mu}\Gamma_{\mu}}{M}\ |\vec{p}=0\ \ ss_3\mu\ \epsilon\rangle$$

$$= \epsilon\ \text{sign}\ \mu\ |pss_3\mu\ \epsilon\rangle \tag{39}$$

From the above equation we see that the vectors $|pss_3\mu\ \epsilon\rangle$ are the positive and negative energy solutions of the "Dirac equation." We want to restrict ourselves to the physical case of only positive energy states $\epsilon = 1$. If we make this restriction, we still have all the solutions of the Dirac equation: the states $|pss_3\mu = \frac{1}{2}\rangle$ are the usual positive energy solutions and the states $|pss_3\mu = -\frac{1}{2}\rangle$ correspond to the negative energy solutions. The usual reinterpretation by the Dirac hole theory is no longer necessary.

The transformation property of the canonical basis under a Lorentz transformation Λ is the usual one

$$U(\Lambda)|pss_3\ \mu\rangle = \sum_{s_3'} |(\Lambda p)ss_3'\mu\rangle\ \mathcal{D}^{s=\frac{1}{2}\ c}_{s_3',\ s_3}(R) \tag{40}$$

where R is the Wigner rotation

$$R = L(\Lambda p)\ \Lambda\ L^{-1}(p) = R(\Lambda, p) \tag{41}$$

The canonical basis is defined in such a manner that the basis vectors and their conjugates are orthogonal.

$$\langle p's's_3'\mu|pss_3\mu\rangle = 2P_0\ \delta^3(\vec{p}'-\vec{p})\ \delta^{\mu'\mu}\ \delta^{s_3's_3}\ \delta^{s's} \tag{42}$$

However, this is not the case for the spinor basis. By expanding the spinor basis in terms of the canonical basis (which is possible since we know the canonical basis bra and ket are orthogonal) we calculate[6)]

$$\langle f^{j'c'}_{j_3'}(p')|f^{jc}_{j_3}(p)\rangle = 2p_0\ \delta^3(\vec{p}'-\vec{p})\,\delta^{c'c}\,\delta^{j'j}\,\frac{1}{m}\,[p_0 - \text{sign}\ c\ \vec{p}\cdot\vec{\sigma}]_{j_3'j_3} \tag{43}$$

The $\vec{\sigma}$ are the usual 2 x 2 Pauli spin matrices. Using the above result, the canonical basis can be expanded in terms of the spinor basis.

$$|pss_3\mu\rangle = \sum_{\substack{j_3, c\\ j_3'}} \int \frac{d^3p'}{2p_o'} |f^{jc}_{j_3}(p')\rangle \frac{1}{m}[p_o' + \text{sign } c\, \vec{p}' \cdot \vec{\sigma}]_{j_3 j_3'} \langle f^{jc}_{j_3'}(p')|pss_3\mu\rangle \tag{44}$$

Evaluating the matrix element yields the result

$$|pss_3\mu\rangle = \sum_{j_3 c} |f^{jc}_{j_3}(p)\rangle \tfrac{1}{2}(1 + i\pi)\, \pi(c)\, \mathcal{D}^{j=\frac{1}{2}c}_{j_3\ s_3}(L^{-1}(p)) \tag{45}$$

where

$$\pi(c) \overset{\text{def}}{=} \begin{cases} 1 & \text{for } c = \frac{3}{2} \\ \text{sign } \mu \overset{\text{def}}{=} \pi & \text{for } c = -\frac{3}{2} \end{cases} \tag{46}$$

and $\mathcal{D}^{j=\frac{1}{2}c}_{j_3\ s_3}(L^{-1}(p))$ is the representation matrix of the "boost." Explicit expressions for $\mathcal{D}^{j=\frac{1}{2},c}_{j_3\ s_3}(L^{-1}(p))$ are tabulated in H. Joos.[7] The quantities

$$U^c_{j_3}(p\sigma\pi) \overset{\text{def}}{=} \tfrac{1}{2}(1 + i\pi)\, \pi(c)\, \mathcal{D}^{j=\frac{1}{2}c}_{j_3\ \sigma}(L^{-1}(p)) \tag{47}$$

which are just the transition coefficients between the canonical basis and the spinor basis will be shown to be the usual Dirac spinors in the next section.

At this point we introduce the conjugate spinor basis

$$\langle \dot{f}^{j'c'}_{j_3'}(p')| = \sum_{\tilde{j}_3} \frac{1}{m}(p_o' + \text{sign } c'\, \vec{p}' \cdot \vec{\sigma})_{j_3', \tilde{j}_3} \langle f^{jc'}_{\tilde{j}_3}(p')| \tag{48}$$

which, of course, satisfies the relation

$$\langle \dot{f}^{j'c'}_{j_3'}(p')|f^{jc}_{j_3}(p)\rangle = 2p_o \delta^3(\vec{p}' - \vec{p})\, \delta^{j'j}\, \delta^{c'c}\, \delta^{j_3' j_3} \tag{49}$$

<u>Section IV.</u>

We define

$$2p_o\delta^3(p'-p)\gamma_{\mu\ BA}^{\ \tilde{c}'\tilde{c}} \overset{\text{def}}{=} \langle \dot{f}^{j\tilde{c}'}_{B}(p')|2\Gamma_\mu|f^{j\tilde{c}}_{A}(p)\rangle \tag{50}$$

and will now verify that the $\gamma_{\mu\,BA}^{\tilde{c}'\tilde{c}}$ are the usual Dirac gamma matrices. As a first step we will show that the defined quantities obey the Dirac equation.

$$\sum_{\tilde{c}A} 2p_o\, \delta^3(p'-p) \frac{p^\mu}{m} \gamma_{\mu\,BA}^{\tilde{c}'\tilde{c}}\, U_A^{\tilde{c}}(p\sigma\pi)$$

$$= \sum_{\tilde{c}A} \frac{p^\mu}{m} \langle f_B^{j\tilde{c}'}(p')|\, 2\Gamma_\mu\, | f_A^{j\tilde{c}}(p)\rangle\, U_A^{\tilde{c}}(p\sigma\pi)$$

$$= \sum_{\tilde{c}A} \langle \dot{f}_B^{j\tilde{c}'}(p')| \frac{2P^\mu\Gamma_\mu}{M} | f_A^{j\tilde{c}}(p)\rangle\, U_A^{\tilde{c}}(p\sigma\pi)$$

From (39) we already know the action of the operator $\frac{2P^\mu\Gamma_\mu}{M}$ on the canonical basis. By expanding the spinor basis in terms of the canonical basis we readily calculate the action of $\frac{2P^\mu\Gamma_\mu}{M}$ on the spinor basis. The result of the calculation is

$$\frac{2P^\mu\Gamma_\mu}{M} | f_A^{j\tilde{c}}(p)\rangle = | \dot{f}_A^{j-\tilde{c}}(p)\rangle \tag{51}$$

Using this result

$$\sum_{\tilde{c}A} 2p_o\, \delta^3(\vec{p}\,'-\vec{p}) \frac{p^\mu}{m} \gamma_{\mu\,BA}^{\tilde{c}'\tilde{c}}\, U_A^{\tilde{c}}(p\sigma\pi)$$

$$= \sum_{\tilde{c}A} \langle \dot{f}_B^{j\tilde{c}'}(p')| \dot{f}_A^{j-\tilde{c}}(p)\rangle\, U_A^{\tilde{c}}(p\sigma\pi)$$

By expressing the vectors $|\dot{f}_{j_3}^{jc}(p)\rangle$ in terms of the spinor basis $|f_{j_3}^{jc}(p)\rangle$, using the expression (43) for the scalar product of two spinor basis vectors, and the explicit expression for $U_A^{\tilde{c}}(p\sigma\pi)$, we calculate

$$= 2p_o\, \delta^3(p'-p)\, \pi\, U_B^{\tilde{c}}(p\sigma\pi) \tag{52}$$

Multiplying both sides of (52) by m and integrating both sides by $\int \frac{d^3p'}{2p_o'}$ we finally obtain

$$\sum_{\tilde{c}A} p^{\mu} \gamma_{\mu\,BA}^{\;\tilde{c}'\tilde{c}} \; U_A^{\tilde{c}}(p\sigma\pi) = \pi m \; U_B^{\tilde{c}'}(p\sigma\pi) \tag{53}$$

If we arrange $U_A^{\tilde{c}}$ as a column matrix and $\gamma_{\mu\,BA}^{\;\tilde{c}'\tilde{c}}$ as a 4 x 4 matrix

$$U(p\sigma\pi) \overset{\text{def}}{=} \begin{bmatrix} U_{1/2}^{3/2} \\ U_{-1/2}^{3/2} \\ U_{1/2}^{-3/2} \\ U_{-1/2}^{-3/2} \end{bmatrix}$$

$$\gamma_{\mu} \overset{\text{def}}{=} \begin{bmatrix} \gamma_{1/2\;\;1/2}^{3/2\;\;3/2} & \gamma_{1/2\;\;-1/2}^{3/2\;\;3/2} & \gamma_{1/2\;\;1/2}^{3/2\;\;-3/2} & \gamma_{1/2\;\;-1/2}^{3/2\;\;-3/2} \\ \gamma_{-1/2\;\;1/2}^{3/2\;\;3/2} & \gamma_{-1/2\;\;-1/2}^{3/2\;\;3/2} & \gamma_{-1/2\;\;1/2}^{3/2\;\;-3/2} & \gamma_{-1/2\;\;-1/2}^{3/2\;\;-3/2} \\ \gamma_{1/2\;\;1/2}^{-3/2\;\;3/2} & \gamma_{1/2\;\;-1/2}^{-3/2\;\;3/2} & \gamma_{1/2\;\;1/2}^{-3/2\;\;-3/2} & \gamma_{1/2\;\;-1/2}^{-3/2\;\;-3/2} \\ \gamma_{-1/2\;\;1/2}^{-3/2\;\;3/2} & \gamma_{-1/2\;\;-1/2}^{-3/2\;\;3/2} & \gamma_{-1/2\;\;1/2}^{-3/2\;\;-3/2} & \gamma_{-1/2\;\;-1/2}^{-3/2\;\;-3/2} \end{bmatrix}$$

Equation (53) may be written

$$p^{\mu}\gamma_{\mu} \; U(p\sigma\pi) = \pi m U(p\sigma\pi) \tag{54}$$

which is simply the Dirac equation. If we define

$$\tilde{U}_{j_3}^{\,c}(p\sigma\pi) = U_{j_3}^{*-c}(p\sigma\pi)$$

we can obtain the following additional relations involving U and $\tilde{U}$:

Adjoint Dirac Equation: $$\tilde{U}(p\sigma\pi)p^{\mu}\gamma_{\mu} = \tilde{U}(p\sigma\pi)\pi m \qquad (55)$$

Orthogonality: $$\tilde{U}(p\sigma'\pi')U(p\sigma\pi) = \pi\delta_{\sigma\sigma'}\delta_{\pi'\pi} \qquad (56)$$

Completeness: $$\sum_{\sigma\pi} \pi\, U_A^{\tilde{c}}(p\sigma\pi)\, U_B^{\tilde{c}'}(p\sigma\pi) = \delta^{\tilde{c}\tilde{c}'}\delta_{AB} \qquad (57)$$

Projection Relation: $$\left(\frac{p^{\mu}\gamma_{\mu}}{m} + \frac{\pi}{2}\right) = \sum_{\sigma} U(p\sigma\pi)\, \tilde{U}(p\sigma\pi) \qquad (58)$$

From Eqs. (54) through (58) we see that $U(p\sigma\pi = 1)$ is the usual positive energy Dirac spinor $U(p,s)$ and $\tilde{U}(p\sigma\pi = 1)$ is the spinor $\bar{U}(p,s)$. Since we have restricted ourselves to positive energies ($\epsilon = 1$), $U(p\sigma\pi = -1)$ is the usual Dirac spinor $V(p,s)$ after it has been reinterpreted by the Dirac hole theory and $\tilde{U}(p\sigma\pi = -1)$ is the usual spinor $\bar{V}(p,s)$ after reinterpretation by the hole theory.

Once again it should be emphasized that the advantage of obtaining the Dirac formalism from such a mathematical structure is that the structure can readily be generalized (by dropping the Dirac representation relation) so that, for example, all half integer or all integer spins are allowed. It seems very probable that such representations will be useful in describing baryons and mesons.

References

1. E. Majorana, Nuovo Cimento 9, 335 (1932).
2. Y. Nambu, Progr. Theoret. Phys. Suppls. 37, 38, 368 (1966).
3. C. Fronsdal, "Relativistic Symmetries," Seminar on High Energy Physics, Trieste (1965), IAEA Vienna; P. Budini and C. Fronsdal, Phys. Rev. Letters 14, 968 (1965).
4. A. O. Barut, Phys. Rev. Letters 20, 893 (1968).
5. A. O. Barut, P. Cordero, and G. C. Ghirardi, Phys. Rev. D1, 536 (1970) and Nuovo Cimento 66A, 36 (1970).
6. A. Böhm and G. B. Mainland, Fortschr. Phys., April (1970) (to appear).
7. H. Joos, Fortschr. Phys. 10, 65 (1962).

$\mathcal{A}_2$†

A. Böhm
Center for Particle Theory
Department of Physics
University of Texas
Austin, Texas 78712

$\mathcal{A}_2$ denotes an algebraic structure that is a generalization of an algebra which I had called $\mathcal{A}_1$.[1] That the algebra $\mathcal{A}_2$ might also have something to do with the A_2-meson is rather accidental, though I have to admit that I was already under Maglic's influence when the possibility of this generalization occurred to me while I was working on a quite different problem. This problem was the infinite dimensional generalization of the 4-dimensional Dirac representation for baryons, which was needed for some other purposes. The essential difference between $\mathcal{A}_1$ and $\mathcal{A}_2$ is that $\mathcal{A}_2$ contains an infinite dimensional generalization of the Dirac representation of γ or SO(3,2), whereas $\mathcal{A}_1$ contains the Majorana representation of SO(3,2) instead.

From the talks of Mainland[2] and Jaffe[3] the basic concepts that I will need should be known, and I shall just briefly remind you of their properties:

The relativistic symmetry γ is essentially the enveloping algebra of the Poincaré group $\mathcal{P}_{P_\mu, L_{\mu\nu}}$ adjoint by a Lorentz-vector operator Γ_μ, which together with the spin part $S_{\mu\nu}$ of the Lorentz group generators $L_{\mu\nu} = M_{\mu\nu} + S_{\mu\nu}$ form the Lie algebra of $SO(3,2)_{S_{\mu\nu}, \Gamma_\mu}$. γ is the associative algebra generated by

$$P_\mu, M = (P_\mu P^\mu)^{\frac{1}{2}}, L_{\mu\nu} = M_{\mu\nu} + S_{\mu\nu}, S_{\mu\nu}, \Gamma_\mu; \quad \nu, \mu = 0,1,2,3$$

in which the multiplication is defined by the relations[10]

---

†Presented at the Symposium on de Sitter and Conformal Groups, University of Colorado, Summer 1970.

$$[P_\mu, P_\nu] = 0 \quad \text{(a)}$$

$$[L_{\mu\nu}, P_\rho] = i(g_{\nu\rho}P_\mu - g_{\mu\rho}P_\nu) \quad \text{(b)}$$

$$[L_{\mu\nu}, L_{\rho\sigma}] = i(g_{\mu\rho}L_{\nu\sigma} + g_{\nu\sigma}L_{\mu\rho} - g_{\mu\sigma}L_{\nu\rho} - g_{\nu\rho}L_{\mu\sigma}) \quad \text{(c)}$$

$$[M_{\mu\nu}, S_{\rho\sigma}] = 0 \quad \text{(d)}$$

$$\tfrac{1}{2}\epsilon_{\mu\nu\rho\sigma}P^\nu M^{\rho\sigma} = 0 \quad \text{(e)}$$

$$[P_\mu, S_{\rho\sigma}] = 0 \; [P_\mu, \Gamma_\nu] = 0 \quad \text{(f)}$$

$$[S_{\mu\nu}, S_{\rho\sigma}] = -i(g_{\mu\rho}S_{\nu\sigma} + g_{\nu\sigma}S_{\mu\rho} - g_{\mu\sigma}S_{\nu\rho} - g_{\nu\rho}S_{\mu\sigma}) \quad \text{(g)}$$

$$[L_{\rho\sigma}, \Gamma_\mu] = [S_{\rho\sigma}, \Gamma_\mu] = i(g_{\sigma\mu}\Gamma_\rho - g_{\rho\mu}\Gamma_\sigma) \quad \text{(h)}$$

$$[\Gamma_\rho, \Gamma_\sigma] = -iS_{\rho\sigma} \quad \text{(i)}$$

where $\mu, \nu, \rho, \sigma = 0,1,2,3$ and $g_{00} = 1$, $g_{11} = g_{22} = g_{33} = -1$. An irrep (irreducible representation) of γ is determined, among others, by the irreducible representation of $SO(3,2)_{S_{\mu\nu},\Gamma_\mu}$ that it contains.

The irreducible representations of SO(3,2) and therefore also the irreps of γ are for our purpose most conveniently characterized by the multiplicity pattern. This is a pattern of n,s which displays the content of irreps of the maximal compact subgroup $SO(2)_{\Gamma_0} \times SO(3)_{S_{\mu\nu}}$, where n = eigenvalue of Γ_0 and $S(S+1)$ = eigenvalue of $\frac{1}{2}S_{ij}S^{ij}$. Examples of such multiplicity patterns have been given by Jaffe.[3)] Fig. 1 shows the multiplicity pattern of the Dirac representation and two of the 4 Majorana representations. And Fig. 2, Fig. 3, and Fig. 4 show the multiplicity pattern of the representations, which we shall consider here, and which we call $(R,\frac{1}{2})$, $(R,0)$, and $(R=2,0)$ respectively (where R is the eigenvalue of the second order Casimir operator). The two irreps in Fig. 4 are the limiting cases of the irrep $(R,0)$ for $R \to 2$. From comparison of the pattern for the $(R,\frac{1}{2})$-representation and the Dirac representation we see already that $(R,\frac{1}{2})$ is in a certain sense an infinite generalization of the Dirac representation and $(R,0)$ is just the integer spin analogue.

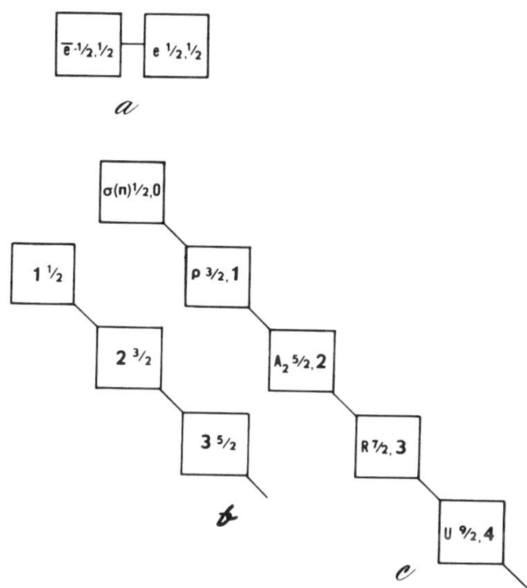

Fig. 1. Multiplicity pattern for the following representations of γ or SO(3,2).

a) Dirac representation

b) Majorana representation with half-integer spin ($k_o = \frac{1}{2}$, $c = 0$)

c) Majorana representation with integer spin ($k_o = 0$, $c = \frac{1}{2}$).

The numbers in the boxes give the values of n,s ; the letters in Figs. 1a and 1c give a possible particle assignment.

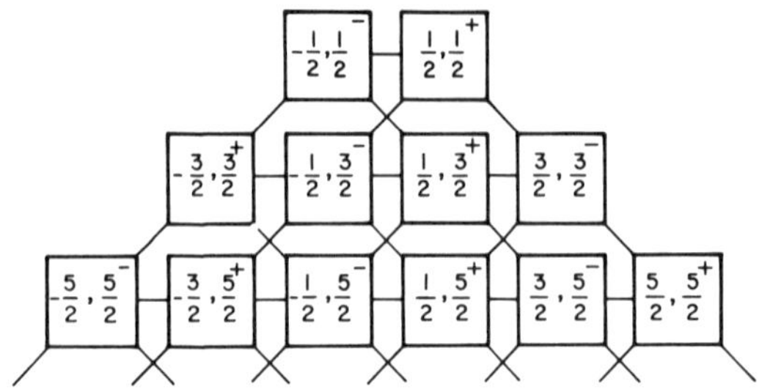

Fig. 2. Multiplicity pattern of the "generalized Dirac" representation (R, ½) of SO(3,2) or γ. The number in the boxes give the values of $\boxed{n, s^P}$.

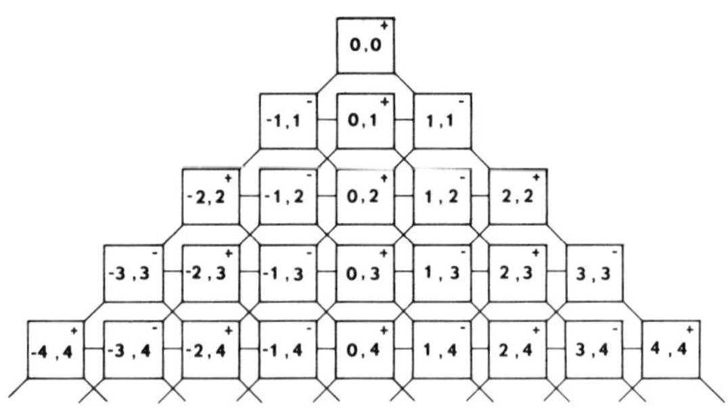

Fig. 3. Multiplicity pattern of the integer-spin representation (R,0).

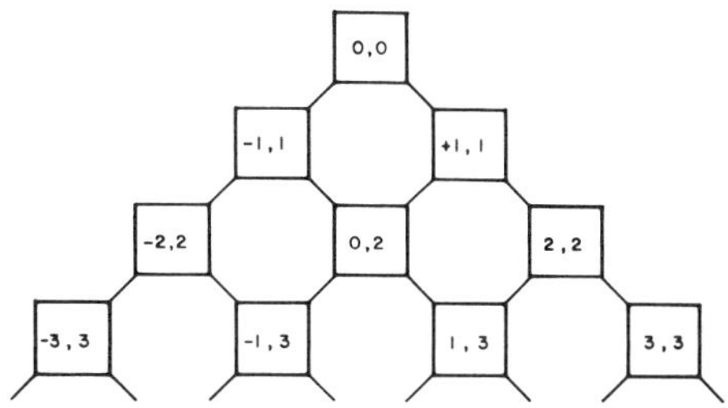

Fig. 4a

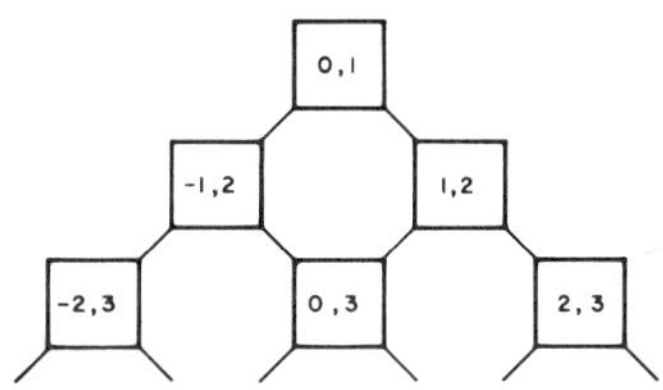

Fig. 4b

Fig. 4. Multiplicity pattern of the representations (R=2,0) (Fig. 4a) and (R=2,1) (Fig. 4b). This is the direct sum of irreducible representations of SO(3,2) that are obtained from the irreducible representation (R,0) in the limiting case R → 2.

To see how physics can be put into these patterns, we have to induce the representations of SO(3,2) to representations of the whole relativistic symmetry γ. This will be done for the representations $(R,\cdot)$ in complete analogy to the case for the 4-dimensional Dirac representation as it was described in the talk by Mainland.[2) The result will then be a representation space, which is the "infinite dimensional" generalization of the space of solutions of the Dirac equation.

Let $\mathcal{H}^{(R,\cdot)}$ denote the irrep space of γ. Then we obtain the canonical basis for $\mathcal{H}^{(R,\cdot)}$ in a completely analogous way to the well known procedure for the Poincaré group:

For the states at rest we take the basis vectors $|p=0,S_3,s,n\rangle$ with the properties

$$\begin{aligned} \vec{S}^2 | \qquad \rangle &= s(s+1) | \qquad \rangle \\ S_{12} | \qquad \rangle &= s_3 \qquad | \qquad \rangle \\ \Gamma_0 | \qquad \rangle &= n \qquad | \qquad \rangle \end{aligned} \tag{1}$$

The operations of $SO(3,2)_{S_{\mu\nu},\Gamma_\mu}$ act only on the indices s_3,s,n and leave $p = 0$ unchanged, and at rest these states correspond to the basis states of SO(3,2) in which SO(3) x SO(2) is diagonal. Then we boost these states into states with momentum p:

$$|p,s_3,s,n\rangle = U(L^{-1}(p))|p=0,s_3,s,n\rangle \tag{2}$$

(where $L(p)p = (m,0,0,0)$, rotation free) and find that these have the usual properties of the canonical basis states of $\mathcal{P}$:

$$\begin{aligned} U^{-1}(L(p))\,\omega_3\,U(L(p))|p,s_3,s,n\rangle &= ms_3|p,s_3,s,n\rangle \\ W|p,s_3,s,n\rangle &= m^2 s(s+1)|p,s_3,s,n\rangle \end{aligned} \tag{3}$$

where

$$\omega_\mu = \tfrac{1}{2}\epsilon_{\mu\nu\rho\sigma} P^\nu L^{\rho\sigma} \quad \text{and} \quad W = -\,\omega_\mu \omega^\mu$$

and the additional property

$$P_\mu \Gamma^\mu |p,s_3,s,n\rangle = m\cdot n|p,s_3,s,n\rangle \tag{4}$$

From this it follows that s and s_3 are the spin and its 3rd component. n is a new quantum number. As the only new object that was adjoint to the enveloping algebra of $\mathcal{P}$, $\epsilon(\mathcal{P})$, is Γ_μ and this commutes with P_μ it is of course clear that m^2 must be an invariant of γ. However Γ^μ_μ does not commute with $\vec{S}^2$ and therefore not with W so that Γ_μ transforms between states with different spin, and the spin is no more an invariant.

For SO(3,2) to each box $\boxed{n,s}$ of the multiplicity pattern corresponds the set of states

$$\{\,|s_3, s, n\rangle \,|\, -s \le s_3 \le s\} \tag{5}$$

For γ we let, because of the above properties, the set of states

$$\{\,|p, s_3, s, n\rangle \,|\, -s \le s_3 \le s,\ P_\mu P^\mu = m^2\} \tag{6}$$

correspond to each box $\boxed{n,s}$ and obtain in this way an extension of the multiplicity pattern of SO(3,2) for γ. But as (6) spans the irrep space of the Poincaré group $\mathcal{H}(m,s)$, we obtain a correspondence between each box of the pattern $\boxed{n,s}$ and an irrep space of $\mathcal{P}$ $\mathcal{H}(m,s,n)$ where n is here an additional label distinguishing between equivalent irrep spaces of $\mathcal{P}$.

So the multiplicity pattern gives us the reduction of the irrep space $\mathcal{H}^{(R,\cdot)}$ of γ with respect to the irreps of the Poincaré group. Now physics has entered into the multiplicity pattern because an irrep space of $\mathcal{P}$ is the mathematical image of an "elementary particle." So each box in the pattern corresponds to an "elementary particle" and the pattern of $(R,\frac{1}{2})$ gives us a spectrum of baryons and the pattern of (R,0) gives us a spectrum of mesons.

Unfortunately this hadron spectrum is still quite unphysical, because it consists of particles which have all the same mass and are distinguished from each other only by their different spins s and by a new additional quantum number n.[4] To obtain a realistic mass spectrum we will have to break the relativistic symmetry γ; this "suitably broken" γ we call $\mathcal{A}_2$.

I cannot give here a detailed description of the properties of $\mathcal{A}_2$ and the reasons for the choice of this structure; this would also lie outside the subject of this conference. I will just let you know the relation that breaks the symmetry and describe its consequences so that you can compare the physical content of these representations $(R,\cdot)$ of γ or SO(3,2) with the experimental data.

We define

$$B_\mu = P_\mu + \Lambda M^{-1} \tfrac{1}{2}\{P^\rho, L_{\rho\mu}\} \tag{7}$$

where $M^2 = P_\mu P^\mu$ and $\Lambda^2 = \lambda_1{}^2 - \lambda_2{}^2 \left(\frac{W}{M^2} - \left(\frac{P_\mu \Gamma^\mu}{M}\right)^2\right)$. λ_1 and λ_2 are two constants whose values are empirically determined to be $\lambda_1{}^2 = (0.30 \pm 0.01)$ BeV^2 and $\lambda_2 = \frac{\lambda_1}{7}$.

Then one can show that $\Lambda^{-1} B_\mu$ and $L_{\mu\nu}$ obey the c.r. of SO(4,1). The second order Casimir operator

$$Q = \Lambda^{-2} B_\mu B^\mu - \tfrac{1}{2} L_{\mu\nu} L^{\mu\nu} \tag{8}$$

commutes because of the construction (7) with the generators P_μ and $L_{\mu\nu}$. We require now that in addition

$$[Q, \Gamma_\mu] = 0 \tag{9}$$

so that the eigenvalue α^2 of Q is an invariant of the whole algebra and characterizes a physical system. This relation is the symmetry breaking relation that gives rise to a non-trivial mass spectrum. In accordance with the O'Raifeartaigh theorem this is not a c.r. between generators but a complicated algebraic relation and the algebra $\mathcal{A}_2$ is not the enveloping algebra of a group. The resulting mass spectrum is:

$$m^2_{(s,n)} = (\lambda_1{}^2 - \lambda_2{}^2 (s(s+1)-n^2))(\alpha^2 - \tfrac{9}{4}) + (\lambda_1{}^2 - \lambda_2{}^2 ((s(s+1)-n^2))s(s+1) \tag{10}$$

where the spectrum of s,n is given by the multiplicity pattern of Fig. 2 and 3. We see that for $\lambda_2{}^2 = 0$ this gives the old rotator spectrum $m^2 = \text{const} + \lambda_1{}^2\, s(s+1)$. In the realistic case $\lambda_2{}^2 \ll \lambda_1{}^2$ (2% of $\lambda_1{}^2$) so that $\lambda_2{}^2$ gives the fine structure splitting between resonances of the same spin.

The comparison of the above predictions for the hadron spectrum are shown in Fig. 5, Fig. 6, Fig. 7 and Fig. 8 for the mesons. (For Fig. 6 the values of the constants $\lambda_1{}^2$ and $\lambda_2{}^2$ were slightly different from the ones given above: $\lambda_1{}^2 = 0.298$ BeV^2, $\lambda_2{}^2 = 0.005$ BeV^2.) Fig. 5 shows the final compilation of the CERN Missing Mass Spectrometer experiment. In the meantime many more resonances have been found at higher masses by the CERN Boson Spectrometer[5] but as our predictions have a big error at those high masses--due to the error of the constants λ_1, λ_2--comparison above the U-mass becomes meaningless as long as one does not know anything about the spin-parity of those higher resonances. It is by now clear that there are at least

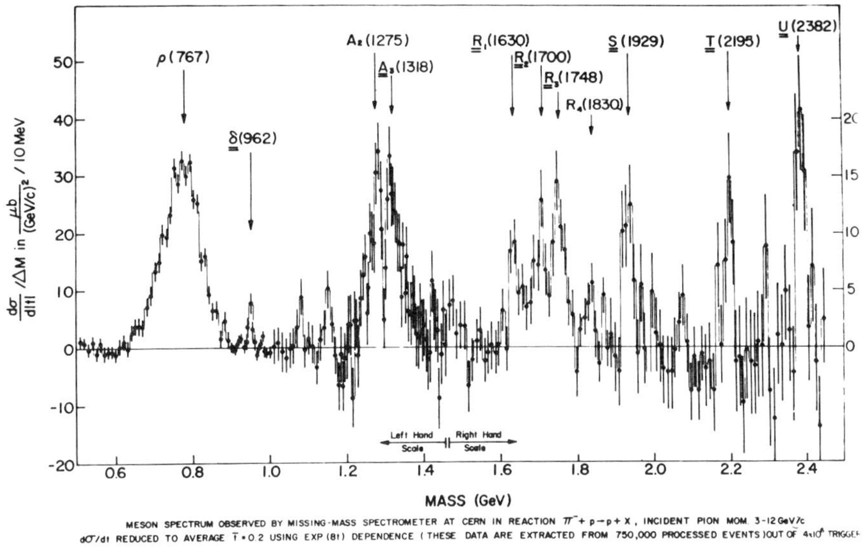

Fig. 5. Meson mass spectrum observed by the Missing Mass Spectrometer.

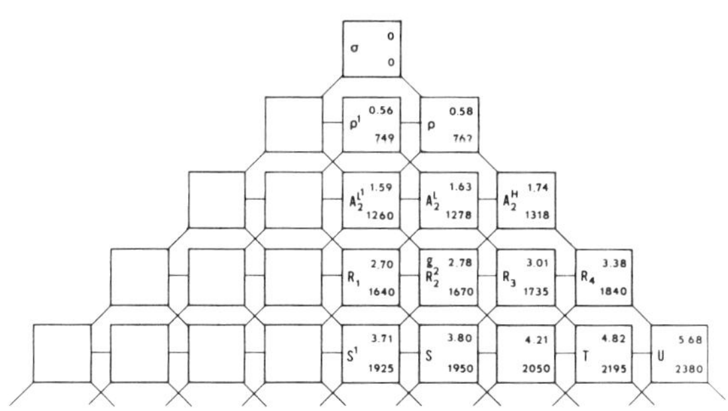

Fig. 6. Multiplicity pattern of $\gamma^{(R,0)}$ with the possible particle assignments and predicted masses. The number in the right upper corner of each box is the predicted mass squared in BeV^2 and the number in the right lower corner is the predicted mass in MeV.

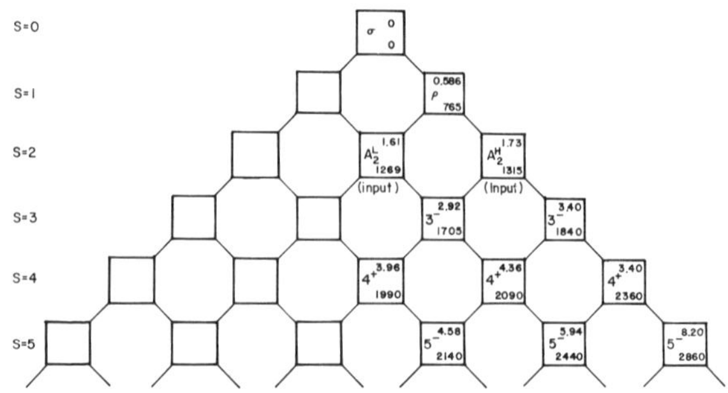

Fig. 7. Predicted particle spectrum for the representation $\gamma^{(R=2,0)}$. The left corner of each box gives the spin-parity, the numbers on the right are the predicted mass squared in BeV^2 and the predicted mass in MeV. For this figure only we have used slightly different values for the symmetry breaking constants: $\lambda_1^2 = 0.298$ BeV^2, $\lambda_2^2 = 0.005$ BeV^2.

$0^+\sigma$

$1^-, \omega$

$2^+, f^\ell$ $\qquad$ $2^+ f$

$3^-, \omega(1965)$ $\qquad$ $3^-, ?$

Fig. 8. The predicted particle spectrum for the I = 0 meson tower with the possible particle assignment. Only the right half of the multiplicity pattern of Fig. 4a has been given here.

two $S^P = 2^+$ mesons in the A_2 region; whether there are also some mesons with other spin parity in this region is not yet clear. We see from Figs. 6 and 7 that this point is well described by the model. We also see that there is almost a one-to-one correspondence between the bumps in the Missing Mass Spectrum and the boxes in the pattern of Fig. 6, the only problem being the ρ'. It should be remarked that the M. M. Spectrum does not contain all I = 1 mesons; e.g. from the backward elastic $\bar{p}p$ cross section it has been inferred[5] that the S is split into two bumps of opposite parity, which is well described by the pattern of Fig. 6, (R,0), but not by the pattern for (R=2,0) of Fig. 7. For I = 0 mesons however the representation (R=2,0) seems to give a better fit (Fig. 8). We remark that the constants λ_1 and λ_2 are universal; therefore we have taken one and the same value for λ_1 and λ_2 for all meson and all baryon towers, only the value of α^2 that characterizes the representation is an adjustable parameter.

For baryons the situation is similar as is shown in the following Fig. 9, Fig. 10 and Fig. 11. Fig. 9 shows the predicted masses-square for nucleon and $I=\frac{1}{2}$ nucleon resonances, Fig. 10 and Fig. 11 show the same for the Σ-resonances and Λ-resonances respectively. The symbol below the value of m^2 gives the partial wave in which a resonance with the right S^P and with a mass in agreement with the predicted value has been found. We see that, except for the $S^P = 3/2^+$ case, the agreement is good.

For the states with negative n the mass formula predicts that they have the same mass as the one with positive n. For the baryons we would interpret these states as the anti-particle states of the states with positive n. In the following we will derive that this is in fact true and the states with quantum number -n are the charge conjugated of the states with quantum number +n. What the interpretation of the states with quantum number -n should be for the mesons, we don't know as yet.

A very nice property of the Dirac representation is that it is also a representation of the discrete operators C, P, T; this is e.g. not true for the Majorana representation where parity leads to representation doubling.

Let me first consider the parity operator U_p. Because of the relation

$$U_p L_{oi} U_p^{-1} = -L_{oi} \qquad i = 1,2,3 \tag{11}$$

one can prove that Γ_μ must fulfill the following relations with U_p:

	n=1/2 P=+	n=3/2 P=-	n=5/2 P=+	n=7/2 P=-	n=9/2 P=+
s = 1/2	0.88 BeV^2 (input)				
s = 3/2	1.67 BeV^2	1.77 BeV^2 D_{13}			
s = 5/2	2.69 BeV^2	2.84 BeV^2 D_{15}	3.12 BeV^2 F_{15}		
s = 7/2	3.68 BeV^2 F_{17}(?)	3.92 BeV^2 G_{17}	4.35 BeV^2 F_{17}(?)	5.02 BeV^2 G_{17}	
s = 9/2	4.07 BeV^2	4.39 BeV^2	5.04 BeV	6.04 BeV	7.36 BeV^2

Fig. 9. The predicted particle spectrum for the nucleon resonances $I = \frac{1}{2}$, $Y = 0$. The numbers are the predicted mass squares in BeV^2, the symbol below the number gives the partial wave in which a resonance in agreement with the predicted mass and spin-parity has been found. Only the right half of the multiplicity pattern of Fig. 2 has been given here.

	n=1/2, P=+	n=3/2, P=-	n=5/2, P=+	n=7/2, P=-
s = 1/2	1.24 (input)			
s = 3/2	2.05	2.13 D_{03}		
s = 5/2	2.98 F_{05}	3.16 D_{05}	3.46 F_{05}	
s = 7/2	4.04 F_{07}	4.28 G_{07}	4.71 F_{07}	5.38 G_{07}
s = 9/2	4.43	4.75	5.40 $\Lambda(2350)9/2^+$	6.40

Fig. 10. Predicted particle spectrum for the Σ resonances.

	n=1/2 P=+	n=3/2 P=-	n=5/2 P=+	n=7/2 P=-
s = 1/2	1.41 input			
s = 3/2	2.20 P_{13}	2.30 D_{13}		
s = 5/2	3.12 F_{15}	3.31 D_{15}	3.63 F_{15}	
s = 7/2	4.21 F_{17}	4.45 G_{17}	4.88	5.55

Fig. 11. Predicted particle spectrum for the Λ resonances.

Either

Case A:	or	Case B:
$U_p \Gamma_o U_p^{-1} = \Gamma_o$		$U_p \Gamma_o U_p^{-1} = -\Gamma_o$
$U_p \Gamma_i U_p^{-1} = -\Gamma_i$		$U_p \Gamma_i U_p^{-1} = \Gamma_i$
(12a)		(12b)

We shall restrict ourselves here to case A, because then $\Gamma_\mu P^\mu$ can be simultaneously diagonalized with U_p, which will not be possible for case B, and is therefore in accordance with the Dirac representation. Then

$$U_P = \eta \, e^{i\pi\Gamma_o} \quad \text{on states at rest}$$

and the phase factor η is chosen such that

for $\gamma(R,\frac{1}{2})$: $\boxed{n = \frac{1}{2},\ s = \frac{1}{2}}$ has P = +1 (nucleon) and

for $\gamma(R,0)$: $\boxed{n = 0,\ s = 0}$ has P = +1 (σ = state with quantum numbers of vacuum)

Then it will turn out that

$$\eta = 1 \quad \text{for } \gamma^{(R,0)} \quad \text{(mesons)} \tag{13a}$$

$$\eta = e^{-i\frac{\pi}{2}} \quad \text{for } \gamma^{(R,\frac{1}{2})} \quad \text{(baryons)} \tag{13b}$$

One obtains that in general parity on the canonical states will be

$$U_p |\underline{p}, s, s_3, n\rangle = (-1)^{[n]} \, |-p, s, s_3, n\rangle$$

where $[n]$ = largest integer which is smaller or equal n. $(-1)^{[n]}$ is given in the upper right corner of the boxes in the figures.

There are 16 extensions of the unitary irreducible representations of the Poincaré group[6] by the discrete operators U_p (space inversion), A_T (time inversion, anti-unitary) and U_c (charge conjugation). They are characterized by the four numbers

$$\pi_c = (U_c U_p)^2, \quad \epsilon_T = A_T^2, \quad \epsilon_I = (U_p A_T)^2, \quad \epsilon_c = (U_c U_p A_T)^2$$

which may independently assume the values +1 and -1. Of these 16 groups only two appear to be relevant for the description of massive particles and one can conclude from some physical arguments[6] that bosons are described by the extension characterized by

$$(\pi_C, (-1)^{2s}\epsilon_T, (-1)^{2s}\epsilon_I, (-1)^{2s}\epsilon_C) = (+,+,+,+) \quad (14)$$

and fermions are described by the extension

$$(\pi_C, (-1)^{2s}\epsilon_T, (-1)^{2s}\epsilon_I, (-1)^{2s}\epsilon_C) = (-,+,+,+) \quad (15)$$

Therefore we have to choose for the baryon representation $\gamma^{(R,\frac{1}{2})}$ an extension by C, P, T for which (15) is valid and for the meson representation $\gamma^{(R,0)}$ an extension for which (14) is valid. From this difference and the difference of the phase factor η for $\gamma^{(R,\frac{1}{2})}$ and $\gamma^{(R,0)}$ given by (13) it follows that the action of the operators U_P, U_C, A_T in $\mathcal{H}^{(R,\frac{1}{2})}$ and $\mathcal{H}^{(R,0)}$ are quite distinct. We cannot give the derivation here but describe only the results:[6)]

For the baryon representation $\gamma^{(R,\frac{1}{2})}$: U_C transforms from states belonging to $\boxed{n,s}$ to states belonging to $\boxed{-n,s}$; this leads to the particle-antiparticle interpretation. A_T does not transform out of the subspace of $\mathcal{H}^{(R,\frac{1}{2})}$ corresponding to $\boxed{n,s}$.

For the meson representation $\gamma^{(R,0)}$: The subspaces corresponding to $\boxed{n,s}$ are U_C eigenspaces with C-parity +1 or -1; so each box $\boxed{n,s}$ corresponds to a meson state with definite C-parity or definite G-parity ($G = (-1)^I C$). A_T transforms from states of $\boxed{n,s}$ into states of $\boxed{-n,s}$; this gives us the interpretation of the states with negative quantum number n. But this also leads us to the unexpected prediction that to every meson state there exists a T-conjugated state, which is distinct from the original state. The question remains open with respect to the physical realizability of these states. As they are T-conjugates of each other they can only be distinguished by observables that do not commute with T and are, therefore, degenerate in all the well known quantum numbers.

References

1. A. Böhm, Phys. Rev. 175, 1767 (1968).
2. B. Mainland, these Proceedings, p. 349.
3. L. Jaffe, these Proceedings, p. 125.
4. The introduction of a new "space-time" quantum number has been suggested before by A. O. Barut and Y. Nambu (see e.g. Proceedings of the 1967 International Conference on Particles and Fields,

Interscience Publishers, Inc., 1968); however their quantum number n differs from our n due to their use of a different representation of SO(3,2).

5. B. Maglic, Rapporteur's Talk at International Conference on Elementary Particles, Lund (1969).
   W. Kienzle, invited paper at the Washington Meeting of the American Physical Society (April 1970), CERN preprint.
6. The derivation of these results is given in A. Böhm, University of Texas preprint CPT-27.

BREAKING OF THE SCALE SYMMETRY AND THE DE SITTER ROTATOR†‡

Akira Inomata
Department of Physics
State University of New York at Albany
Albany, New York 12203

Introduction

In this talk I would like to discuss a specific breaking scheme of the symmetry under scale transformations and its possible relation to a hadron model. In the very beginning of this symposium, Professor Barut pointed out the three topical branches of application of the De Sitter and conformal groups; namely, symmetries of space-time in the large, the scale invariance problem in strong interactions, and the theory of dynamical groups for composite systems. These are by no means mutually exclusive. What follows is an example in which all come together one way or another.

Construction of field theory in space-time of higher symmetries has been attempted by a number of authors with the hope that the divergence difficulty may be resolved in a natural way.[1] A trivial question may arise, however, as to where such symmetries are supposed to manifest themselves in space-time. The Poincaré symmetry is not suited for the observed curved structure of the universe in the cosmological scale. Modification of the symmetry is certainly needed for the large scale. The De Sitter symmetry, which allows for the possibility of an expanding universe, would perhaps be a better approximation. Nevertheless, in local phenomena, the Poincaré symmetry is very precise at least up to the dimension of 10^{-13} cm. The divergence seems to occur when the same symmetry is extended to the extreme dimension ($\sim 10^{-33}$ cm in QED). If space-time is homogeneous everywhere and in every scale, the equations of motion may be described in terms of appropriate global coordinates. It would then be expected that the space-time structure in the large will directly affect the singular character in the small dimension. On the

†Presented at the Symposium on De Sitter and Conformal Groups, University of Colorado, Summer 1970.
‡Supported in part by the Research Foundation of SUNY.

other hand, one might think of a hierarchical structure of the universe. Suppose the De Sitter approximation is good in the large. Then the same could be considered true in the small. In any event, the structure of our universe is not so simple as to be described by a single set of coordinates; it is homogeneous only on the average in the cosmological scale. It would not be surprising that space-time of the micro-domain smaller than 10^{-13} cm is rather different from the world of special relativity.

As is well known, the group SO(4,1) resolves the mass degeneracy inherent in the Poincaré group,[2] the resultant mass spectrum being characterized by a contraction parameter λ,

$$m^2 = m_o^2 + \lambda^2 J(J+1) \quad . \tag{1}$$

The algebra of SO(4,1) reduces to that of the Poincaré group as λ tends to zero. If the SO(4,1) is the space-time symmetry, the Rydberg energy λ in Eq. (1) should be related to the radius of space-time by $\lambda = 1/R$ with $\hbar = c = 1$. Use of the radius of the cosmological universe ($\sim 10^{27}$ cm) yields a value for λ too small to be responsible for any physically conceivable mass splittings. To be compatible with the empirical hadron mass spectrum, the radius R must be of the order of 10^{-13} cm. This can hardly be interpreted as an effect from the space-time structure in the large; this would rather suggest that we either consider the micro-domain relevant to the hadron strongly curved, or give up the idea of counting the SO(4,1) group as a space-time symmetry group. The group SO(4,1) has been treated as a dynamical group in flat space-time.[3] The dynamical group for a composite system is in general to contain as its maximal compact subgroup a symmetry group from which degeneracy of energy results, and as its limiting noncompact group, a kinematical group which describes the composite system as a point particle. Since the SO(4,1) group is not a symmetry group of space-time in this case, the parameter λ in Eq. (1) remains to be determined phenomenologically. A feature of the dynamical group theory is that the equations of motion are realized on the infinite component basis. Under certain circumstances, however, it would be established that the infinite component theory defined in flat space-time is equivalent to the curved space-time formulation describing the small domain for a composite system.

Let us now look at the mass spectrum (1) from the symmetry breaking point of view. In the broken SU(3) scheme, breaking of the symmetry takes place in a rather simple way; very strong interactions are invariant under SU(3) and medium strong interactions break SU(3) symmetry, yet remain invariant under SU(2). Schematically,

$$\text{Broken } SU(3) \supset SU(2) \quad . \tag{2}$$

In a similar fashion, we consider the scheme in which the conformal group SO(4,2) breaks down into SO(4,1);

$$\text{Broken } SO(4,2) \supset SO(4,1) \quad . \tag{3}$$

It has been known that conformal symmetry is the maximal space-time symmetry of massless free field equations. Apparently, the presence of nonvanishing masses breaks such invariance. In turn, its breaking could be considered to have contributed to their presence. As the broken internal symmetry (2) has resulted in the Gell-Mann-Okubo mass formula, the mass spectrum (1) may reasonably be regarded as a consequence of the broken space-time symmetry (3). If the SO(4,2) symmetry breaks down in such a way that the Poincaré symmetry is always preserved, the mass splitting (1) does not occur; the mass spectrum remains to be degenerate. In the scheme we consider, the conformal symmetry reduces to the De Sitter symmetry by breaking and the De Sitter symmetry becomes the Poincaré symmetry by contraction; i.e., in the limit when the mass splitting disappears. In this respect, it is essential to ascribe the spectrum generating group SO(4,1) to the De Sitter structure of the micro-domain for a hadron.[4)]

Breaking of Scale Symmetry

Let us start with the conformal group on flat space-time,[5)] which consists of the Lorentz transformations L, the translations T, the dilations or the scale transformations D, and the special conformal transformations K. This group has the SO(4,2) structure

$$[J_{ab}, J_{cd}] = i(\eta_{ad}J_{bc} - \eta_{bd}J_{ac} + \eta_{bc}J_{ad} - \eta_{ac}J_{bd}) \tag{4}$$

where $\eta_{11} = \eta_{22} = \eta_{33} = -\eta_{44} = \eta_{55} = -\eta_{66} = 1$ and $\eta_{ab} = 0$ for $a \neq b$ ($a,b,c,\ldots = 1,2,3,4,5$, and 6). The four dimensional transformations L, T, D and K are generated by J_{ij} ($i,j = 1,2,3$, and 4), $P_j = \frac{1}{2}\lambda(J_{5j} + J_{6j})$, $D = J_{56}$ and $K_j = \frac{1}{2}(J_{5j} - J_{6j})$, respectively. The λ appearing in the definition of P_j is the contraction parameter.

With the usual realization $P_j = -i\partial_j$ and $J_{ij} = x_iP_j - x_jP_i + S_{ij}$, we have[6)]

$$D = x^jP_j + i\ell \tag{5}$$

$$K = \tfrac{1}{4}\lambda(x^2P_j + 2x^iJ_{ij} + 2ix_j\ell) \tag{6}$$

where ℓ is the scale dimension of the fields on which these operators work. The local currents associated with the dilations D and the special conformal transformations K can be expressed in terms of the improved stress-energy tensor θ_{ij} of Callan, Coleman and Jackiw as[7)]

$$D_k = x^j \theta_{jk} \tag{7}$$

$$K_{ij} = (2x_i x^k - \delta_i^{\ k} x^2)\,\theta_{kj} \tag{8}$$

provided that the Lagrangian $L(\varphi_A, \partial_i \varphi_A)$ is so chosen that

$$\sum_A \left(\ell_A \frac{\partial L}{\partial \partial_i \varphi_A} \varphi_A + \frac{\partial L}{\partial \partial_j \varphi_A} S_{ij} \varphi_A \right) = \partial^i \sigma_{ij} \tag{9}$$

where σ_{ij} is some tensor. The divergences of these currents are given by[7)]

$$\partial^i D_j = \theta^j_{\ j} \tag{10}$$

$$\partial^j K_{ij} = 2x_i \theta^j_{\ j} \quad . \tag{11}$$

Apparently, with the choice (9), breaking of scale symmetry necessarily implies failure of conformal symmetry; the symmetries generated by D and K are simultaneously broken. In other words, breaking of conformal symmetry is as minimal as is induced by that of scale symmetry. This is the minimal breaking scheme considered by Mack and Salam in a slightly less general form.[6)] Since the symmetries generated by J_{ij} and P_j can be retained under the minimal breaking condition (9), the Poincaré symmetry may be preserved;

$$\text{Broken } SO(4,2) \supset \text{Poincaré} \tag{12}$$

the scheme on which most attention has been focused.

There is an alternative scheme in which SO(4,2) symmetry is broken under the condition (9). The alternative is that translational symmetry, in addition to D and K symmetries, ceases to be good, while the symmetries generated by J_{ij} and $J_{5j} = (P_j + \lambda K_j)/\lambda$ are preserved. This is indeed the aforementioned scheme (3), which results in the mass spectrum (1). The second scheme is as minimal as the first in the sense that satisfying the condition (9) it admits the same number of good generators as those of the first. The first allows the 10 parameter Poincaré group and the second the 10 parameter De Sitter group. In the second scheme, translation invariance does not exist, so that space-time is no longer flat. The symmetry generated by J_{5j} requires a uniform deformation of space-time.

Under SO(4,2), a sphere S(4,2) = SO(4,2)/SO(4,1) remains invariant. The isotropy group SO(4,1) of SO(4,2) may be considered as a group of rotations about the sixth axis on the surface S(4,2). Now the symmetry is broken in the directions of J_{6j} and J_{56} but the symmetry about the sixth axis is retained. The sphere S(4,2) is thus deformed so as to be, for instance, an onion shaped surface with the sixth axis as the symmetry axis. Each slice of the onion cut perpendicular to the symmetry axis indicates a De Sitter space-time. The subgroup SO(4,1) carries a sphere S(4,1) = SO(4,1)/SO(3,1) into itself. The Lorentz group SO(3,1) may be taken as a group of rotations about the fifth axis. Then J_{5j} generates the displacements of the fifth axis on the surface S(4,1). The stereographic projection induces on the sphere the conformally flat metric[8)]

$$g_{\mu\nu} = \phi^2 \delta_\mu^{\ i} \delta_\nu^{\ i} \eta_{ij} \tag{13}$$

where

$$\phi = (1 + \tfrac{1}{4}\lambda^2 x^2)^{-1} . \tag{14}$$

It is easy to show that the space-time with the metric (13) is of constant curvature

$$R_{\mu\nu\rho\sigma} = \lambda^2 (g_{\mu\rho} g_{\nu\sigma} - g_{\mu\sigma} g_{\nu\rho}) . \tag{15}$$

The divergences of local currents previously defined in flat space-time can be re-expressed in terms of conformally flat coordinates. However, insofar as local properties are concerned, such modification is unnecessary. So much for the minimal breaking of scale symmetry.

General Relativistic Considerations

As is seen in Eqs. (10) and (11), conformal symmetry is minimally broken by the presence of fields whose stress-energy tensor is of nonvanishing trace. Since the breaking scheme we are concerned with requires a deformation of space-time, the stress-energy tensor with a nonvanishing trace may have to be considered to serve as the source to the geometry. A general tensor equation of the lowest rank linking a conserved matter distribution with the space-time geometry is in fact the Einstein equation in general relativity:

$$G_{\mu\nu} \equiv R_{\mu\nu} - \tfrac{1}{2} g_{\mu\nu} R = \varkappa T_{\mu\nu} \quad , \tag{16}$$

where $T_{\mu\nu}$ is the usual symmetric stress-energy tensor and the term corresponding to the cosmological term in the theory of gravitation is ignored from the Machian view.[9] Callan, Coleman and Jackiw have modified the Einstein equation (16) in the form[7]

$$R_{\mu\nu} - \tfrac{1}{2} g_{\mu\nu} R = \varkappa (1 - \varkappa \Phi)^{-1} \, \theta_{\mu\nu} \tag{17}$$

with the improved stress-energy tensor

$$\theta_{\mu\nu} = T_{\mu\nu} + \nabla_{\nu} \nabla_{\mu} \Phi - g_{\mu\nu} \nabla^{\lambda} \nabla_{\lambda} \Phi \tag{18}$$

where Φ is a function of scalar fields. In the absence of scalar fields, Φ vanishes and the modified equation (17) coincides with the Einstein equation (16).

As a result of the broken SO(4,2), we have obtained the De Sitter geometry (15). For this geometry, the Einstein tensor in Eq. (16) takes a specific form,

$$G_{\mu\nu} = - 3\lambda^2 g_{\mu\nu} \; . \tag{19}$$

Thus, if Einstein's formulation is appropriate for linking any matter distribution to a geometry or space-time, the matter fields allowed in the geometry (15) must have the stress-energy tensor

$$\theta_{\mu\nu} = - (1 - \varkappa \Phi) \, \rho g_{\mu\nu} \tag{20}$$

or

$$T_{\mu\nu} = -\rho \, g_{\mu\nu} \tag{21}$$

where

$$\varkappa \rho = 3\lambda^2 \; . \tag{22}$$

The divergence of the local dilation current is now given by

$$\partial^{i} D_{i} = - 4(1 - \varkappa \Phi) \rho \tag{23}$$

which does not vanish unless the matter distribution disappears. This indicates that the symmetry breaking agents in the present scheme are not the scalar fields but some other fields which have the stress-energy tensor of the form (21) altogether. Since the scalar fields are quite free from space-time geometry, they can be independent agents of breaking scale symmetry even in the limit where SO(4,1) reduces to the Poincaré. Such scalar fields could therefore be thought of as the Nambu-Goldstone bosons associated with a spontaneous breakdown of scale symmetry. As space-time is curved, however, conformal symmetry is broken as a consequence of matter concentration, responding in a sense to Mach's idea.

Let us now play a cabalistic game with Eq. (22). Take appropriate values for the coupling $\varkappa$, the matter density ρ, and the Rydberg energy λ or the corresponding radius $R = 1/\lambda$. Then find combinations of interest. The following is a list of some combinations which seem to make sense a bit:

Universe:

$\varkappa = \varkappa_G$, $R \sim 10^{27}$ cm, $\rho \sim 10^{-28}\,\mathrm{g/cm^3}$

Neutron stars:

$\varkappa = \varkappa_G$, $R \sim 10^{9}$ cm, $\rho \sim 10^{15}\,\mathrm{g/cm^3}$

H-atom:

$\varkappa = 10^{37}\varkappa_G$, $R \sim 10^{-5}$ cm, ($\lambda \sim$ Rydberg Energy R_H) $\rho \sim 1\,\mathrm{g/cm^3}$

Hadrons:

$\varkappa = 10^{39}\varkappa_G$, $R \sim 10^{-14}$ cm ($\lambda \sim 0.5$ Bev) $\rho \sim 10^{15}\,\mathrm{g/cm^3}$,

Here $\varkappa_G$ is the gravitational coupling constant. In the example of the H-atom, the coupling $\varkappa$ is taken to be of the order of the electromagnetic coupling. The radius R differs from the Bohr radius and does not have a particular physical meaning, but its inverse gives the Rydberg energy R_H consistent with the observed value. It may be noted that the De Sitter symmetry is not the space-time symmetry of the H-atom. It is perhaps an effective radius of a geometry which the H-atom would form as a relativistic composite system. In the case of hadrons, the coupling $\varkappa$ is taken to be of strong interaction and the mass distribution is of the nuclear density. The Rydberg energy λ in the mass formula (1) then becomes ~ 0.5 Bev which is a desired result. If the results of this type of games are taken seriously, use of the Einstein equation (16) should not be limited to the gravitational processes. The Einstein equation would have to be looked over from a more flexible stand.

Model for the De Sitter Matter

Historically, the De Sitter universe is believed to be an empty space-time just as the Minkowski space-time provides only a framework of the world. The reason for this is as follows. Matter distributing in the universe may reasonably be approximated by an ideal gas of the form

$$T_{\mu\nu} = p\, g_{\mu\nu} - (p+\rho)\, u_\mu u_\nu \tag{24}$$

where p is the pressure and u_μ is the four-velocity. In order to obtain matter of the De Sitter type (21), we must assume that

$$p+\rho = 0 . \tag{25}$$

Since $p \geq 0$ and $\rho \geq 0$ must be required from the physical grounds, Eq. (25) leads to $p = \rho = 0$. Consequently, we have $T_{\mu\nu}$ vanishing. The Maxwell field is another reasonable field for filling in the universe, but cannot be of the De Sitter type. Furthermore, if the cosmological term is introduced in the Einstein equation (16), no matter distribution is actually needed for the De Sitter geometry (15). Thus one is led to a belief that there is no reasonable matter in the De Sitter universe.

Despite this old belief, the symmetry breaking scheme we have considered urges us to hunt out a De Sitter matter.[10] Let us then consider the case of the Dirac field. The stress-energy tensor is now given by

$$T_{\mu\nu} = \tfrac{1}{4}(\bar{\psi}\gamma_\mu \nabla_\nu \psi - \nabla_\nu \bar{\psi}\gamma_\mu \psi + \bar{\psi}\gamma_\nu \nabla_\mu \psi - \nabla_\mu \bar{\psi}\gamma_\nu \psi) \tag{26}$$

where

$$\gamma_\mu \gamma_\nu + \gamma_\nu \gamma_\mu = 2g_{\mu\nu} . \tag{27}$$

The Dirac equation

$$\gamma^\mu \nabla_\mu \psi + m\psi = 0 \tag{28}$$

admits a solution ψ satisfying the condition

$$\nabla_\mu \psi = -\tfrac{1}{4} m \gamma_\mu \psi . \tag{29}$$

Since the adjoint field $\bar{\psi}$ obeys

$$\nabla_\mu \bar{\psi} = \tfrac{1}{4} m \bar{\psi} \gamma_\mu , \tag{30}$$

it is straightforward to show that the scalar bilinear $\bar{\psi}\psi$ is constant. Under the conditions (29) and (30), the stress-energy tensor (26) reduces indeed to the De Sitter form (21) with

$$\rho = \tfrac{1}{4} m \bar{\psi}\psi . \tag{31}$$

Thus we have found a matter that can serve as a source to the De Sitter geometry. As m, the mass of the field ψ, tends to zero, the matter density ρ vanishes and the dilation current conserves. In this respect, the Dirac field ψ under the condition (29) might be considered as a Nambu-Goldstone fermion associated with a spontaneous breakdown of conformal symmetry. However, since space-time is not flat in the presence of the field and hence the vacuum state cannot be well defined, such an interpretation remains open.

The integrability condition of Eq. (29),

$$(\nabla_\nu \nabla_\mu - \nabla_\mu \nabla_\nu)\psi = \tfrac{1}{4} R_{\mu\nu\rho\sigma} \gamma^\rho \gamma^\sigma \psi , \tag{32}$$

together with the given geometry (15) yields the relation

$$m = 2\lambda \tag{33}$$

which implies that the mass of the Dirac field filling up the De Sitter micro-domain is of the order of 1 Bev. Recall that the De Sitter space-time of the radius $R = 1/\lambda$ corresponds to a slice of the broken SO(4,2) onion. Because of the mass splitting (1), each De Sitter space-time must be characterized by one of the values $\lambda^2 J(J+1)$ replacing λ^2.

Breaking of the De Sitter Symmetry

So far we have assumed that the SO(4,1) symmetry realized as a result of broken SO(4,2) is exact. As the mass formula (1) ignores mass differences among particles of the same spin, the De Sitter symmetry is insufficient for the hadron model. How one can relate the internal symmetry with this external symmetry is an open question. An ambitious program would be to seek the source of the fine structure of the mass spectrum out of the broken symmetry mechanism as Böhm did in the purely algebraic aspect.[11] For this purpose, the SO(4,1) model is obviously too rigid. Moreover, if the micro-domain is of the exact De Sitter structure, space-time has to have a double structure; the De Sitter feature inside and the Minkowskian character outside. There is a singular surface which cuts off any communication between the inside and outside worlds. Fluctuation of some kind in the matter distribution near the surface would circumvent this barrier. In these contexts, breaking of SO(4,1) seems necessary.

To see the behavior of the singular surface, let us again make use of the ideal gas model (24) which is this time allowed to have a negative pressure. Suppose the distribution is spherically symmetric in the spatial region of radius r_0. Then we adopt the following solutions: For $r > r_0$,

$$ds^2 = -\left(1 - \frac{a}{r}\right)^{-1} dr^2 - r^2 (d\theta^2 + \sin^2\theta d\varphi^2) + \left(1 - \frac{a}{r}\right) dt^2 \tag{34}$$

where $p = 0$, $\rho = 0$ and $a = \frac{\varkappa m}{4\pi}$ is Schwarzschild's radius; and for $r < r_0$,

$$ds^2 = -(1 - \lambda^2 r^2)^{-1} dr^2 - r^2 (d\theta^2 + \sin^2\theta d\varphi^2) + [A - B\sqrt{1 - \lambda^2 r^2}]^2 dt^2 \tag{35}$$

with

$$\varkappa p = \lambda^2 \left[\frac{3B\sqrt{1 - \lambda^2 r^2} - A}{A - B\sqrt{1 - \lambda^2 r^2}}\right] \tag{36}$$

$$\varkappa \rho = 3\lambda^2 \tag{37}$$

where A and B are constants to be determined by boundary conditions. Making a connection of these solutions at the boundary surface $r = r_0$, we obtain for $r_0 > a$,

$$a = \lambda^2 r_0^3 , \tag{38}$$

$$(A - B\sqrt{1 - \lambda^2 r_0^2})^2 = 1 - a/r_0 . \tag{39}$$

With another condition $p(r=r_0) = 0$, A and B can be determined. As a result, the inside solution (35) is fixed in the form

$$ds^2 = -(1 - \lambda^2 r^2)^{-1} dr^2 - r^2 (d\theta^2 + \sin^2\theta d\varphi^2) + \\ + \tfrac{1}{4}(3\sqrt{1 - \lambda^2 r_0^2} - \sqrt{1 - \lambda^2 r^2})^2 dt^2 \tag{40}$$

with

$$\varkappa p = \lambda^2 \left[\frac{3\sqrt{1 - \lambda^2 r^2} - 3\sqrt{1 - \lambda^2 r_o^2}}{3\sqrt{1 - \lambda^2 r_o^2} - \sqrt{1 - \lambda^2 r^2}} \right] \quad . \tag{41}$$

The De Sitter solution is

$$ds^2 = -(1 - \lambda^2 r^2)^{-1} dr^2 - r^2 (d\theta^2 + \sin^2\theta d\varphi^2) + (1 - \lambda^2 r^2)\, dt^2 \tag{42}$$

with

$$\varkappa p = -\, 3\lambda^2 \tag{43}$$

which is a limiting case of Eq. (40) where the distribution radius r_o approaches Schwarzschild's radius a. Thus the exact De Sitter symmetry imposes a singular surface at $r = r_o$.

In order to avoid the singularity, we wish to suppose that the radius r_o is extended outside Schwarzschild's radius a; that is, we wish to modify the De Sitter solution minimally. Since the requirement (25) must be satisfied, the pressure inside has to be negative. Thus we look for an approximate solution for which the negative pressure is possible. In fact, Eq. (40) admits a negative pressure solution if we chose r_o such that

$$a < r_o < \frac{9}{8} a \quad . \tag{44}$$

The solution of this type may be considered as a broken SO(4,1) solution. Further details will be given on other occasions.

References

1. See, for instance, papers of Börner, Castell, Ingraham, Kastrup and others presented at this symposium.
2. A. O. Barut and A. Böhm, Phys. Rev. 139, B1107 (1965); P. Roman, Non-Compact Groups in Particle Physics, edited by Y. Chow (Benjamin, New York, 1966), p. 89.
3. See, for instance, A. O. Barut, Lectures in Theoretical Physics Vol. XB, edited by A. O. Barut and W. E. Brittin (Gordon and Breach, New York, 1968), p. 377.
4. A. Inomata and D. Peak, Prog. Theor. Phys. 42, 134 (1969).
5. T. Fulton, R. Rohrlich and L. Witten, Rev. Mod. Phys. 34, 442 (1962).
6. G. Mack and A. Salam, Ann. Phys. (N.Y.) 53, 174 (1969).
7. C. G. Callan, S. Coleman and R. Jackiw, MIT preprint CTP 713 (1969).

8. F. Gürsey, Group Theoretical Concepts and Methods in Elementary Particle Physics (Gordon and Breach, New York, 1962), p. 365.
9. A. Inomata, Prog. Theor. Phys. 39, 1370 (1968).
10. A. Inomata, Nuovo Cimento 46, 132 (1966).
11. A. Böhm, a paper presented at this symposium.

LORENTZ INVARIANT ALGEBRAIZATION
OF VERTEX FUNCTIONS.
DYNAMICAL GROUP $SO(3,1) \otimes SO(4,3)$†‡

Milan Noga
Department of Physics
Purdue University
Lafayette, Indiana 47907

I. Introduction

The hypothesis that the dynamics of a given quantum mechanical system can be completely described by some dynamical group as well as by the Schrödinger equation has been verified for almost all interesting and important quantum mechanical problems. In quantum mechanics we postulate a Hamiltonian $\mathcal{H}$ which is usually a complicated differential operator and then a solution to the Schrödinger equation

$$\mathcal{H}\psi_n = E_n \psi_n \tag{I.1}$$

determines the energy levels E_n and the set of the quantum numbers n of a given quantum mechanical system, which is completely described by the wavefunctions ψ_n. In the approach using dynamical groups one starts from a chosen dynamical group G and phenomenologically identifies its generators with operators of physical observables rather than postulating the Hamiltonian. In addition, the quantum mechanical wavefunctions ψ_n are assumed to form a basis for a single unitary irreducible representation of the group in question and in such a way measurable physical quantities can be straightforwardly calculated. The same idea was consequently generalized and used in particle physics.

Over the last few years the relativistic framework of dynamical groups proposed by Barut[1)] has been successfully applied to strong

†Presented at the Symposium on De Sitter and Conformal Groups, University of Colorado, Summer 1970.
‡Work supported in part by the U. S. Atomic Energy Commission.

decays of meson[2] and baryon[3] resonances as well as to the study of mass spectral and form factors of hadrons.[4] The essential assumptions made in such studies may be summarized as follows:

a) Hadron states are assigned to unitary irreducible representations of some noncompact group G which contains the Lorentz group as a subgroup (in order to guarantee relativistic invariance of the theory). A priori, suitable candidates for G are, for example, the groups SO(3,1), SO(3,2), SO(4,2), SL(2,C), SL(6,C), etc. The ultimate selection, however, is only to be dictated by results which agree with experiment.

b) Once the group G has been selected then its generators (which are self-adjoint operators in the Hilbert space of physical states) are phenomenologically identified with physical observables such as, for example, momentum, angular momentum, electromagnetic current, etc. In this way matrix elements representing measurable quantities may then be easily calculated by group theoretical considerations.

It is amazing that physical consequences following from the above set of assumptions agree quite well with experiments for suitable choices of the group G.

In fact, even the simplest possible dynamical group, SO(3,1), has been able to describe very well the pion-baryon decay rates of many resonances[3] by making use of only two free parameters which are an effective coupling constant g and an eigenvalue ν of one of the Casimir operators of the SO(3,1) group. Of course, the parameter ν is adjusted phenomenologically by requiring a best fit to the experimental data. However, if one wishes to avoid the freedom in the choice of ν one is then naturally led to the study of larger dynamical groups such as, for example, SO(4,2), which was proposed by Barut and Tripathy.[5] This group has received a great deal of attention in the series of excellent papers by Barut et al,[6] Nambu,[7] and Yao.[8]

Although all the calculations mentioned above show good agreement with experiment, it is of course not at all clear whether other choices for G might be more suitable. Furthermore, if one really expects the study of a given dynamical group to be physically meaningful one should be able to derive its Lie algebra from general physical assumptions.

It is the purpose of this talk to show that the dynamical group which describes the hadronic world may be rigorously derived starting from the following usually accepted physical hypotheses:

a) Relativistic and isotopic invariance of the theory,

b) Validity of the Lehmann-Symanzik-Zimmerman (LSZ) reduction technique,[9]

c) Either an effective interaction Lagrangian $\mathcal{L}_I$ of the type[10)] $\mathcal{L}_I = (F_\pi)^{-1} A_\mu^\alpha(x)\, \partial^\mu \Phi^\alpha(x)$ (where $F_\pi \simeq 190$ MeV is the pion decay amplitude, $A_\mu^\alpha(x)$ is the axial vector current, $\Phi^\alpha(x)$ is the pion field, and $\alpha = 1,2,3$ and $\mu = 0,1,2,3$ are isovector and space-time indices respectively) or the validity of PCAC (i.e. the assumption that the soft pion technique may be employed whenever $(F_\pi m_\pi^2)^{-1} \partial^\mu A_\mu^\alpha(x)$ is chosen as an interpolating pion field),[11)]

d) Validity of the usual equal-time commutators between axial charges,[12)] and

e) Absence of exotic states having isospin I = 2.

From the above assumptions it is then easy to reduce the dynamical problem of calculating decay amplitudes involving pions to the study of representations of a certain noncompact group. In fact, we show in the next two sections that the above assumptions lead us to conclude that hadron states form unitary representations of the noncompact group SO(3,1) ⊗ SO(4,3). In addition we show that the physical interpretation of the generators of this group is unique and unambiguous and that the relativistic transition amplitude is written as a sum of matrix elements of a certain class of generators of the group in question.

II. Reduction of the Dynamical Problem to the Algebra of Matrix Elements

We start by considering a general pion transition process

$$a(p) \to b(p') + \pi(q,\alpha) \quad , \qquad \text{(II.1)}$$

where a(p) and b(p′) denote arbitrary hadron states with momenta p and p′ respectively while $\pi(q,\alpha)$ denotes a pion with momentum q and isospin index α. The S-matrix for this process is defined by

$$^{in}\langle b\underset{\sim}{p}'; \underset{\sim}{q}\alpha | S | a\underset{\sim}{p}\rangle^{in} = {}^{out}\langle b\underset{\sim}{p}'; \underset{\sim}{q}\alpha | a\underset{\sim}{p}\rangle^{in} \quad , \qquad \text{(II.2)}$$

(where a and b denote all other quantum numbers) and is related to the invariant Feynman amplitude $M_\alpha(p',q;\, p)$ by

$$^{in}\langle b\underset{\sim}{p}', \underset{\sim}{q}\alpha | S | a\underset{\sim}{p}\rangle^{in} = -\,(2\pi)^4\, \delta^4(p'+q-p)\,(2\pi)^{-9/2}$$

$$\times\, (8q^o p^o p'^o)^{-1/2} M_\alpha(p' q;\, p) \quad . \qquad \text{(II.3)}$$

By use of LSZ reduction technique[9)] one may then write

$$ {}^{in}\langle b\underset{\sim}{p}'; \underset{\sim}{q}\alpha | S | a\underset{\sim}{p}\rangle^{in} = -\frac{i}{(2\pi)^{3/2}} \frac{1}{(2p^o)^{1/2}} \int d^4x\, e^{-iqx} (\Box - m_\pi^2) \langle b\underset{\sim}{p}' | \Phi^\alpha(x) | a\underset{\sim}{p}\rangle , \quad (II.4) $$

where the states $|a\underset{\sim}{p}\rangle^{in}$ and $|b\underset{\sim}{p}'\rangle^{in}$ have been normalized to

$$ {}^{in}\langle b\underset{\sim}{p}' | a\underset{\sim}{p}\rangle^{in} = \delta_{ab}\, \delta^3(\underset{\sim}{p} - \underset{\sim}{p}') . \quad (II.5) $$

We can then proceed further either by making use of an effective interaction Lagrangian[10] of the type

$$ \mathcal{L}_I = (F_\pi)^{-1} A_\mu^\alpha(x)\, \partial^\mu \Phi^\alpha(x) \quad (II.6) $$

and its corresponding equations of motion

$$ (\Box - m_\pi^2)\Phi^\alpha(x) = -(F_\pi)^{-1}\partial^\mu A_\mu^\alpha(x) \quad (II.7) $$

or PCAC[11] (i.e. $(F_\pi m_\pi^2)^{-1}\partial^\mu A_\mu^\alpha(x)$ is chosen as an interpolating pion field and the soft pion technique is then employed) and rewrite Eq. (II.4) in the following form

$$ {}^{in}\langle b\underset{\sim}{p}', \underset{\sim}{q}\alpha | S | a\underset{\sim}{p}\rangle^{in} = F_\pi^{-1} \frac{(2\pi)^{5/2}}{(2q^o)^{1/2}}\, \delta^4(p'+q-p) \times (p-p')^\mu \langle b\underset{\sim}{p}' | A_\mu^\alpha(0) | a\underset{\sim}{p}\rangle . \quad (II.8) $$

(In the derivation of the last equation we have also used translation invariance, i.e.

$$ \langle b\underset{\sim}{p}' | A_\mu^\alpha(x) | a\underset{\sim}{p}\rangle = \exp[-i(p'-p)x]\, \langle b\underset{\sim}{p}' | A_\mu^\alpha(0) | a\underset{\sim}{p}\rangle). $$

Comparing Eq. (II.8) with (II.3) we note that the invariant Feynman amplitude $M_\alpha(p'q; p)$ may be written as

$$ M_\alpha(p', q; p) = F_\pi^{-1} (2\pi)^3 (4p^o p'^o)^{1/2} (p-p')^\mu \langle b\underset{\sim}{p}' | A_\mu^\alpha(0) | a\underset{\sim}{p}\rangle . \quad (II.9) $$

From the above equation it is now evident that $M_\alpha(p',q;\,p)$ may be obtained by calculating the matrix elements of $A^\alpha_\mu(0)$ between two hadron states $|b\underset{\sim}{p}'\rangle$ and $|a\underset{\sim}{p}\rangle$. It should also be mentioned at this point that the state $|a\underset{\sim}{p}\rangle$ representing a hadron of momentum $\underset{\sim}{p}$ may be obtained from the state $|a\rangle$ at rest by means of a homogeneous Lorentz transformation, i.e.

$$|a\underset{\sim}{p}\rangle = e^{i\underset{\sim}{\xi}\underset{\sim}{M}}|a\rangle \quad , \qquad \text{(II.10)}$$

where $\underset{\sim}{M}$ denotes the boost operator and $\underset{\sim}{\xi}$ is a vector in the direction of $\underset{\sim}{p}$ with magnitude given by

$$\tanh|\underset{\sim}{\xi}| = \frac{|\underset{\sim}{p}|}{p^o} \quad , \qquad \text{(II.11)}$$

Use of Eq. (II.10) simplifies extremely the calculations of the matrix element $\langle b\underset{\sim}{p}'|A^\alpha_\mu(0)|a\underset{\sim}{p}\rangle$. In fact, since the invariant amplitude (II.9) is Lorentz invariant, we may assume without loss of generality that the initial state $|a\underset{\sim}{p}\rangle$ is at rest. The final state $|b\underset{\sim}{p}'\rangle$ is then obtained by boosting the state $|b\rangle$ at rest to momentum $\underset{\sim}{p}'$. The calculation of the invariant amplitude is then reduced to the determination of matrix elements of the type

$$\langle b|\, e^{-i\underset{\sim}{\xi}\underset{\sim}{M}}\, A^\alpha_\mu(0)|a\rangle \ . \qquad \text{(II.12)}$$

Since the matrices of finite Lorentz transformations

$$B_{ba} \equiv \langle b|\, e^{-i\underset{\sim}{\xi}\underset{\sim}{M}}|a\rangle \qquad \text{(II.13)}$$

can be found in the literature[13] the invariant Feynman amplitude may be written as

$$M_\alpha(p',q;\,p) = F_\pi^{-1}\,(2\pi)^3\,\left(4p^o p'^o\right)^{\frac{1}{2}}(p-p')^\mu\, B_{bn}\langle n|A^\alpha_\mu(0)|a\rangle$$

and the problem of determining it is then reduced to the determination of the matrix elements of $A^\alpha_\mu(0)$ between two states at rest, i.e. to the calculation of

$$\langle b|A^\alpha_\mu(0)|a\rangle \ . \qquad \text{(II.14)}$$

In order to determine these matrix elements we start by assuming the validity of the usual equal-time commutation relations between axial charges,[12] i.e.

$$\left[\int d^3x\, A_0^{\alpha}(\underset{\sim}{x},t)\; \int d^3y\, A_0^{\beta}(\underset{\sim}{y},t)\right] = i\,\epsilon^{\alpha\beta\gamma}\, I^{\gamma}\;, \qquad \text{(II.15)}$$

where I^{γ} denotes the generator of isospin transformations.

Let us next consider the matrix element of the commutator (II.15) between hadron states $|b\underset{\sim}{p}'\rangle$ and $|a\underset{\sim}{p}\rangle$. We obtain

$$\langle b\underset{\sim}{p}'|\left[\int d^3x\, A_0(\underset{\sim}{x},0),\; \int d^3y\, A_0^{\beta}(\underset{\sim}{y},0)\right]|a\underset{\sim}{p}\rangle = i\,\epsilon^{\alpha\beta\gamma}(I^{\gamma})_{ba}\,\delta^3(\underset{\sim}{p}-\underset{\sim}{p}'). \qquad \text{(II.16)}$$

To evaluate the left-hand side of the last equation one inserts a complete set of intermediate states $|n\underset{\sim}{p}_n\rangle$ and uses translational invariance to carry out the spatial and momentum $\underset{\sim}{p}_n$ integrations. This yields

$$(2\pi)^6 \sum_n \{\langle b\underset{\sim}{p}|A_0^{\alpha}(0)|n\underset{\sim}{p}\rangle\langle n\underset{\sim}{p}|A_0^{\beta}(0)|a\underset{\sim}{p}\rangle - \langle b\underset{\sim}{p}|A_0^{\beta}(0)|n\underset{\sim}{p}\rangle\langle n\underset{\sim}{p}|A_0^{\alpha}(0)|a\underset{\sim}{p}\rangle\}$$

$$= i\,\epsilon^{\alpha\beta\gamma}(I^{\gamma})_{ba}\;, \qquad \text{(II.17)}$$

where we have cancelled a common factor of $\delta^3(\underset{\sim}{p}-\underset{\sim}{p}')$ on both sides. It should also be stressed that the above relation can only be derived for states $|b\underset{\sim}{p}\rangle$ and $|a\underset{\sim}{p}\rangle$ with the same momentum $\underset{\sim}{p}$. Therefore, we shall restrict ourselves to hadron states at rest without loss of generality.

To proceed further let us define three matrices $x_0^{\alpha}(\alpha = 1,2,3)$ by

$$(x_0^{\alpha})_{bn} = \langle b|A_0^{\alpha}(0)|n\rangle\; (2\pi)^3\;, \qquad \text{(II.18)}$$

where b and n denote the bth row and nth column of the matrix x_0^{α}. With this notation Eq. (II.17) may then be written in matrix form as

$$x_0^{\alpha}x_0^{\beta} - x_0^{\beta}x_0^{\alpha} = i\,\epsilon^{\alpha\beta\gamma}I^{\gamma}\;, \qquad \text{(II.19)}$$

where also I^{γ} is a matrix. Introducing the usual abbreviation

$$\left[x_0^{\alpha},\, x_0^{\beta}\right] \equiv x_0^{\alpha}x_0^{\beta} - x_0^{\beta}x_0^{\alpha}\;, \qquad \text{(II.20)}$$

Eqs. (II.17) and (II.19) may then be written as

$$\left[x_0^\alpha, x_0^\beta\right] = i\,\epsilon^{\alpha\beta\gamma} I^\gamma . \tag{II.21}$$

The above equation will play an important role in the rest of our discussion. In fact, we proceed in the next section to make use of this equation along with Lorentz invariance and absence of exotic states to study relations among the matrix elements $\langle b|A_\mu^\alpha(0)|a\rangle$ and show that they form a closed algebra which is isomorphic to the Lie algebra of the group $SO(3,1) \otimes SO(4,3)$.

III. Derivation of the Dynamical Algebra

Any reasonable theory describing strong interactions of pions with hadrons must be Lorentz and isotopic spin invariant. This then implies that the invariance symmetry group K is the direct product of the isospin group $SU(2)_I$ and the Lorentz group $SO(3,1)$, i.e. $K = SU(2)_I \otimes SO(3,1)$. Clearly the group K is generated by the following Lie algebra

$$[I^\alpha, I^\beta] = i\,\epsilon^{\alpha\beta\gamma} I^\gamma , \tag{III.1}$$

$$[J_{\mu\nu}, J_{\rho\sigma}] = i(g_{\nu\rho} J_{\mu\sigma} - g_{\nu\sigma} J_{\mu\rho} - g_{\mu\rho} J_{\nu\sigma} + g_{\mu\sigma} J_{\nu\rho}) , \tag{III.2}$$

and

$$[I^\alpha, J_{\mu\nu}] = 0 , \tag{III.3}$$

where $\alpha,\beta,\gamma = 1,2,3$ are isospin indices, $\mu,\nu,\rho,\sigma = 1,2,3,0$ are space-time indices, and I^α and $J_{\mu\nu}$ are the generators of the groups $SU(2)_I$ and $SO(3,1)$ respectively. The metric tensor $g_{\mu\nu}$ is defined by

$$g_{00} = 1,\ g_{11} = g_{22} = g_{33} = -1,\ g_{\mu\nu} = 0 \ \text{if}\ \mu \neq \nu .$$

The most important operator in our theory is the axial vector current $A_\mu^\alpha(0)$ taken at the origin of a reference frame. This operator is, of course, an isovector and a Lorentz four-vector and therefore obeys the following set of commutation relations

$$[I^\alpha, A_\mu^\beta(0)] = i\,\epsilon^{\alpha\beta\gamma} A_\mu^\gamma(0) \tag{III.4a}$$

and

$$[J_{\mu\nu}, A^{\alpha}_{\rho}(0)] = i(g_{\nu\rho}A^{\alpha}_{\mu}(0) - g_{\mu\rho}A^{\alpha}_{\nu}(0)) \; . \tag{III.4b}$$

We next consider the sum rules obtained by taking matrix elements of the commutators (III.1)-(III.4) between two hadron states $|f\rangle$ and $|i\rangle$ at rest. Thus we define twelve matrices x^{α}_{μ} by

$$(x^{\alpha}_{\mu})_{fi} = \langle f|A^{\alpha}_{\mu}(0)|i\rangle \; (2\pi)^3 \; , \tag{III.5}$$

where the subscripts f and i denote the fth row and ith column of the matrix x^{α}_{μ}. The matrix relations following from the commutators (III.4) then take the form

$$[I^{\alpha}, x^{\beta}_{\mu}] = i\,\epsilon^{\alpha\beta\gamma}x^{\gamma}_{\mu} \tag{III.6}$$

and

$$[J_{\mu\nu}, x^{\alpha}_{\rho}] = i(g_{\nu\rho}x^{\alpha}_{\mu} - g_{\mu\rho}x^{\alpha}_{\nu}) \; , \tag{III.7}$$

where we have used the abbreviation defined in Eq. (II.20). It should also be mentioned that the symbols I^{α} and $J_{\mu\nu}$ occurring in the last two equations are not operators but matrices. However, we have used the same symbols for the linear operators I^{α} and $J_{\mu\nu}$ as for their algebraical realizations since these matrices will, of course, satisfy the same algebraic relations as those given by Eqs. (III.1)-(III.3). To avoid any confusion we stress that from now any commutator which will be derived must be understood as a matrix relation.

If we were now able to construct uniquely the set of twelve matrices x^{α}_{μ} then we would be finished and the invariant Feynman amplitude (II.9) would then be uniquely determined by

$$M_{\alpha}(p',q;\, p) = F_{\pi}^{-1}\,(4p^{o}p^{o\prime})(p-p')^{\mu}\,B\,x^{\alpha}_{\mu}$$

where the matrices B and x^{α}_{μ} are defined by Eqs. (II.13) and (III.5) respectively. To proceed further in this construction we next write the most general form for the commutators $[x^{\alpha}_{\mu}, x^{\beta}_{\nu}]$. In order to do this we note that this quantity is antisymmetric with respect to the interchange of pairs of indices $(\alpha\mu)$ and $(\beta\nu)$. Therefore the most general decomposition of this commutator is:

$$[x^{\alpha}_{\mu}, x^{\beta}_{\nu}] = i\,\epsilon^{\alpha\beta\gamma}Y^{\gamma}_{\{\mu\nu\}} + i\,Z^{\{\alpha\beta\}}_{[\mu\nu]} \; , \tag{III.8}$$

where $Y^{\gamma}_{\{\mu\nu\}}$ and $Z^{\{\alpha\beta\}}_{[\mu\nu]}$ are matrices and the symbols $[\cdot\cdot]$ and $\{\cdot\cdot\}$ are abbreviations for antisymmetricity and symmetricity in the corresponding pairs of indices. It is now a simple matter to prove that $Y^{\gamma}_{\{\mu\nu\}}$ is an isovector and symmetric Lorentz tensor while $Z^{\{\alpha\beta\}}_{[\mu\nu]}$ is a reducible symmetric isotensor and an antisymmetric Lorentz tensor. This proof can be found in Appendix A.

The decomposition of the commutator $[x^{\alpha}_{\mu}, x^{\beta}_{\nu}]$ given by Eq (III.8) reminds us very strongly of the algebraic structure of Weinberg's superconvergence conditions[10] analyzed in a series of papers.[14),15),16)] In these works the left hand size of Eq. (III.8) is interpreted as the s and u-channel contributions to a superconvergence sum rule while the right hand side correspond to t-channel meson exchange contributions. In accordance with this philosophy the matrices $Z^{\{\alpha\beta\}}_{[\mu\nu]}$ are only related to the exchange of mesons with isospin I = 0,2 since $Z^{\{\alpha\beta\}}_{[\mu\nu]}$ is a symmetric isotensor. It is usually assumed that isospin two states (which belong to the class of so-called exotic states) do not exist. Therefore we require that the part of $Z^{\{\alpha\beta\}}_{[\mu\nu]}$ which transforms under $SU(2)_I$ as an irreducible symmetric tensor with I = 2 must vanish. This implies immediately that $Z^{\{\alpha\beta\}}_{[\mu\nu]}$ is only an isospin scalar, namely,

$$Z^{\{\alpha\beta\}}_{[\mu\nu]} = - \delta^{\alpha\beta}\, T_{\mu\nu}\,, \tag{III.9}$$

where the minus sign is only a convention and $T_{\mu\nu}$ is a matrix which transforms as an antisymmetric Lorentz tensor and obeys the following set of matrix relations:

$$[J_{\mu\nu}, T_{\rho\sigma}] = i(g_{\nu\rho}T_{\mu\sigma} - g_{\nu\sigma}T_{\mu\rho} - g_{\mu\rho}T_{\nu\sigma} + g_{\mu\sigma}T_{\nu\rho}) \tag{III.10}$$

and

$$[I^{\alpha}, T_{\mu\nu}] = 0. \tag{III.11}$$

We stress that the application of Eq. (III.8) to the time components

$$[x^{\alpha}_{0}, x^{\beta}_{0}] = i\,\epsilon^{\alpha\beta\gamma}\, Y^{\gamma}_{\{00\}} \tag{III.12a}$$

must give the same result as Eq. (II.21), i.e.

$$[x_0^\alpha, x_0^\beta] = i\,\epsilon^{\alpha\beta\gamma}\, I^\gamma \quad , \tag{III.12b}$$

and which follow from the equal-time commutator algebra which the axial vector charges satisfy.[12)]

Comparing the last two equations we obtain

$$Y_{\{00\}}^{\gamma} = I^\gamma \, . \tag{III.12c}$$

Since I^γ is a Lorentz scalar it follows immediately (by making use of the commutators $[J_{\mu\nu}, Y_{\{00\}}^{\gamma}] = 0$; see Appendix A) that $A_{\{\mu\nu\}}^{\gamma}$ is a symmetric invariant Lorentz tensor and therefore we conclude

$$Y_{\{\mu\nu\}}^{\gamma} = g_{\mu\nu}\, I^\gamma \quad . \tag{III.13}$$

Using the results given by Eqs. (III.9) and (III.13) we rewrite the important relation (III.8) as follows:

$$[x_\mu^\alpha, x_\nu^\beta] = i g_{\mu\nu}\, \epsilon^{\alpha\beta\gamma}\, I^\gamma - i\delta^{\alpha\beta}\, T_{\mu\nu} \quad . \tag{III.14}$$

From the last equation it is now simple to express $T_{\mu\nu}$ in terms of the x_μ^α's and make use of the Jacobi identity to determine the commutators $[T_{\mu\nu}, x_\rho^\alpha]$ and $[T_{\mu\nu}, T_{\rho\sigma}]$. These calculations can be found in Appendix B, where the following results are derived:

$$[T_{\mu\nu}, x_\rho^\alpha] = i(g_{\nu\rho} x_\mu^\alpha - g_{\mu\rho} x_\nu^\alpha) \tag{III.15}$$

and

$$[T_{\mu\nu}, T_{\rho\sigma}] = i(g_{\nu\rho} T_{\mu\sigma} - g_{\nu\sigma} T_{\mu\rho} - g_{\mu\rho} T_{\nu\sigma} + g_{\mu\sigma} T_{\nu\rho}) \, . \tag{III.16}$$

The commutators (III.1)-(III.3), (III.6), (III.7), (III.10), (III.11) and (III.14)-(III.16) show that the 27 matrices I^α, $J_{\mu\nu}$, $T_{\mu\nu}$ and x_μ^α form a closed algebra which may be identical to the Lie algebra of some dynamical group G. If we are able to find the structure of this group G then our dynamical problem will be completely reduced to the study of unitary representations of this group.

In order to find the structure of G we find it convenient to introduce 6 matrices $F_{\mu\nu}$ defined as follows:

$$F_{\mu\nu} = J_{\mu\nu} - T_{\mu\nu} \quad . \tag{III.17}$$

It is then a simple matter to verify that the matrices $F_{\mu\nu}$ commute with all the 21 matrices I^{α}, $T_{\mu\nu}$ and x^{α}_{μ}, and that they satisfy the Lie algebra of the SO(3,1) group, namely,

$$[F_{\mu\nu}, F_{\rho\sigma}] = i(g_{\nu\rho}F_{\mu\sigma} - g_{\nu\sigma}F_{\mu\rho} - g_{\mu\rho}F_{\nu\sigma} + g_{\mu\sigma}F_{\nu\rho}). \tag{III.18}$$

This implies that the group G is the direct product of SO(3,1) with a group G_0 which is generated by the Lie algebra given by commutators (III.1), (III.6), (III.11), and (III.14)-(III.16). Thus the problem is now reduced to finding the group structure of G_0. This can be done quite easily if we define a metric tensor g_{ab} for

$$a,b = \mu,\nu,\rho,\sigma \ldots = 1,2,3,0 \qquad \text{and}$$

$$a,b = \alpha,\beta,\gamma \ldots = 5,6,7 \qquad \text{by}$$

$$g_{11} = g_{22} = g_{33} = -1,$$

$$g_{00} = g_{55} = g_{66} = g_{77} = +1, \text{ and}$$

$$g_{ab} = 0 \quad \text{if } a \neq b, \tag{III.19}$$

and introduce in addition matrices

$$L_{ab} = -L_{ba} \quad \text{defined by}$$

$$L_{\alpha\beta} \equiv -\epsilon^{\alpha\beta\gamma} I^{\gamma},$$

$$L_{\alpha\mu} \equiv x^{\alpha}_{\mu} , \tag{III.20}$$

and

$$L_{\mu\nu} \equiv T_{\mu\nu} \quad .$$

With the above definitions the commutators (III.1), (III.6), (III.11) and (III.14)-(III.16) may then be compactly rewritten in the form

$$[L_{ab}, L_{cd}] = i(g_{bc}L_{ad} - g_{bd}L_{ac} - g_{ac}L_{bd} + g_{ad}L_{bc}). \tag{III.21}$$

The above commutation relations define the well known Lie algebra of the noncompact rotational group SO(4,3).

To conclude this section we would like to stress once again that the dynamical problem of determining the Feynman invariant amplitude for processes involving pions has been completely reduced to the study of the algebra of the noncompact group $SO(3,1) \otimes SO(4,3)$. Since operators representing physical observables operate on the Hilbert space of physical states this then implies that hadron states must form a representation space of the dynamical algebra of observables, i.e. of the Lie algebra of the group $SO(3,1) \otimes SO(4,3)$. From this it then follows that any unitary (reducible or irreducible) representation of this group may correspond to possible physical states. Of course, there is no reason at all to demand that physical states transform according to unitary irreducible representations of this group, since the required Lie algebra relations are also fulfilled if one considers unitary reducible representations.

IV. Connection With Dynamical Groups Proposed By Barut

We have proved that matrix elements of physical observables form the closed algebra of a dynamical group which combines in a nontrivial way internal (isospin) symmetry with space-time symmetry. Originally the dynamical groups proposed by Barut et al[1),6)] were only restricted to the external (space-time) properties of hadrons while later these groups were combined with internal symmetries by taking their direct products.[2),3)]

We would next like to discuss what happens if we restrict ourselves to matrix elements of physical observables describing external properties of hadrons, i.e. to sets of hadrons with the same internal quantum numbers. This is equivalent to considering hadron families with the same third component of isospin and thus implies that we rule out all matrices Γ^{α} connected with internal symmetries as well as the matrices x^1_{μ} and x^2_{μ} which change the charges of the hadrons under consideration. Thus we shall only deal now with the 16 matrices $J_{\mu\nu}$, $T_{\mu\nu}$ and $X^3_{\mu} \equiv \Gamma_{\mu}$. It is then simple to verify that they form a closed algebra which is identical with the Lie algebra of the group $SO(3,1) \otimes SO(3,2)$. This result tells us that hadron states with the same third component of isospin must transform according to unitary (reducible or irreducible) representations of this group.

The dynamical group SO(3,2) was proposed by Barut, Corrigan, and Kleinert[4)] in order to calculate mass spectra and electromagnetic form factors of hadrons. In their framework hadron states are assumed to transform according to unitary irreducible representations of this group and the matrix Γ_μ introduced above plays the role of their so-called algebraic current. They then consider one class of representations of the group SO(3,2), which, of course, are also representations of the group SO(3,1) ⊗ SO(3,2) which we have derived here by identifying the matrices $T_{\mu\nu}$ with the matrices $J_{\mu\nu}$. Thus we have shown that the assumptions made by the preceding authors on the basis of an excellent physical intuition can in fact be uniquely derived making use of usually accepted dynamical assumptions.

V. Summary and Conclusions

Several dynamical models for the description of hadron states which lead naturally to relations identical to the algebra of certain Lie groups have been proposed over the last few years. Among them we start by mentioning the popular Chew static bootstrap model,[17] which was completely reworded in group theoretic language by Cook, Goebel, and Sakita.[18] Next we mention the work of Capps[19] who has shown under fairly general assumptions that if one saturates superconvergence sum rules with single particle states one is naturally led to models in which hadron states are associated with unitary representations of certain Lie groups. More recently, algebraic superconvergence conditions for the forward scattering of massless pions with hadrons have been derived by Weinberg[10),20] making use of the effective chiral Lagrangian formalism.

All the preceding treatments led to the conclusion that hadron states form a basis for unitary representations of certain Lie groups. On the other hand, in the framework of dynamical groups one usually makes the ad hoc assumption that hadron states form unitary irreducible representations of some noncompact group. Since this approach has been rather successful one is then led to conjecture that these dynamical groups might in fact be derived from generally accepted physical assumptions. We have shown that this is actually the case. In fact, we have derived relations identical to the Lie algebra of the group SO(3,1) ⊗ SO(4,3) merely by assuming isospin and Lorentz invariance, usual equal-time commutator algebra between axial charges, absence of exotic states, and either an effective interaction Lagrangian or PCAC. Since physical observables are self-adjoint operators in the Hilbert space $\mathcal{H}$ of hadronic physical states it then follows that $\mathcal{H}$ is the representation space of the Lie group SO(3,1) ⊗ SO(4,3). Thus hadron states must form a basis for unitary (irreducible or reducible) representations of this group, which is a nontrivial combination of the isospin group SU(2) with the Lorentz group SO(3,1).

The generalization to larger internal symmetry groups (for example SU(3)) is straightforward and may be done along the lines discussed in this paper.

The dynamical groups proposed by Barut and his collaborators were SO(3,1), SO(3,2)and SO(4,2), which are all subgroups of SO(3,1) ⊗ SO(4,3) so that all representations of the latter are also reducible representations of the former groups. If we only restrict ourselves to the external properties of hadrons we have found that hadron states with the same third components of isospin are classified according to unitary representations of the group SO(3,1) ⊗ SO(3,2). Note that the group SO(3,2) is exactly the one proposed by Barut et al[4] in their calculations of electromagnetic form factors and mass spectra of physical states.

To conclude this discussion we stress that the dynamical calculation of the pion-hadron vertex function was reduced to a set of algebraic relations which turned out to be the same as the Lie algebra of the group SO(3,1) ⊗ SO(4,3). Finally, it should also be mentioned that an algebraic treatment to the dynamical problem of pion-hadron coupling constants has also been extensively developed in a series of papers by Sugawara who makes use of the LSZ reduction technique and the assumption that the dispersive part of the three point function may be completely saturated by single particle intermediate states.

Acknowledgments

I would like to express my gratitude to Dr. Jose Katz who has collaborated with me on this subject. I wish also to thank Professors R. H. Capps, S. P. Rosen and M. Sugawara for discussions and their kind hospitality at Purdue.

Appendix A

In this appendix we show that the matrices $Y^{\gamma}_{\{\mu\nu\}}$ and $Z^{\{\alpha\beta\}}_{[\mu\nu]}$ introduced in Eq. (III.8) transform as tensors under Lorentz and isospin transformations. We start with the matrix relation (III.8), which is of the form

$$[x^{\alpha}_{\mu}, x^{\beta}_{\nu}] = i\,\epsilon^{\alpha\beta\gamma}\, Y^{\gamma}_{\{\mu\nu\}} + iZ^{\{\alpha\beta\}}_{[\mu\nu]} \quad . \tag{A.1}$$

Our first step is to express $Y^{\gamma}_{\{\mu\nu\}}$ in terms of x^{α}_{μ}. This can simply be done and one obtains the following result

$$Y^{\gamma}_{\{\mu\nu\}} = -\frac{i}{2}\,\epsilon^{\alpha\beta\gamma}\,[x^{\alpha}_{\mu}, x^{\beta}_{\nu}] \quad . \tag{A.2}$$

The commutator

$$[J_{\rho\sigma}, Y^{\gamma}_{\{\mu\nu\}}] = -\frac{i}{2}\,\epsilon^{\alpha\beta\gamma}[J_{\rho\sigma}, [x^{\alpha}_{\mu}, x^{\beta}_{\nu}]] \qquad (A.3)$$

can then be rewritten by making use of the Jacobi identity as

$$[J_{\rho\sigma}, Y^{\gamma}_{\{\mu\nu\}} = \frac{i}{2}\,\epsilon^{\alpha\beta\gamma}\{[x^{\beta}_{\nu}, [J_{\rho\sigma}, x^{\alpha}_{\mu}]] + [x^{\alpha}_{\mu}, [x^{\beta}_{\nu}, J_{\rho\sigma}]]\}. \qquad (A.4)$$

Carrying out the algebraic reduction using the commutation relations (III.7) and (III.8) we finally obtain the result

$$[J_{\rho\sigma}, Y^{\gamma}_{\{\mu\nu\}}] = i(g_{\sigma\mu}Y^{\gamma}_{\{\rho\nu\}} + g_{\sigma\nu}Y^{\gamma}_{\{\rho\mu\}} - g_{\rho\mu}Y^{\gamma}_{\{\sigma\nu\}} - g_{\rho\nu}Y^{\gamma}_{\{\sigma\mu\}}) . \qquad (A.5)$$

From the above equation we see that the matrices $Y^{\gamma}_{\{\mu\nu\}}$ transform as a symmetric Lorentz tensor. The same procedure can be used to prove that $Y^{\gamma}_{\{\mu\nu\}}$ transforms as an isovector while $Z^{\{\alpha\beta\}}_{[\mu\nu]}$ transforms as a symmetric isotensor and an antisymmetric Lorentz tensor.

Appendix B

The purpose of this appendix is to derive the commutators $[T_{\mu\nu}, x^{\alpha}_{\rho}]$ and $[T_{\mu\nu}, T_{\rho\sigma}]$. We start from relation (III.14) and obtain

$$T_{\mu\nu} = \frac{i}{3}[x^{\alpha}_{\mu}, x^{\alpha}_{\nu}] \quad . \qquad (B.1)$$

By making use of the above equation we may then write

$$[T_{\mu\nu}, x^{\beta}_{\rho}] = -\frac{i}{3}[x^{\beta}_{\rho}, [x^{\alpha}_{\mu}, x^{\alpha}_{\nu}]] \quad . \qquad (B.2)$$

We next apply the Jacobi identity to the double commutator given above and obtain

$$[T_{\mu\nu}, x^{\beta}_{\rho}] = \frac{i}{3}\{[x^{\alpha}_{\nu}, [x^{\beta}_{\rho}, x^{\alpha}_{\mu}]] + [x^{\alpha}_{\mu}, [x^{\alpha}_{\nu}, x^{\beta}_{\rho}]]\} \quad . \qquad (B.3)$$

Making use of Eq. (III.14) we then carry out the algebraic reduction of the double commutators on the right hand side of Eq. (B.3). This yields the result

$$[T_{\mu\nu}, x^\beta_\rho] = \frac{2i}{3}\{g_{\nu\rho}x^\beta_\mu - g_{\mu\rho}x^\beta_\nu\} - \frac{1}{3}\{[T_{\rho\mu}, x^\beta_\nu] + [T_{\nu\rho}, x^\beta_\mu]\} . \tag{B.4}$$

The preceding commutator is then used to calculate the sum $[T_{\rho\mu}, x^\beta_\nu] + [T_{\nu\rho}, x^\beta_\mu]$. After some simple algebra we obtain

$$[T_{\rho\mu}, x^\beta_\nu] + [T_{\nu\rho}, x^\beta_\mu] = \frac{i}{2}\{g_{\rho\mu}x^\beta_\nu - g_{\rho\nu}x^\beta_\mu\} - \frac{1}{2}[T_{\mu\nu}, x^\beta_\rho]. \tag{B.5}$$

Inserting the last result into Eq. (B.4) we then obtain the relation

$$[T_{\mu\nu}, x^\beta_\rho] = i(g_{\nu\rho}x^\beta_\mu - g_{\mu\rho}x^\beta_\nu) , \tag{B.6}$$

which has been used in Section III.

We can now proceed further and calculate

$$[T_{\mu\nu}, T_{\rho\sigma}] = \frac{i}{3}[T_{\mu\nu}, [x^\alpha_\rho, x^\alpha_\sigma]] . \tag{B.7}$$

In order to do this we once again make use of the Jacobi identity for the double commutator and obtain upon using Eq. (B.6) the result

$$[T_{\mu\nu}, T_{\rho\sigma}] = \frac{1}{3}\{g_{\nu\rho}[x^\alpha_\sigma, x^\alpha_\mu] - g_{\mu\rho}[x^\alpha_\sigma, x^\alpha_\nu] + g_{\sigma\mu}[x^\alpha_\rho, x^\alpha_\nu] - g_{\nu\sigma}[x^\alpha_\rho, x^\alpha_\mu]\}. \tag{B.8}$$

Combining the above relation with Eq. (B.1) we then find

$$[T_{\mu\nu}, T_{\rho\sigma}] = i(g_{\nu\rho}T_{\mu\sigma} - g_{\nu\sigma}T_{\mu\rho} - g_{\mu\rho}T_{\nu\sigma} + g_{\mu\sigma}T_{\nu\rho}) . \tag{B.9}$$

Thus the matrices $T_{\mu\nu}$ form a closed algebra identical to the Lie algebra of the group SO(3,1). Relation (B.6) then tells us that the matrices x^α_μ transform as four-vectors with respect to the group in question.

References

1. A. O. Barut, Conference on Symmetry Principles at High Energy (W. H. Freeman and Co., San Francisco, 1964); Phys. Rev. 135, B839 (1964); ibid. 156, 1538 (1967); Proceedings of the First Coral Gables Conference on Symmetry Principles at High Energy (W. H. Freeman and Co., San Francisco, 1965).
2. A. O. Barut and K. C. Tripathy, Phys. Rev. Letters 19, 918 (1967).

3. A. O. Barut and H. Kleinert, Phys. Rev. Letters 18, 745 (1967); H. Kleinert, ibid. 18, 1027 (1967).
4. A. O. Barut and H. Kleinert, Phys. Rev. 156, 1546 (1967); ibid. 161, 1464 (1967); A. O. Barut, D. Corrigan, and H. Kleinert, ibid. 167, 1527 (1968).
5. A. O. Barut and K. C. Tripathy, Phys. Rev. Letters 19, 1081 (1967).
6. A. O. Barut, D. Corrigan and H. Kleinert, Phys. Rev. Letters 20, 167 (1968); A. O. Barut and A. Baiguni, Phys. Rev. 184, 1342 (1969); A. O. Barut, P. Cordero, and G. C. Ghirari, Phys. Rev. 1, D536 (1970).
7. Y. Nambu, Phys. Rev. 160, 1171 (1967).
8. T. Yao, University of Pittsburgh preprint.
9. H. Lehman, K. Symanzik, and W. Zimmerman, Nuovo Cimento 1, 205 (1955).
10. S. Weinberg, Phys. Rev. 177, 2604 (1969).
11. M. Gell-Mann and M. Lévy, Nuovo Cimento 16, 705 (1960).
12. M. Gell-Mann, Phys. Rev. 125, 1067 (1962).
13. S. Ström, Arkiv Fysik 29, 467 (1965); ibid. 33, 465 (1966).
14. M. Noga and C. Cronström, Nucl. Phys. B15, 61 (1970); Phys. Rev. 1D, 2414 (1970).
15. A. McDonald, Phys. Rev. D1, 721 (1970).
16. L. R. Ram Mohan, Phys. Rev. (to be published).
17. G. F. Chew, Phys. Rev. Letters 9, 233 (1962).
18. T. Cook, C. J. Goebel, and B. Sakita, Phys. Rev. Letters 15, 35 (1965).
19. R. H. Capps, Phys. Rev. 168, 1731 (1968); 171, 1591 (1968).
20. S. Weinberg, Phys. Rev. Letters 22, 1023 (1969); Proceedings of the 14th International Conference on High Energy Physics, Vienna 1968, p. 253.
21. M. Sugawara, in Particle Physics, edited by Paul Urban (Springer-Verlag, Berlin, 1969), p. 310; Phys. Rev. Letters 24, 691 (1970).

Section C: Mass Zero Particles

MASS ZERO PARTICLES IN DE SITTER SPACE†

Gerhard Börner‡
Max-Planck-Institut für Physik und Astrophysik
Munich, Germany

Introduction

If in Einstein's equation with a zero energy-momentum tensor we allow a nonzero cosmological constant Λ, then we find as a solution the curved but empty De Sitter spacetime. There are essentially two models with different global topology, which can best be visualized as hypersurfaces in a flat 5-dimensional Minkowski space M_5:

a) $\xi_0^2 - \xi_1^2 - \xi_2^2 - \xi_3^2 - \xi_4^2 = -R^2 \qquad (\Lambda \equiv \frac{3}{R^2})$

The group SO(1,4) acts transitively on this manifold.

b) $\xi_0^2 - \xi_1^2 - \xi_2^2 - \xi_3^2 + \xi_4^2 = -R^2$

This space has SO(2,3) as a transitively acting symmetry group.

In the following we shall be concerned with model a).[1]

Obviously the infinitesimal generators of SO(1,4) are the operators of angular momentum J_{ab} in M_5. There are two Casimir operators

$$I_1 = -\frac{1}{2R^2} J_{ab}\, \eta^{ac}\, \eta^{bd}\, J_{cd}$$

$$I_2 = +v^a \eta_{ab} v^b, \qquad v^a = \epsilon^{abcde} J_{bc} J_{de}$$

(η^{ab}: metric-, ϵ^{abcde} completely antisymmetric tensor in M_5)

†Presented at the Symposium on De Sitter and Conformal Groups, University of Colorado, Summer 1970.

‡Address from Sept. 1970 to June 1971: City College of C.U.N.Y., New York, NY 10031.

The contraction $R \to \infty$ gives back the generators of the Poincaré group. I_1 and I_2 go over into the operators of mass and spin respectively.[1] In De Sitter space we characterize a particle by its eigenvalues with respect to I_1 and I_2, which in view of the above mentioned correspondence to the flat space quantities we might call again "mass" and "spin."

I. Spin Zero Particles

A. Klein-Gordon Equation

Under $x' = g(x)$ a scalar field transforms as follows

$$\varphi'(x') = \varphi(x) \tag{1}$$

Introducing a convenient coordinate system on a):[1]

$$ds^2 = \lambda^{-2}(R^2 d\lambda^2 - (d\underline{y})^2)^* \tag{2}$$

$$-\infty < \lambda < \infty \qquad \lambda \neq 0$$

$$-\infty < y^\alpha < \infty$$

we calculate the generators using (1) and get[1]

$$I_1 = -\frac{\lambda^2}{R^2}\partial_\lambda\partial_\lambda + \frac{2\lambda}{R^2}\partial_\lambda + \lambda^2\partial_\lambda\partial_\lambda$$

$$I_2 = 0$$

The field equation is

$$I_1\varphi = \varkappa^2\varphi \tag{3}$$

We solve this equation as in Ref. 1 under the condition that I_1 must be a self-adjoint operator defined on the Hilbert space of square-integrable functions with respect to the invariant measure on a). Thus we obtain irreducible representations of the De Sitter group which are decompositions of the quasiregular representation of the De Sitter group on a), defined on this same Hilbert space.

---

*Here $\xi^0 = \frac{R}{2}(\lambda - \lambda^{-1}) - \frac{1}{2R}y^2\lambda^{-1}$; $\xi^{1,2,3} = \lambda^{-1}y^{1,2,3}$; $\xi^4 = -\frac{R}{2}(\lambda + \lambda^{-1}) + \frac{1}{2R}y^2\lambda^{-1}$. For $R \to \infty$ we require $\lambda \sim (1 + \frac{x^0}{R})$, $\underline{y} \to \underline{y}$ then $ds^2 \to (dx^0)^2 - (d\underline{y})^2$.

The spectrum of eigenvalues $R^2\varkappa^2 = 9/4 - \nu^2$ consists of a continuous

$$\nu = i\rho, \qquad 0 < \rho < \infty \tag{4}$$

and a discrete part

$$\nu = (n + \tfrac{1}{2})^2, \quad n = 0,1,2,\ldots \tag{5}$$

(this is the so-called "most degenerate" spectrum[2]).

For $R \to \infty$ a whole bunch of eigenvalues $\varkappa^2$ goes to zero, as if the curvature of spacetime has removed a degeneracy present in flat spacetime.

To each value of $\varkappa^2$ belongs a definite eigenfunction:[1),4)]

$$\varphi_{\rho\pm} = N_\rho \lambda^2 e^{i\underset{\sim}{k}\cdot\underset{\sim}{y}} \{\theta(\lambda) \pm \theta(-\lambda)\} \begin{Bmatrix} \mathrm{Im}\ (e^{i\pi/4}\ h_{-\frac{1}{2}+i\rho}(|\underset{\sim}{k}\lambda|R)) \\ \mathrm{Re}\ (\qquad\quad '' \qquad\quad) \end{Bmatrix} \tag{6}$$

$N\rho$: normalization factor, h_ν: spherical Hankelfunction (± indicates the $\mathfrak{F}$-parity:[1] $\varphi_{\rho\pm}(-\lambda, \underset{\sim}{y}) = \pm\varphi_{\rho\pm}(\lambda, \underset{\sim}{y})$; where $\mathfrak{F}: \lambda \to -\lambda$, $\underset{\sim}{y} \to \underset{\sim}{y}$ is the discrete symmetry transforming antipodic points of the De Sitter space into each other, commuting with every group element)

$$\varphi_n = N_n \lambda^2 e^{i\underset{\sim}{k}\cdot\underset{\sim}{y}}\ j_n(\lambda|\underset{\sim}{k}|R) \tag{6}$$

$$\mathfrak{F}\varphi_n = (-)^n \varphi_n$$

We have two completeness relations,[4] one for each system of functions with a definite $\mathfrak{F}$-parity:

$$\int d^3k \left\{ \sum_{n} \varphi^*_{\substack{2n\\2n+1}}(\lambda', \underset{\sim}{y}')\, \varphi_{\substack{2n\\2n+1}}(\lambda, \underset{\sim}{y}) + \int d\rho\, \varphi^*_{\rho\pm}(\lambda', \underset{\sim}{y}')\, \varphi_{\rho\pm}(\lambda, \underset{\sim}{y}) \right\}$$

$$= \frac{\lambda^4}{R}\{\delta(\lambda-\lambda') \pm \delta(\lambda+\lambda')\}\, \delta^3(y-\underset{\sim}{y}') \tag{7}$$

B. Quantum Theory of the Scalar Field

The commutation function $[\varphi(x), \varphi(y)]$ for real fields must be an invariant two-point function which is

i) a solution of the homogeneous equation (3)

ii) constructed from $\varphi_{\rho+}$, $\varphi_{\rho-}$, φ_n

iii) antisymmetric

iv) causal, i.e. zero for spacelike distances

$$s \equiv (\lambda\lambda')^{-1}\{R^2(\lambda-\lambda')^2 - (\underset{\sim}{y} - \underset{\sim}{y}')^2\}$$

For a fixed value of $\varkappa^2$ out of the continuous spectrum we can construct a function meeting all the requirements[1]

$$\Delta(\lambda, \underset{\sim}{y}; \lambda', \underset{\sim}{y}') = i\int d^3k(\overset{*}{\varphi}_{\rho+}(\lambda', \underset{\sim}{y}')\varphi_{\rho-}(\lambda, \underset{\sim}{y}) - \varphi_{\rho+}(\lambda, \underset{\sim}{y})\overset{*}{\varphi}_{\rho-}(\lambda', \underset{\sim}{y}'))\coth\pi\rho$$

$$\Delta_{|\lambda=\lambda'} = 0 \qquad (\frac{\lambda}{R}\partial_\lambda\Delta)_{|\lambda=\lambda'} = \lambda^3\delta^3(\underset{\sim}{y} - \underset{\sim}{y}') \tag{8}$$

We see that this function gives rise to a canonical quantization of the free scalar fields in De Sitter space. The above construction made use of the fact that to each value of ρ there exist two eigenfunctions of $\mathfrak{J}$-parity +1 resp. -1. In the case of the discrete spectrum the $\mathfrak{J}$-parity of the eigenfunctions is $(-)^n$, i.e. the eigenvalue of the Casimir operator determines the $\mathfrak{J}$-parity. Therefore we have only one possible candidate for an invariant two-point function solving (3), and it turns out that this is the acausal function. Let us look more closely at the state $\varkappa^2 = \frac{2}{R^2}$ $(n = 0)$. This is the conformal invariant state, because the equation $I_1\varphi = \frac{2}{R^2}\varphi$ can be transformed into

$$\frac{\partial_\lambda\partial_\lambda}{R^2} - \partial_\alpha\partial_\alpha \quad \varphi = 0 \tag{9}$$

The equation is formally conformal invariant, but the solutions of (9) in De Sitter space are

$$\varphi_o = \lambda^2 e^{ik\cdot y}\frac{\sqrt{2|\underset{\sim}{k}|}}{(2\pi)^2} j_o(\lambda|\underset{\sim}{k}|R) \tag{10}$$

which do not form complete basis for an irreducible representation of the conformal group SO(2,4). The invariant function constructed from (10) is $\sim \frac{1}{s}$, the causal function $D \sim \epsilon\delta(s)$ cannot be obtained.

We have to conclude that the conformal invariant solution in De Sitter space does not possess a causal commutation function, and can therefore not be interpreted as a particle. Since it seems reasonable to ascribe mass zero to the conformal invariant state, which does not feel the curvature of spacetime, we may state that there are no spin zero, mass zero particles in De Sitter space. The condition for the physical state space is $\varkappa^2 > 9/4R^2$.

C. Goldstone Particle

There is no Goldstone particle in (1+4) De Sitter space. A formal transcription of the original (not quite rigorous) proof of the Goldstone theorem into De Sitter space[3] indicates that the additional state appearing in case of a spontaneous symmetry breakdown has to have a mass $\varkappa^2 = 0$ (n=1). This state belongs to the discrete spectrum, has only the acausal invariant two-point function, and can therefore not be interpreted as a particle.

It may be interesting to note that the conformal invariant state and the Goldstone state have different quantum numbers.

II. Photons

A. Transformation Properties of Fields with Spin

To define the transformation properties of fields with spin[1] we use the fact that SO(1,3) is a subgroup of SO(1,4) and that it is the stability group of the point $x^0 = (0,0,0,0,-R)$ (in 5-dim. M_5 coord.). Defining $g_x \in SO(1,4)$ as the group element transforming x into x^0, we find that to each group element g $(x' = g(x))$, there corresponds a rotation around x^0: $g_x^{-1} g\, g_x$. The fields are representations of this group $\{g_x^{-1} g\, g_x\}$ which leaves x^0 fixed, i.e. the homogeneous Lorentz group:

$$\psi'(x') = T(g_x^{-1} g\, g_x)\, \psi(x) \tag{11}$$

To determine the infinitesimal generators for fields with spin, we have to add the local variation to the displacement operators derived for the scalar field. These spin operators are uniquely determined by writing (11) for an infinitesimal transformation,[1] and we find for the generators

$$B_o = \overset{o}{B}_o \qquad\qquad (B_\mu = \frac{1}{R} J_{4\mu})$$

$$B_\alpha = -i\partial_\alpha + \frac{1}{R} J_{o\alpha}$$

$$J_{o\alpha} = L_{o\alpha} + \lambda S_{o\alpha} + \frac{y^\beta}{R} S_{\beta\alpha}$$

$$J_{\alpha\beta} = L_{\alpha\beta} + S_{\alpha\beta} \qquad\qquad \alpha,\beta = 1,2,3 \tag{12}$$

Here $\overset{o}{B}_\mu$, $L_{\mu\nu}$ are the generators represented on a space of scalar functions. The spin matrices $S_{\mu\nu}$ satisfy the commutation relation

$$i[S_{\mu\nu}, S_{x\lambda}] = \eta_{\mu x} S_{\lambda\nu} - \eta_{\nu x} S_{\lambda\mu} + \eta_{\lambda\nu} S_{x\mu} - \eta_{\lambda\mu} S_{x\nu} \tag{13}$$

Fields of spin 1 transform like vectors under the rotation group. Together with (13) this determines $S_{\mu\nu}$ to be

$$(S_{\alpha\beta})_{\rho\nu} = -i(\eta_{\alpha\nu}\eta_{\beta\rho} - \eta_{\alpha\rho}\eta_{\beta\nu}) \qquad \rho,\nu = 0,..,3$$

$$(S_{o\alpha})_{\rho\nu} = -i(\eta_{\alpha\nu}\eta_{o\rho} + \eta_{\alpha\rho}\eta_{o\gamma}) \qquad \alpha,\beta = 1,2,3$$

$$\eta = (+---) \tag{14}$$

Then

$$S_{12}^2 + S_{23}^2 + S_{31}^2 = \begin{pmatrix} 2 & 0 & 0 & 0 \\ 0 & 2 & 0 & 0 \\ 0 & 0 & 2 & 0 \\ 0 & 0 & 0 & 0 \end{pmatrix}$$

B. Field Equations for Spin 1

Using (12) and (14) we calculate the Casimir operator I_1 and end up with the field equations

$$(I_1)_\mu{}^\nu A_\nu = \varkappa^2 A_\mu$$

$$(\overset{o}{I}_1 + \frac{2}{R^2} + \varkappa^2)A_\alpha = \frac{2\lambda}{R} \partial_\alpha A_o$$

$$(\overset{o}{I}_1 + \varkappa^2)A_o = \frac{2\lambda}{R} \partial_\alpha A_\alpha \tag{15}$$

where

$$\overset{o}{I}_1 = \frac{\lambda^2}{R^2} \partial_\lambda \partial_\lambda - \frac{2\lambda}{R^2} \partial_\lambda - \lambda^2 \partial_\alpha \partial_\alpha$$

For the second Casimir operator I_2 we obtain

$$I_2 = 2I_1 + B \tag{16}$$

where $B_\mu{}^\nu A_\nu = 0$ is equivalent to

$$\frac{\lambda}{R} \partial_\lambda A_o - \frac{3}{R} A_o - \lambda \partial_\alpha A_\alpha = 0 \tag{17}$$

(17) is the Lorentz condition generalized to De Sitter space.

To solve (15) and (17) we make the following ansatz:

$$A_\alpha = e^{i\underset{\sim}{k}\cdot\underset{\sim}{y}} \left(a_\alpha^\perp H_1(w) + \frac{ik_\alpha}{|\underset{\sim}{k}|} H_2(w) \right)$$

$$A_o = H_3(w) \qquad\qquad w = |\underset{\sim}{k}|\lambda R \tag{18}$$

where $a_\alpha^\perp \cdot k_\alpha = 0$, $(\underset{\sim}{a}^\perp)^2 = 1$ (we omit the explicit construction of $a_\alpha^\perp$).

Inserting into (15) we get three equations

$$\left(\partial_w \partial_w - \frac{2}{w}\partial_w + 1 + \frac{\varkappa^2 R^2 + 2}{w^2}\right) H_1(w) = 0 \tag{19}$$

$$\left(\partial_w \partial_w - \frac{2}{w}\partial_w + 1 + \frac{\varkappa^2 R^2 + 2}{w^2}\right) H_2(w) = \frac{2}{w} H_3(w) \tag{20}$$

$$\left(\partial_w \partial_w - \frac{2}{w}\partial_w + 1 + \frac{\varkappa^2 R^2}{w^2}\right) H_3(w) = -\frac{2}{w} H_2(w) \tag{21}$$

The general solutions of (19) can immediately be written down

$$H_1(w) = N_\mu^1 \, w^{3/2} (J_\mu(w) + c_\mu^1 J_{-\mu}(w))$$

$$\mu^2 = \tfrac{1}{4} - \varkappa^2 R^2 \tag{22}$$

J_μ: Bessel function, N_μ: Normalization factor

(20) and (21) can be solved by the ansatz

$$H_3(w) = w^\rho J_\mu(w) \qquad\qquad H_2(w) = w^\nu (\alpha + \beta w \partial_w) J_\mu(w) \tag{23}$$

$$\rho = \nu + 1, \quad \nu = 3/2, \quad \alpha = \tfrac{1}{2}, \quad \beta = -1; \quad \mu^2 = \tfrac{1}{4} - \varkappa^2 R^2 \tag{24}$$

So the general solutions of (15) have the form

$$A_\alpha = e^{i\underset{\sim}{k}\cdot\underset{\sim}{y}} w^{3/2} \left(a_\alpha^\perp N_\mu^1 \{J_\mu + c_\mu^1 J_{-\mu}\} + \frac{ik_\alpha}{|\underset{\sim}{k}|} N_\mu^2 (\tfrac{1}{2} - w\partial_w)\{J_\mu + c_\mu^2 J_{-\mu}\}\right)$$

$$A_o = e^{i\underset{\sim}{k}\cdot\underset{\sim}{y}} w^{5/2} N_\mu^3 (J_\mu + c_\mu^3 J_{-\mu}) \tag{25}$$

Inserting (25) into (17) we obtain the additional conditions

$$N_\mu^3 = N_\mu^2 \qquad\qquad c_\mu^3 = c_\mu^2 \tag{26}$$

C. Spectrum and Eigenfunctions

Let us make an attempt to proceed in analogy to the scalar case. The main difficulty initially is the indefinite norm. We are interested in solutions for which

$$0 < \int A_\mu^\dagger \cdot A_\mu \, d\Omega < \infty$$

with the measure $d\Omega$ being the invariant measure on the De Sitter hypersurface induced from M_5. In addition we require these solutions to form a manifold on which I_1 and I_2 are simultaneously "self-adjoint." Just as in the case of the scalar field these conditions determine a spectrum of $\varkappa^2$ and give a definite form for the eigenfunctions. By investigating formally the self-adjointness condition we find that there is a continuous spectrum $\frac{1}{4} < \varkappa^2 R^2 < \infty$, where to each value x^2 of the Casimir operator correspond two eigenfunctions of opposite $\mathfrak{F}$-parity just as in the scalar case:

$$R^2 \varkappa^2 = \rho^2 - \tfrac{1}{4} \qquad 0 < \rho < \infty$$

$$\rho \pm A_\alpha(w) = \lambda^{3/2} e^{i\underset{\sim}{k}\cdot\mathcal{Y}} (\theta(\lambda) \pm \theta(-\lambda)) \left\{ (N'_\rho a_\alpha^\perp + \frac{ik_\alpha}{|\underset{\sim}{k}|} N_\rho^2 (\tfrac{1}{2} - w\partial_w)) \times \right.$$

$$\left. \times \begin{Bmatrix} \mathrm{Im}(e^{i\pi/4} H_{i\rho}(|w|)) \\ \mathrm{Re}(\quad '' \quad) \end{Bmatrix} \right\}$$

$$\rho \pm A_0(w) = \lambda^{5/2} e^{i\underset{\sim}{k}\cdot\mathcal{Y}} N_\rho^2 (\theta(\lambda) \pm \theta(-\lambda)) \begin{Bmatrix} \mathrm{Im}(H_{i\rho}(|w|)\, e^{i\pi/4}) \\ \mathrm{Re}(\quad '' \quad) \end{Bmatrix} \qquad (27)$$

where $H_{i\rho}$: Hankelfunction of first kind; $\theta(x) = \begin{matrix} +1 & x > 0 \\ -1 & x < 0 \end{matrix}$

And there is a discrete spectrum, where the eigenvalue of the Casimir operator I_1 determines the $\mathfrak{F}$-parity of the functions:

$$R^2 \varkappa^2 = \tfrac{1}{4} - (n+\tfrac{1}{2})^2 \qquad n=0,1,2,\ldots$$

$$n \pm A_\alpha(w) = \tfrac{1}{2}(1 \pm (-)^n)\lambda^{3/2}\, e^{i\underset{\sim}{k}\cdot \underset{\sim}{y}} \left\{ (N'_n a^{\perp}_\alpha + \frac{ik_\alpha}{|k|} N^2_n (\tfrac{1}{2} - w\partial_w))J_{n+\frac{1}{2}}(w) \right\}$$

$$n \pm A_o(w) = \tfrac{1}{2}(1 \pm (-)^n)\lambda^{5/2}\, e^{i\underset{\sim}{k}\cdot \underset{\sim}{y}}\, N^2_n J_{n+\frac{1}{2}}(w) \tag{28}$$

These functions can of course be only "improper" eigensolutions. By looking at the integrals

$$\int (\rho_1 \pm A_\mu(w_1))^\dagger \cdot (\rho_2 \pm A_\mu(w_2)) \frac{R}{\lambda^4}\, d\lambda\, d^3y = \delta^3(\underset{\sim}{k_1} - \underset{\sim}{k_2})\,\delta(\rho_1 - \rho_2) \tag{29}$$

and

$$\int n_1 A^\dagger_\mu(w_1) \cdot n_2 A_\mu(w_2) \frac{R}{\lambda^4}\, d\lambda\, d^3y = \delta_{n_1 n_2}\, \delta^3(\underset{\sim}{k_1} - \underset{\sim}{k_2}) \tag{30}$$

we find that only the normalization factors N_ρ of the continuous part of the spectrum can be determined as finite quantities, while the discrete spectrum does not allow the determination of finite factors N_n, because the norm of these functions is infinite. The solutions (28) do not belong to our "Hilbert space." The physical state space is built only upon the solutions (27), with the spectrum condition $\varkappa^2 > \frac{1}{4}R^2$.

This is an indication that wavefunctions with infinitely many spin components are needed to give rise to unitary representations of the symmetry group for $R^2\varkappa^2 = \frac{1}{4} - (n+\frac{1}{2})^2$, just as it is the case in Minkowski space for imaginary mass. We see that the discrete part of the spectrum again corresponds to the imaginary mass states of Minkowski space as in the case of spin zero.

D. Conformal Invariance, Gauge Invariance

The conformal invariant state, invariant under the transformation of a covering group of the conformal group SO(2,4), is contained in the discrete part of the spectrum: It is the state with $\varkappa^2 = 0$: With the help of (17) we can bring (15) for $\varkappa^2 = 0$ into the form ($A_\mu \to \lambda A_\mu$)

$$\left(\frac{\partial_\lambda \partial_\lambda}{R^2} - \partial_\beta \partial_\beta\right) A_\alpha = -\partial_\alpha \partial_\beta A_\beta + \partial_\alpha \frac{\partial_\lambda}{R} A_o$$

$$\left(\frac{\partial_\lambda \partial_\lambda}{R^2} - \partial_\beta \partial_\beta\right) A_o = \frac{1}{R^2}\partial_\lambda \partial_\lambda A_o - \frac{\partial_\lambda}{R}\partial_\beta A_\beta \tag{31}$$

which explicitly shows the conformal invariance. Note that the term $\partial_\mu(\partial_\nu A_\nu)$ does not vanish, because the auxiliary condition still does not have the usual conform invariant form, but reads

$$\frac{\lambda}{R}\partial_\lambda A_o - \frac{2}{R}A_o - \lambda\,\partial_\alpha A_\alpha = 0 \tag{32}$$

From (31) it is clear that $\varkappa^2 = 0$ gives also the field invariant under gauge transformations. Indeed, the substitution $A_\mu \to A_\mu + \partial_\mu f$ $(\partial_o \equiv 1/R\ \partial_\lambda)$ leaves (31) invariant and (32) gives for f the condition

$$\left(\frac{\lambda^2}{R^2}\partial_\lambda\partial_\lambda - \lambda^2\partial_\alpha\partial_\alpha - \frac{2\lambda}{R^2}\partial_\lambda\right)f = 0 \tag{33}$$

which is, by the way, not the conformal invariant scalar equation.

Our previous investigations have shown that the state with $\varkappa^2 = 0$ does not belong to the space of physical states for which $\varkappa^2 > \frac{1}{4}R^2$. So the states for $\varkappa^2 = 0$ cannot be interpreted as particles, the usual characterization of the photon as the conformal invariant and gauge invariant particle cannot be kept up in (1+4) De Sitter space.

The remaining possible candidates for the photon come from the continuous spectrum $\varkappa^2 > \frac{1}{4}R^2$. As has been shown for the scalar field a causal commutation function can be obtained, if for each value of the Casimir operator there are two eigenfunctions of opposite $\mathfrak{F}$-parity. The arguments given there hold true in the spin 1 case too. In fact, the calculations are almost identical, so that we can omit them here.

One can now arbitrarily choose a state with $\varkappa^2 = \frac{\alpha}{R^2}$, $\alpha > \frac{1}{4}$ which gives zero for $R \to \infty$. This undesirable arbitrariness carries in it also some inconsistency: The light cone is as usual described by the classical propagation of lightwaves, corresponding to the conformal invariant value $\varkappa^2 = 0$. Now we have the paradox situation that the quantum particles cannot travel along their classical path, that photons do not run along null geodesics.

III. Spin $\frac{1}{2}$ Fields

A similar analysis as in Sec. II can be carried out for the spin $\frac{1}{2}$ fields.[1] The second order Casimir operator is

$$I_1^{\frac{1}{2}} = \overset{o}{I}_1 - \frac{3}{4R^2} + \lambda\sigma^\alpha\partial_\alpha\cdot\gamma_5 \tag{34}$$

$$\sigma^{\alpha}\text{: Pauli matrices} \qquad \gamma_5 = \begin{pmatrix} I & 0 \\ 0 & -I \end{pmatrix}$$

$I_1^{\frac{1}{2}}$ can be factorized into

$$I_1^{\frac{1}{2}} = (i\gamma^{o} \frac{\lambda}{R}\partial_{\lambda} + i\lambda\gamma^{\alpha}\partial_{\alpha} + i\gamma^{o}\frac{3}{2R})^2 + \frac{3}{2R^2} \tag{35}$$

Hence

$$(I_1^{\frac{1}{2}} + \varkappa^2)\,\psi = 0 \tag{36}$$

is equivalent to the De Sitter invariant Dirac equation

$$(i\gamma^{o} \frac{\lambda}{R}\partial_{\lambda} + i\lambda\gamma^{\alpha}\partial_{\alpha} + i\gamma^{o}\frac{3}{2R})\psi = m\psi \tag{37}$$

with

$$\varkappa^2 = m^2 + \frac{3}{2R^2} \tag{38}$$

for physical states $\varkappa^2 > \frac{3}{2R^2}$. So the conformal invariant state $m^2 = 0$, or $\varkappa^2 = \frac{3}{2R^2}$, again does not belong to the physical state space, but its mass is the lower bound for spin $\frac{1}{2}$ particle masses.

Acknowledgment

It is a pleasure to thank Dr. L. Castell for valuable comments and many enlightening discussions.

References

1. G. Börner and H. P. Dürr, Il Nuovo Cimento 64A, 669 (1969).
2. N. Limič, J. Niederle, and R. Raczka, J. Math. Phys. 7, 1861 (1966).
3. G. Börner, Prog. Theor. Phys. 43, no. 1, 244 (1970). L. Castell, Il Nuovo Cimento 61A, 585 (1969).
4. L. Castell, Jour. Math. Phys. 11, 1669 (1970).

DE SITTER ALGEBRAIC APPROACH TO ELECTROMAGNETIC INTERACTIONS†‡

B. J. Dalton
Department of Physics
Florida State University
Tallahassee, Florida

I. Introduction

Here, we discuss a new (group-theoretic) approach to electromagnetic interactions. The method used assumes that the interacting particles are each separately described by a De Sitter momentum space as discussed in the author's first lecture at this symposium and in a recent publication.[1] Our aim is to give a reason, from group theoretic arguments, why the particular force pattern

$$\vec{F} = q\vec{E} + q\vec{V} \times \vec{B} \tag{1}$$

should occur in nature. The second term can, of course, be obtained from the first by transformations from the rest frame where the first term holds. Conventionally, one could take this first term as an a priori fundamental property of the electric field and charge or one can construct an appropriate Lagrangian (or Hamiltonian) and reobtain the experimental form (1) from extremum equations.

Our procedure for describing the interaction patterns is to study the composite group representation for the two particles and how it changes as the two extended particles move closer together (or apart) in the background X-space. As will be seen below, the particular nature of force patterns such as (1) is determined (in this picture) by "how" the composite representation makes a transition between the uncoupled and vector-coupled limits as the two extended particles come together.

†Presented at the Symposium on De Sitter and Conformal Groups, University of Colorado, Summer 1970.
‡Research sponsored in part by the National Science Foundation Grant No. GP-7901.

The two geometries describing the two finite particles are not part of the background Minkowski space. We use small symbols with prime and double prime to label the components for the two particle geometries, i.e. (p', x'), (p'', x'') etc., and a large symbol X to label the background space. Large P' and P'' symbols are also used to describe the two average (observable) momentum spaces for the two particles. The averaging is over the finite extent of the two systems. The centers for the finite particles have "positions" in the background space, and an x-space interval in the particle's geometry may correspond to some interval in the background X-space. However, we will not need to explicitly give a mapping between intervals in these two spaces in order to obtain force patterns.

II. Two Extreme Limits for the Group Representation

We first describe two extreme limits for the composite group representation for the two particle-geometries. The first is the uncoupled limit (independent particle limit). This limit corresponds to the case where the two particles have a very large (or infinite) separation in the background X space. The two geometric particles are described independently of one another in this limit. We have for the prime system the eigenvalue equations.[1]

$$I_2' \ \Psi' = \frac{J_{ab}' J_{ab}'}{2} \ \Psi' \tag{2}$$

$$I_4' \ \Psi' = W_a' \ W_a' \ \Psi' \tag{3}$$

$$\mathfrak{m}_{54}' \ \Psi' = J_{54}' \ \Psi' \tag{4}$$

$$\mathfrak{m}_{12}' \ \Psi' = J_{12}' \ \Psi' \tag{5}$$

and for the double-prime system, the equations

$$I_2'' \ \Psi'' = \frac{J_{ab}'' J_{ab}''}{2} \ \Psi'' \tag{6}$$

$$I_4'' \ \Psi'' = W_a'' \ W_a'' \ \Psi'' \tag{7}$$

$$\mathfrak{m}_{54}'' \ \Psi'' = J_{54}'' \ \Psi'' \tag{8}$$

$$\mathfrak{m}_{12}'' \ \Psi'' = J_{12}'' \ \Psi'' \tag{9}$$

In keeping with the physical interpretation of the first lecture, we refer to J'_{54}, J''_{54} as the charge generators and J'_{12}, J''_{12} as the spin-projection generators. In this uncoupled limit, the charge and spin projections for the individual systems are good quantum (labeling) numbers. The wavefunction $\Psi(UC)$ for the composite representation in this limit is written as a kronecker product

$$\Psi(UC) = \Psi'_{\alpha}(p')\,\Psi''_{\beta}(p'') \tag{10}$$

where α and β represent the labeling numbers for a given solution and UC refers to this uncoupled limit.

The second limit corresponds to the case where the two geometric particles have come together in the background X-space to form one system. We represent this limit by the vector-coupled representation with composite generators given by

$$J_{ab} = J'_{ab} + J''_{ab} \tag{11}$$

This corresponds to the conditions

$$\theta_{ab} \equiv \theta'_{ab} = \theta''_{ab} \tag{12}$$

i.e. the two systems transform together in the same parameter space $\{\theta_{ab}\}$. We have the following eigenvalue equations for this case

$$I_2\Psi(VC) = \frac{J_{ab}J_{ab}}{2}\,\Psi(VC)$$

$$= \left[\frac{J'_{ab}J'_{ab}}{2} + \frac{J''_{ab}J''_{ab}}{2} + J'_{ab}J''_{ab}\right]\Psi(VC) \tag{13}$$

$$I_4\Psi(VC) = W_a W_a\,\Psi(VC) \tag{14}$$

$$\mathfrak{m}_{54}\Psi(VC) = J_{54}\,\Psi(VC) \tag{15}$$

$$\mathfrak{m}_{12}\Psi(VC) = J_{12}\,\Psi(VC) \tag{16}$$

Here, the condition

$$\mathfrak{m}_{12} = \mathfrak{m}'_{12} + \mathfrak{m}''_{12} \tag{17}$$

corresponds to spin projection addition, and the condition

$$m_{54} = m'_{54} + m''_{54} \tag{18}$$

corresponds to addition of electric charge in this picture. This interpretation is in keeping with the interpretation in the first lecture of J'_{54} and J''_{54} as the charge generators. In this vector-coupled (VC) limit, we write the wavefunction Ψ(VC) in the form

$$\Psi(\mathrm{VC}) = \sum_{\alpha\beta} C_{\alpha\beta}\, \Psi'_{\alpha}(p')\ \Psi''_{\beta}(p'') \tag{19}$$

where

$$\alpha \equiv (m'_{54},\ m'_{12}), \qquad \beta \equiv (m''_{54},\ m''_{12}) \tag{20}$$

The sum over m_{54} is replaced by an integral for the 4+1 group. Here, we make no attempt to evaluate the generalized Clebsch-Gordan coefficients $C_{\alpha\beta}$, but just treat them as constants. In this limit, we cannot say that m'_{54} and m'_{12} are good quantum numbers, i.e. the prime system does not have a good charge and spin projection.

III. Interaction Representation

We assume that as the two particles are approaching one another in the background X-space, the composite representation is undergoing a transition from the uncoupled to the vector-coupled limit. In this in-between case, neither the uncoupled, or the vector-coupled representation is valid. We assume that the wavefunction for this transition region can be written, however, in the form

$$\Psi(\mathrm{IR}) = \sum_{\alpha\beta} T_{\alpha\beta}(X)\, \Psi'_{\alpha}(p')\, \Psi''_{\beta}(p'') \tag{21}$$

where the coefficients $T_{\alpha\beta}(X)$ are functions of the separation of the two particles in the background X-space. As the two systems come completely together in the background space, we require that

$$T_{\alpha\beta}(X) \to C_{\alpha\beta}, \qquad \Psi(\mathrm{IR}) \to \Psi(\mathrm{VC}) \tag{22}$$

so that the representation reaches the vector-coupled limit.

To obtain equations for the transition region, we assume that the interaction representation corresponds to the path (set of equations) which minimizes the structural difference between the uncoupled and vector-coupled limits. This minimum structure-difference approach is somewhat analogous to the extremum path approach of general relativity, and the Lagrangian approach of classical mechanics and field theory. The two limiting representations differ from each other in several structural aspects. For instance, the labeling numbers m'_{54}, and m'_{12} are not good labeling numbers for the vector-coupled limit, but m_{54} is. One structural difference measure between the two limiting representations then, is the value for the commutator

$$\left[\frac{J_{ab}J_{ab}}{2}, J'_{54}\right] \tag{23}$$

since $J_{ab}J_{ab}$ is diagonal in one limit and J'_{54} is diagonal in the other. We assume that the coefficients $T_{\alpha\beta}(X)$ in (21) are chosen to minimize the structure-difference measure

$$\left|\bar{\Psi}(IR)\left[\frac{J_{ab}J_{ab}}{2}, J'_{54}\right]\Psi(IR)\right| \tag{24}$$

To see the patterns which are contained in these structure-difference measures, consider the commutators

$$I_{\mu} \equiv \left[\frac{J_{ab}J_{ab}}{2}, J'_{5\mu}\right] = 2i\left[J''_{5b}J'_{b\mu} - J'_{5b}J''_{b\mu}\right] \tag{25}$$

where we have used (13) and carried through the indicated commutation relations using

$$\left[J'_{ab}, J'_{cd}\right] = i\left[\delta_{ac}J'_{bd} - \delta_{ad}J'_{bc} - \delta_{bc}J'_{ad} + \delta_{bd}J'_{ac}\right] \tag{26}$$

Consider the commutation measure

$$\left|\bar{\Psi}(IR)\, I_{\mu}\, \Psi(IR)\right| = \left|\sum_{\substack{\alpha\beta\\ \gamma\delta}} \bar{T}_{\alpha\beta}\, T_{\gamma\delta}\, \langle I_{\mu}\rangle^{\alpha\beta}_{\gamma\delta}\right| \tag{27}$$

where

$$\langle I_{\mu}\rangle^{\alpha\beta}_{\gamma\delta} \equiv \bar{\Psi}'_{\alpha}\,\bar{\Psi}''_{\beta}\left[J''_{5b}J'_{b\mu} - J'_{5b}J''_{b\mu}\right]\Psi'_{\gamma}\,\Psi''_{\delta} \tag{28}$$

To obtain an x′ and x″ space version of these structure-difference measures, we take Fourier transformations over the particle momentum spaces p′ and p″. Since the coefficients $T_{\alpha\beta}(X)$ are independent of p′ and p″, we obtain the form

$$\sum_{\substack{\alpha\beta \\ \gamma\delta}} \bar{T}_{\alpha\beta} T_{\gamma\delta} F_{\mu}(\alpha\beta; \gamma\delta) \tag{29}$$

where

$$\frac{F_{\mu}}{2i} = \int d\tau' \int d\tau'' \exp(ip'_{\nu} x'_{\nu}) \exp(ip''_{\rho} x''_{\rho}) \langle I_{\mu} \rangle^{\alpha\beta}_{\gamma\delta} \tag{30}$$

To bring some of the terms in (30) into recognizable form, consider the diagonal terms of (30), $(\gamma,\delta) = (\alpha,\beta)$ for the scalar representation where

$$J'_{ab} = \ell'_{ab}, \qquad J''_{ab} = \ell''_{ab} \tag{31}$$

For this case, (28) becomes

$$\langle I_{\mu} \rangle^{\alpha\beta}_{\gamma\delta} = \bar{\Psi}'_{\alpha} \bar{\Psi}''_{\beta} \left[\ell''_{5b} \ell'_{b\mu} - \ell'_{5b} \ell''_{b\mu} \right] \Psi'_{\alpha} \Psi''_{\beta} \tag{32}$$

Using (32) in (30) and the interpretation (and definitions for the electromagnetic current densities and field tensors) given in the first lecture, we have

$$\frac{F_{\mu}}{2i} = 2i \left[M' \underset{\sim}{J}''_{b}(x'') F'_{b\mu}(x') - M'' \underset{\sim}{J}'_{b}(x') F''_{b\mu}(x'') \right] \tag{33}$$

where M′ and M″ are the respective radii of curvature for the two De Sitter momentum space geometries.[1] In (33) $\underset{\sim}{J}'_{\mu}$ represents only that part of the current-densities that involves the generators $\ell'_{5\mu}$, i.e.

$$\underset{\sim}{J}'_{\mu}(x') = \int d\tau' \exp(ip'_{\nu} x'_{\nu}) \bar{\Psi}'_{\alpha} \ell'_{5\mu} \Psi'_{\alpha} \tag{34}$$

The physical patterns of (33) are easier to recognize in conventional vector form. We define the vector components in the usual way, i.e.

$$E'_k = iF'_{4k}, \quad B'_k = F'_{ij} \quad \text{cyclic } 1,2,3 \tag{35}$$

$$E''_k = iF''_{4k}, \quad B''_k = F''_{ij} \quad \text{cyclic } 1,2,3 \tag{36}$$

For $\mu = 4$, (35) and (36) in (33) gives

$$\frac{F_4}{2i} = -2\left[M' \vec{J}'' \cdot \vec{E}' - M'' \vec{J}' \cdot \vec{E}'' \right] \tag{37}$$

For $\mu = 1,2,3$ we combine the three terms of (33) into a single vector

$$\frac{\vec{F}}{2i} \equiv \frac{1}{2i}\left[\hat{e}_1 F_1 + \hat{e}_2 F_2 + \hat{e}_3 F_3 \right]$$

$$= -2i\left[M' (i J''_4 \vec{E}' + \vec{J}'' \times \vec{B}') \right.$$

$$\left. - M'' (i J'_4 \vec{E}'' + \vec{J}' \times \vec{B}'') \right] \tag{38}$$

Comparison of the form (38) with (1) indicates that (38) is the usual total electromagnetic force density minus the self-force terms

$$i J'_4 \vec{E}', \quad i J''_4 \vec{E}'', \quad \vec{J}' \times \vec{B}', \quad \vec{J}'' \times \vec{B}'' \tag{39}$$

Likewise, no self terms like

$$\vec{J}' \cdot \vec{E}' \quad \text{and} \quad \vec{J}'' \cdot \vec{E}'' \tag{40}$$

occur in (37). From our earlier arguments, we can only require a minimum on the measure (27) for $\mu = 4$. This involves the terms (37), but not those of (38).

We recall that the above terms are only the diagonal parts of the structure-difference measure (29). The off diagonal terms involve states of different charge and spin. This suggests that one should perhaps try and relate these terms to charge and spin exchange, between the two particles. An exchange interpretation is supported in another way by the form (and nonzero value) of the quantities $F_\mu(\alpha, \beta; \gamma, \delta)$ in (30). Inspection of the diagonal terms (37) and (38) indicates that the quantities F_μ measure the deviation away from the action-reaction symmetry between the two geometric systems in the transition region. One very suggestive interpretation then, involves associating the quantities F_μ with exchange impulse densities between the two particle geometries.

For the non-scalar representations, one will obtain patterns identical to the above, but with the orbital generators replaced by the total generators J'_{ab} etc. The patterns in the structure-difference measures then involve intrinsic (spin) terms as well as cross terms between the orbital and spin parts. One cross term of interest arises in the form

$$\vec{B}' \times (\vec{J}'')^I \quad \text{and} \quad \vec{B}'' \times (\vec{J}')^I \tag{41}$$

where $(\vec{J}')^I$ represents the intrinsic part of the current density as discussed in the first lecture. From experiment, one has the (magnetic field)-(magnetic dipole) force form

$$\vec{B} \times \vec{\mu} \tag{42}$$

Comparison of (41) and (42) would suggest that one try and interpret the intrinsic part of the currents (which arise from spin operators S_{5i}) in terms of magnetic dipoles. Analysis of the above patterns in terms of geometric inversion behavior (e.g. $p'_\mu \to -p'_\mu$, $M' \to -M'$) would be a step towards finding possible physical interpretations for the extra terms. The structure-difference measure, involving J'_{54} and the fourth order Casimir invariant, should be minimized simultaneously with (24).

IV. Concluding Remarks

The above algebraic approach to electromagnetic interactions has three important characteristics. First, the same field distributions (e.g. $F'_{\mu\nu}(x')$) which occur in the free (non-interacting) case also occur in the force patterns. The interaction then (in this picture) does not give rise to the functional form (like $\frac{1}{r^2}$) for the electromagnetic field tensors. Rather, the interaction force patterns are involved in structure-difference measures which we minimize. By minimizing these measures, we are attempting to keep the free particle structure (field distributions) intact, as much as possible, while approaching the vector-coupled representation. The underlying physical assumption is that the change of relative motion (i.e. forces) between particles is nature's way of minimizing the difference between two mutually incompatible structures; that of the individual particles on the one hand, and that of an "in-unison" structure (described here by the vector-coupled limit) on the other. The second characteristic is that one does not have any self force terms in the structure-difference patterns. Third, the particular force patterns which result depend explicitly upon the structure of the free particles (i.e. upon which generators are diagonal). If one took an alternate set (e.g. a linear combination) of the generators and diagonalized in this set, the interaction-force patterns derived, as well as the field distributions, would be different.

Acknowledgments

Professors S. Edwards and D. Robson are gratefully acknowledged for many helpful discussions concerning the nature of interaction forces. I wish to thank Professor A. O. Barut for the invitation to attend this informal and friendly Boulder symposium.

Reference

1. B. J. Dalton, Phys. Letters 32B, 211 (1970).

ELECTRODYNAMICS AS PROPERTIES OF DE SITTER MOMENTUM SPACE†‡

B. J. Dalton
Department of Physics
Florida State University
Tallahassee, Florida

In this paper, we consider the possibility of describing an elementary electromagnetic system such as a photon, or an electron as a geometry. By this, we mean to relate particle and field entities (e.g. currents and field tensors) to pure geometric quantities such as group generators of motion. The aim is <u>not</u> to read in Maxwell's equations relating the currents and field tensors, but to find a geometry which gives rise to these equations.

In this geometric picture, the particle is not viewed as a local deviation away from (twist in) the background Minkowski space. Rather, the particle is viewed as a geometry (little universe) itself which is independent of (pinched off from) the background space. The center of the particle's geometry is assumed, however, to occupy some position on the background space, but just where, does not matter to the particle's structure.

In order to obtain non-trivial structure, one must use for the particle's geometry a metric significantly different from the Minkowski one. In general relativity, deviations of the metric away from the Minkowski picture are usually important over large intervals. For elementary particles, however, one has large intervals in momentum space. Because of this, we study the particle's structure as a Riemann geometry in a momentum space p, rather than in an X space.[1)]

We wish to consider a momentum space whose invariant line element admits a Lorentz transformation subgroup. One well known line element with this property is that of the conformally flat De Sitter space[2),3)]

---

†Presented at the Symposium on De Sitter and Conformal Groups, University of Colorado, Summer 1970.
‡Research sponsored in part by the National Science Foundation Grant No. GP-7901.

$$- dq^2 = Q^2 dp_\mu \, dp_\mu \tag{1}$$

$$Q = \left(1 + \frac{p_\mu p_\mu}{4M^2}\right)^{-1} \tag{2}$$

where M is a constant which will appear in a mass spectrum discussed later in this lecture. The ten generators of motion which keep (1) invariant are

$$\ell_{\mu\nu} = - i\left(p_\mu \frac{\partial}{\partial p_\nu} - p_\nu \frac{\partial}{\partial p_\mu}\right) \tag{3}$$

$$\ell_{5\mu} = - iMQ^{-1} \frac{\partial}{\partial p_\mu} + \frac{p_\nu}{2M} \ell_{\mu\nu} \tag{4}$$

This model corresponds to a free (non-interacting) extended (finite) particle. For each Lorentz frame, the particle is spread out over the subset of points U_o which are reached from a given point by the four non-Lorentz transformations whose generators are defined in (4). Later in this paper, we average the little p_μ over U_o, i.e. over the particle's extension in the little p space for each Lorentz frame. These average quantities are assumed to give a measure for the observable momentum components P_μ (indicated by large P_μ symbols), i.e.

$$P_\mu = \langle p_\mu \rangle_{U_o} \tag{5}$$

The set of points $\{P_\mu, \mu=1-4\}$ are reached from one another by the Lorentz subgroup of transformations. In this way, we obtain for the free "finite" particle a 4-momentum space which is spanned by the Lorentz group of transformations.

In a recent article,[3] it was shown that equations identical to those of electrodynamics are realized as relations between the above ten generators (3) and (4). There, ten p-space densities were defined by

$$K_\mu = \bar{\Psi} \ell_{5\mu} \Psi + iM \bar{\Psi} Q^{-1} \frac{\partial \Psi}{\partial p_\mu} \tag{6}$$

$$K_{\mu\nu} = - \frac{i}{2M} \bar{\Psi} \ell_{\mu\nu} \Psi \tag{7}$$

where Ψ is an eigenfunction of the two Casimir invariant operators and a choice of two commuting generators. $\bar{\Psi}\Psi$ is a De Sitter scalar in a given representation. Using (3) and (4) in (6) and (7), one can show that the densities K_μ and $K_{\mu\nu}$ satisfy the relations

$$K_\mu = i\, p_\nu K_{\mu\nu} \tag{8}$$

$$p_\rho K_{\mu\nu} + p_\nu K_{\rho\mu} + p_\mu K_{\nu\rho} = 0, \qquad \rho \neq \mu \neq \nu \tag{9}$$

$$p_\mu K_\mu = 0 \tag{10}$$

Corresponding functions in a little x-space for the particle's geometry are introduced via a Fourier transformation, i.e.

$$J_\mu(x) = \int d\tau\, \exp(i\, p_\rho x_\rho)\, K_\mu(p) \tag{11}$$

$$F_{\mu\nu}(x) = \int d\tau\, \exp(i\, p_\rho x_\rho)\, K_{\mu\nu}(p) \tag{12}$$

where $d\tau \equiv Q^4\, dp_1, dp_2\, dp_3\, dp_4$ is the invariant volume element. This little x-space for the particle's geometry should not be confused with the background Minkowski X-space in which one may have translations. Using (11) and (12), equations (8), (9), and (10) become

$$J_\mu(x) = \frac{\partial F_{\mu\nu}(x)}{\partial x_\nu} \tag{13}$$

$$\frac{\partial F_{\mu\nu}}{\partial x_\rho} + \frac{\partial F_{\rho\mu}}{\partial x_\nu} + \frac{\partial F_{\nu\rho}}{\partial x_\mu} = 0, \quad \rho \neq \mu \neq \nu \tag{14}$$

$$\frac{\partial J_\mu(x)}{\partial x_\mu} = 0 \tag{15}$$

respectively under the Fourier transformation. Equations (13) and (14) are identical in form to Maxwell's equations of electrodynamics, and (15) corresponds to the charge conservation, or continuity equation. In order to obtain these equations in terms of the large X background space, one must assume some mapping between the large X and small x spaces. This topic will be treated by the author in a future publication.

For non-scalar representations of the De Sitter group, one has ten intrinsic spin operators, or matrices $S_{\mu\nu}$ and $S_{5\mu}$ corresponding to the ten orbital generators $\ell_{\mu\nu}$ and $\ell_{5\mu}$. The orbital generators appear in the current densities and field tensors above. The orbital and spin operators usually appear on a parallel basis as parts of a "total" spin operator, i.e.,

$$J_{\mu\nu} = \ell_{\mu\nu} + S_{\mu\nu}, \quad J_{5\mu} = \ell_{5\mu} + S_{5\mu} \tag{16}$$

The idea of having intrinsic contributions to the current densities and field tensors then occurs. Following in analogy with (6), (7), (11) and (12), we define intrinsic components by

$$J_\mu^{\ I}(x) = \int d\tau \exp(i\, p_\rho x_\rho)\, \bar{\Psi}\, S_{5\mu}\, \Psi \tag{17}$$

$$F_{\mu\nu}^{\ I}(x) = \frac{i}{2M} \int d\tau \exp(i\, p_\rho x_\rho)\, \bar{\Psi}\, S_{\mu\nu}\, \Psi \tag{18}$$

If the $S_{\mu\nu}$ and $S_{5\mu}$ are just linear representation matrices (or operators) in the regular sense (e.g. like in the Dirac[4] case, $S_{\mu\nu} = \frac{i}{2}\gamma_\mu\gamma_\nu$, $S_{5\nu} = \frac{i}{2}\gamma_5\gamma_\nu$) then one has no inhomogeneous field equations for the intrinsic components in analogy to (13). However, consider the case in which one has a linear representation for the Lorentz subgroup, i.e. the $S_{\mu\nu}$ satisfy

$$[S_{\mu\nu}, S_{\lambda\rho}] = i[\delta_{\mu\lambda} S_{\nu\rho} - \delta_{\mu\rho} S_{\nu\lambda} - \delta_{\nu\lambda} S_{\mu\rho} + \delta_{\nu\rho} S_{\mu\lambda}] \tag{19}$$

If we define the four operators $S_{5\mu}$ by[5]

$$S_{5\mu} = \frac{p_\nu S_{\mu\nu}}{2M} \tag{20}$$

then one can show that the ten "total" generators $J_{\mu\nu}$ and $J_{5\mu}$ satisfy the usual commutation rule

$$[J_{ab}, J_{cd}] = i[\delta_{ac} J_{bd} - \delta_{ad} J_{bc} - \delta_{bc} J_{ad} + \delta_{bd} J_{ac}] \tag{21}$$

In this case, one has inhomogeneous field equations for the intrinsic components also, i.e.

$$J_\mu^{\ I}(x) = \frac{\partial F_{\mu\nu}^{\ I}(x)}{\partial x_\nu} \tag{22}$$

where (20) has been used in (17). For this latter case, one has a continuity equation for the intrinsic 4-currents, i.e.

$$\frac{\partial J_\mu^I(x)}{\partial x_\mu} = 0 \tag{23}$$

so that the intrinsic charge is conserved independent of the orbital charge. Using (16), one may form "total" current densities and field tensors. However, only in the latter case will the total quantities fully obey the inhomogeneous field equations. For the intrinsic field components, the usual homogeneous field equations (14) do not hold, i.e. one has magnetic charge and current densities J_α(mag.) defined by

$$J_\alpha(\text{mag.}) = \frac{\partial F_{\mu\nu}^I}{\partial x_\rho} + \frac{\partial F_{\rho\mu}^I}{\partial x_\nu} + \frac{\partial F_{\nu\rho}^I}{\partial x_\mu} \tag{24}$$

It should be made clear that these magnetic charge-current densities above arise only from the spin part and are defined by (24) in analogy to (13). The inhomogeneous equations (13) are due to the nonlinear (with respect to the p_μ) transformations whose generators have the particular form given in (4). However, the intrinsic current densities in (22) with $S_{5\mu}$ given in (20) are similar in form to the magnetic charge densities (24).

Let us consider for example a particular 2-dimensional representation of the latter type discussed above.[5)]

$$S_{\mu\nu} = \frac{i}{4}(\bar{\sigma}_\mu \sigma_\nu - \bar{\sigma}_\nu \sigma_\mu) \tag{25}$$

$$S_{5\mu} = \frac{p_\nu}{2M} S_{\mu\nu}$$

$$\bar{\sigma}_i = -\sigma_i, \quad \bar{\sigma}_4 = \sigma_4, \quad \sigma_4^2 = -1 \tag{26}$$

The magnetic charge density J_4 (mag.) defined by (24) is given by

$$J_4(\text{mag.}) = -\frac{1}{4M}\frac{\partial}{\partial x_i}\int d\tau\, \bar{\Psi}\sigma_i \Psi \exp(i p_\mu x_\mu) \tag{27}$$

in this representation. The intrinsic part of the charge density defined in (22) becomes

$$J_4^{\,I}(x) = \frac{i}{4M}\,\frac{\partial}{\partial x_i}\int d\tau\,\bar{\Psi}\,\sigma_i\,\Psi\,\exp(i\,p_\mu\,x_\mu) \tag{28}$$

Thus, the intrinsic part of the charge density is just proportional to the magnetic charge density (in this particular representation), i.e.

$$J_4^{\,I}(x) = -\,i\,J_4\,(\text{mag.}) \tag{29}$$

Since we have operators (group generators) appearing in the charge-current densities, one might ask if the nature of charge could not be explained in terms of the De Sitter eigenvalue spectrum. The operator J_{54} appears in the total charge density. Consider a representation for which J_{54} is diagonal along with J_{12}, the spin projection operator, i.e.

$$J_{54}\,\Psi = \mathfrak{m}_{54}\,\Psi \tag{30}$$

$$J_{12}\,\Psi = \mathfrak{m}_{12}\,\Psi \tag{31}$$

The eigenvalue $\mathfrak{m}_{54}$ can take both positive and negative values. The positive and negative nature of charge could possibly be explained in this manner. However, before this possibility can be realized, one must satisfy another physical condition. One has (for large r) a $\frac{1}{r^2}$ electric field for charged particles. Since the wavefunction Ψ defined above depends upon the labeling numbers $\mathfrak{m}_{54}$, $\mathfrak{m}_{12}$ and the two Casimir invariants I_2 and I_4, the x-space form for the field likewise depends (via (7) and (12)) upon these labeling numbers. A particular representation (and magnitude of $\mathfrak{m}_{54}$) must be found which will give the proper $\frac{1}{r^2}$ distribution for large r. If a solution Ψ (representation) can be found which satisfies this condition, then one is left with the problem of finding a physical interpretation for the other solutions (e.g. for different values of $\mathfrak{m}_{54}$). By using the continuity equation, one could probably show that a number of these solutions do not have stable charge distributions in the Lorentz rest frame, and hence are non-physical.

Next, we discuss the average momentum components P_μ which we define by

$$P_\mu = \langle p_\mu \rangle = \int_{U_o} d\tau\,\bar{\Psi}\,p_\mu\,\Psi \tag{32}$$

The subset U_o over which we integrate is chosen to satisfy the two conditions

$$U_o \supset \{p_\mu = 0, \quad \mu = 1\text{-}4\} \tag{33}$$

$$I \text{ is transitive to } U_o \tag{34}$$

where I represents the non-Lorentz transformations whose generators are defined in (4). These average components have the following properties

$$\text{Under I:} \quad U_o \rightarrow U_o, \quad \langle p_\mu \rangle \rightarrow \langle p_\mu \rangle \tag{35}$$

Under L ≡ Lorentz Subgroup:

$$p_\mu{}' = T_{\mu\nu}\, p_\nu \;, \quad \langle p_\mu{}' \rangle = T_{\mu\nu} \langle p_\nu \rangle \tag{36}$$

In the subset U_o, the components p and p_4 have the following limits

$$p^2 \equiv p_1{}^2 + p_2{}^2 + p_3{}^2 \tag{37}$$

$$4 + 1 \text{ case, Real M,} \quad \begin{array}{l} 0 \leq |p| \leq M \\ 0 \leq |p_4| \leq \infty \end{array} \tag{38}$$

$$3 + 2 \text{ case, Imaginary M,} \quad \begin{array}{l} 0 \leq |p| \leq \infty \\ 0 \leq |p_4| \leq |M| \end{array} \tag{39}$$

The cutoff in p for the 4+1 case may be of physical interest with regard to possible singularities in conventional electrodynamics as the radius goes to zero.

Because of the definition (32) and the choice of U_o, the components P_μ do not change under the non-Lorentz transformations, but transform like a Lorentz 4-vector under the Lorentz subgroup. The Lorentz rest frame is defined to be the case for which

$$\langle p_i \rangle = 0, \quad i = 1,2,3$$

$$\langle p_4 \rangle \neq 0$$

$$\equiv \text{Rest Mass} \tag{40}$$

For any representation for which one can form a scalar (and have finite integrals) one can show from dimension analysis that the rest energy has the form (for diagonal J_{54} and J_{12})

$$\langle p_4 \rangle = |M| \; F(m_{12}, m_{54}, I_2, I_4) \tag{41}$$

where the function F depends upon the particular representation. The radius of curvative M appears as an overall constant in the mass spectrum (41).

In summary, one has in this picture one means by which to introduce non-compact Lie groups in describing elementary particles. For the De Sitter geometry, one has a realization of Maxwell type equations for each representation (in both the 3+2 and 4+1 cases) of the De Sitter group for which one can form a scalar. By averaging over the internal momentum subspace U_0, one can obtain final momentum components whose space is spanned by the Lorentz group. For this free (non-interacting) finite particle model, the mass spectrum (41) results from the integration over the finite extent of the particles and depends upon the labeling numbers for the representation considered. The composition and interaction of two geometric particles with each separately described as above is discussed in the author's second lecture presented at this symposium.

References

1. Possibilities of generalizing the momentum space have been considered by H. S. Snyder, Phys. Rev. 71, 38 (1947); V. G. Kadyshevski, Soviet Physics (JETP) 14, 1340 (1961); Proceedings of the International Conference on High-Energy Physics (CERN, 1962); Yu A. Gol'Fand, Soviet Physics (JETP) 10, 356 (1960); 16, 184 (1963); A. O. Barut, P. Budini and C. Fronsdal, Proc. Roy. Soc. London A291, 106 (1966); G. Cicogna, Nuovo Cimento XLIIA, 656 (1966).
2. The x-space version of this space is discussed in a review article by F. Gürsey, in Group Theoretical Concepts and Methods in Elementary Particle Physics, edited by F. Gürsey (Gordon and Breach Science Publishers, Inc., New York, 1965), p. 365.
3. B. J. Dalton, Phys. Letters 32B, 211 (1970).
4. P. A. M. Dirac, Ann. Math. 36, 675 (1935).
5. This type of representation is discussed by G. Börner and H. P. Dürr, "Classical and Quantum Fields in De Sitter Space," preprint, Münich, June 1969.

THE PARTICIPANTS

Carl Andersen*	College of William and Mary
A. O. Barut*	University of Colorado
Johan G. F. Belinfante*	Carnegie-Mellon University
Gerhard Berendt	Freie Universität Berlin
A. Böhm*	University of Texas
Gerhard Börner*	Max-Planck Institute
W. E. Brittin	University of Colorado
L. Castell*	Max-Planck Institute
Prakash Chand	Colorado State University
Franklin F. K. Cheung*	Notre Dame
Sung C. Chung	West Virginia Tech
Herbert C. Corben	Cleveland State University
Christofer Cronström*	Institute for Advanced Study
B. J. Dalton*	Florida State University
Harald O. Dogliani	Kirtland Air Force Base
G. Domokos	Johns Hopkins University
Susan Kovesi-Domokos	Johns Hopkins University
Joseph Dreitlein	University of Colorado
Ismail Hakki Duru	University of Colorado
J. Elhadad	Faculté des Sciences - Marseille
Jason Ellis	University of Texas at Arlington
Kathryn Ellis	North Texas State University
Gordon N. Fleming	Pennsylvania State University
Stephen A. Fulling	Princeton University
F. Gerjuoy	University of Pittsburgh
Alberto Giovannini	Istituto Di Fisica Teorica - Torino
L. Girardello	University of Colorado
Peter Haberler*	Duke University
Gerhard E. Hahne	Stanford University
Rolf B. Haugen	University of Colorado
Wayne J. Holman III*	Middle East Technical University
R. L. Ingraham*	New Mexico State University
Akira Inomata*	SUNY at Albany
R. Jackiw*	Massachusetts Inst. of Tech.
Leon Jaffe*	University of Texas
Herbert Jehle	George Washington University
Thomas F. Jordan	University of Pittsburgh
David W. Joseph	University of Nebraska
H. A. Kastrup*	Universität München
D. Y. Kim	University of Saskatchewan

*Lecturers

H. Kleinert	Freie Universität Berlin
Kaiser S. Kunz	New Mexico State University
Neil W. Macfadyen*	Carnegie-Mellon University
K. T. Mahanthappa	University of Colorado
G. Bruce Mainland*	University of Texas
Shimon Malin	Colgate University
K. Mariwalla*	Arya-Mehr University
Shigeyuki Miyashita	University of Colorado
Milan Noga*	Purdue University
B. Pirrung*	Inst. f. Th. Physik Marburg
Kenneth Rafanelli	Cleveland State University
Budh Ram	New Mexico State University
Howard R. Reiss	The American University
Sebastián Salamó	University of Colorado
Tetsuo Sawada	University of Oregon
Fritz Schwarz*	University of Texas
C. S. Shukre	University of Pittsburgh
Staffan Ström*	University of Texas
George Thomas	University of Western Ontario
Paul Winternitz*	Carnegie-Mellon University
B. G. Wybourne	University of Canterbury
Walter Wyss	University of Colorado
Tsu Yao*	University of Pittsburgh